P9-CAR-606

CLEANSING
THE DOORS OF PERCEPTION

If the doors of perception were cleansed, everything would appear to man as it is, infinite.

—William Blake

CLEANSING THE DOORS OF PERCEPTION

The Religious Significance of Entheogenic Plants and Chemicals

HUSTON SMITH

JEREMY P. TARCHER/PUTNAM
a member of
Penguin Putnam Inc.
NEW YORK

Most Tarcher/Putnam books are available at special quantity discounts for bulk purchases for sales promotions, premiums, fund-raising, and educational needs. Special books or book excerpts also can be created to fit specific needs.
For details, write Putnam Special Markets, 375 Hudson Street, New York, NY 10014.

Jeremy P. Tarcher/Putnam
a member of
Penguin Putnam Inc.
375 Hudson Street
New York, NY 10014
www.penguinputnam.com

The substances discussed in this book are not without risks. Further, most of them are illegal under controlled-substance laws in the United States and elsewhere, even for religious use. Neither the publisher nor the author accepts liability for harms of any kind arising in connection with these substances.

Published simultaneously in Canada

Library of Congress Cataloging-in-Publication Data

Smith, Huston.
Cleansing the doors of perception: the religious signifiance of entheogenic plants and chemicals/Huston Smith.
p. cm.
ISBN 1-58542-034-4
1. Hallucinogenic drugs and religious experience. I. Title.
BL65.D7 S55 2000 99-087639
291.4'2—dc21

CSP Entheogen Project Series, No. 5
This book also appears in a limited edition, ISBN 1-889725-03-x,
published by the Council on Spiritual Practices (www.csp.org).

Printed in the United States of America
10 9 8 7 6 5 4 3 2 1

This book is printed on acid-free paper. ♾

Book design by Martino Mardersteig

To the memory of
Walter Pahnke,
and to his wife, Eva,
and their children, Kristin, David, and Jonathan

The mescaline experience is without any question the most extraordinary and significant experience available to human beings this side of the Beatific Vision. To be shaken out of the ruts of ordinary perception, to be shown for a few timeless hours the outer and inner worlds, not as they appear to an animal obsessed with survival or to a human being obsessed with words and notions, but as they are apprehended, directly and unconditionally, by Mind at Large – this is an experience of inestimable value to anyone.

—Aldous Huxley

The spiritual crisis pervading all spheres of Western industrial society can be remedied only by a change in our world view. We shall have to shift from the materialistic, dualistic belief that people and their environment are separate, toward a new consciousness of an all-encompassing reality which embraces the experiencing ego, a reality in which people feel their oneness with animate nature and all of creation.

Everything that can contribute to such a fundamental alteration in our perception of reality must therefore command earnest attention. Foremost among such approaches are the various methods of meditation, which aim to deepen the consciousness of reality by way of a total mystical experience. Another important, but still controversial, path to the same goal is the use of the consciousness-altering properties of psychopharmaceuticals.

—Albert Hofmann

Ecstasy! In common parlance ecstasy is fun. But ecstasy is not fun. Your very soul is seized and shaken until it tingles. After all, who will choose to feel undiluted awe? The unknowing vulgar abuse the word; we must recapture its full and terrifying sense.

—R. Gordon Wasson

Our normal waking consciousness . . . is but one special type of consciousness, whilst all about it, parted from it by the filmiest of screens, there lie potential forms of consciousness entirely different. We may go through life without suspecting their existence; but apply the requisite stimulus, and at a touch they are there in all their completeness. . . . No account of the universe in its totality can be final which leaves these other forms of consciousness quite disregarded. How to regard them is the question,–for they are so discontinuous with ordinary consciousness. Yet they may determine attitudes though they cannot furnish formulas, and open a region though they fail to give a map. At any rate, they forbid a premature closing of our accounts with reality. Looking back on my own experiences [with nitrous oxide] they all converge toward a kind of insight to which I cannot help ascribing some metaphysical significance.

—William James

The greatest of blessings come to us through madness, when it is sent as a gift of the gods. Heaven-sent madness is superior to man-made sanity.

—Plato, in the *Phaedrus*

CONTENTS

APPENDICES

ACKNOWLEDGMENTS

I am indebted to the Council on Spiritual Practices (www.csp.org) for asking me to gather into a single volume the essays I have written on the entheogens. Without this request it would not have occurred to me to do that.

The Council has supported the project from beginning to end, and has my sincerest thanks.

PREFACE

Is it possible today, in the climate of fear created by the war on drugs, to write a book on the entheogens with the informed objectivity of Aldous Huxley's *The Doors of Perception,* the understanding that Albert Hofmann accorded the topic in *LSD: My Problem Child,* the expertise Gordon Wasson brought to it in his *SOMA*, and the open-mindedness with which William James approached the subject in *The Varieties of Religious Experience?* And is the reading public ready for such a book?

I do not know the answers to those questions, but I find myself wanting to put them to the test. My reasons are theoretical rather than adversarial, for I am more philosopher than activist. It is true that, though this book is being published as a free-standing volume in its own right, it is also listed as number five in a series of books on the entheogens – virtually nonaddictive drugs that seem to harbor spiritual potentials – that the Council on Spiritual Practices is issuing. I am comfortable with this, for not only did that Council instigate this book by asking me to pull its essays together; I support its objectives, which include working cautiously toward carving out a space where serious students of the entheogens can pursue their interests carefully and lawfully. I was fortunate in being able to do that under the umbrella of Harvard University's 1960–63 research program before it careened off-course, and it is only fair to do what I can to accord others the same opportunity.

This said, however, I come back to my concerns here being philosophical rather than programmatic. During the semester that Aldous Huxley was at M.I.T., he remarked in the course of a seminar that nothing was more curious, and to his way of thinking more important, than the role that mind-altering plants and chemicals have played in human history. Add to that William James's point that no account of the universe in its totality can be taken as final if it ignores extraordinary experiences of the sort he himself encountered through the use of nitrous oxide. This entire book can be seen as an extended meditation on those two ideas.

As for the other two parties that I mentioned in my opening paragraph (and who join Huxley and James on the frontispiece of this book), I will defer until chapter 4 the story of the summer I spent working with Gordon Wasson on his claim that India's sacramental plant, soma, was a psychoactive mushroom. Albert Hofmann, the discoverer of LSD, I include not only for the judiciousness of his discussion of his problem child, but for a personal reason as well. A friend of mine who visited him in Switzerland had occasion to mention my book *Forgotten Truth,* which outlines the metaphysical position – roughly the Great Chain of Being – that my entheogenic encounters enabled me to experience. When my friend returned from that visit and told me that Hofmann had expressed interest in my book, I sent him a copy. The letter I received in reply opened by saying, "No other book in the last years has meant more to me than your *Forgotten Truth.* My experience and awareness of reality and its different aspects correspond completely with your view. The reward I got by studying your book was to find my insights, which are those of a natural scientist, a layman in philosophy, confirmed and expanded more fully by a professional in this field."

The essays in this book span almost forty years. I have edited them liberally, excising repetitions and passages I no longer consider important. Each essay is introduced by a statement that notes the occasion for which it was written and locates it on the trajectory of the book as a whole. My intent has been to produce a work that touches on the major facets of its enigmatic subject as seen through the eyes of someone (myself) who, given my age, may have thought and written more about it than anyone else alive.

Nomenclature has been a problem. I never use the word "hallucinogen" because error is built into its definition – *Webster's New Universal Unabridged Dictionary* defines "hallucination" as "(1) the apparent perception of sights, sounds, etc., that are not actually present [which] may occur in certain mental disorders; (2) the imaginary object apparently seen, heard, etc." The word "psychedelic" is etymologically innocuous, literally meaning "mind-manifesting," but it is dated, tagged to "the psychedelic sixties" when recreational use of drugs took over, and thus clearly inappropriate when speaking of shamans, Eleusis, and the Native American Church. We need a word that designates virtually nonaddictive mind-altering substances that are approached seriously and reverently, and the word "entheogens" does just that. It is not without problems of its own, for etymologically it suggests "God-containing," whereas "God-enabling" would be more accurate – Aldous Huxley told me never to say that chemicals *cause* visionary experiences; say that they *occasion* them. I retain "psychedelic" in the early essays of this book which were written when it was the going word, but, thereafter, I follow the lead of Wasson, Hofmann, Richard Schultes, and other pioneers in concluding that "entheogens" is the appropriate word for mind-changing substances when they are taken sacramentally.

HUSTON SMITH
Berkeley, California

INTRODUCTION

This book opens with a description of my first entheogen experience, and – because objective understanding of these substances is the book's primary aim – this leaves me with no alternative but to talk about myself; for there is no direct line from chemical brain states to the experiences they occasion. Invariably the psychological makeup of the subject (his "set," as investigators call it) acts as a filter, as do the circumstances surrounding the ingestion, its "setting." The reader needs to be aware of this, for whether I am reporting experiences I have had or registering conclusions that I have reached on the tricky issues this book takes up, I (the book's author) am inevitably present in its pages for establishing the angle from which the subject is viewed. The object of this Introduction is to make that angle clear, and it can be relatively brief, for only things that bear on the book's subject need be included.

Set

As far back as I can remember, my recollections have a metaphysical tinge to them and bodings of ultimacy. My parents were Protestant missionaries in rural China, and as disease was rampant it is not surprising that my earliest memory is of a burning fever during which I was rationed to a teaspoonful of boiled water every forty-five minutes because that was as much as my stomach could keep down. Those intolerably long waits, the desperate thirst! *Dukkha,* life is suffering, the first of the Buddha's Four Noble Truths. In a second illness (a year or so later) my fever reached the proportions of delirium, and I experienced my body as distended and filling the bedroom like a balloon. Metaphysics stalks such memories, for suffering raises the question of why it occurs, and delirium (by showing that the world can appear in radically different guises) raises the question of which guise is real. Did Chuang Tzu dream that he was a butterfly, or was his "waking" state a butterfly's dream?

When thoughts begin to enter my memories, they too have metaphysical edges. Whether the memory is of waking on a wintry night to the unbelievable splendor of a star-strewn sky – seen as clearly from the Yangtze delta as if I had been on Mount Everest, for there was no atmospheric pollution in rural China then – or of the agonized screams of a neighbor who was dying of meningitis, or of our cook's report at breakfast that an infant had been left on our doorstep during the night in its parents' desperate hope that we would adopt and feed it, my earliest memories all carry overtones of life, death, and the things that matter most. Happy memories tumble in quickly – dew between the toes of my bare feet at the start of an incomparably beautiful summer's day is one that remains surprisingly vivid – but as far as I can make out these come from shallower strata of memory. When I came upon the title of de Chirico's painting, "Nostalgia for the Infinite," it presented itself as the signature for my life.

In due course my childhood ruminations took on the outlines of the Protestantism that I assimilated from my parents *cum lacte,* as the Romans said: with my mother's milk. I am grateful for this heritage, whose details have changed but whose premises have not, for to have been launched on life with the assurance that we are in caring hands and that in gratitude for that fact we should try to bear one another's burdens, strikes me as a priceless inheritance.

As we were the only Westerners in our town, my parents were my only Caucasian role models, so I grew up assuming that missionaries were what American children grew up to be. This had me coming to the United States for college thinking it was to acquire the credentials that would take me back to China in my parents' footsteps, but I had not reckoned on the West's dynamism. America was exciting, to the point that within two weeks I had shelved all thoughts of returning to China. No vocational crisis occurred, however, for instead of being a missionary I would be a minister. That resolve lasted for two years when, in my junior year in college – Central Methodist College in Fayette, Missouri – ideas took over and I decided to teach rather than preach. Both vocational shifts occurred smoothly, for throughout them my eye remained on the Big Picture. I wanted to know the final nature of things: reality's deepest structure and what follows from that structure for maximizing the human potential.

While I was in college, theology presented itself as the window onto the Big Picture, but in graduate school at the University of Chicago, philosophy displaced it for seeming to have the wider purview. As for which of philosophy's metaphysical systems was true, naturalism seemed at first to be the obvious choice. Naturalism was tailor-made for science, and science was what distinguished the West from sleepy traditional civilizations. I had caught glimpses of its prowess back in China as I watched townsfolk line up for the smallpox vaccinations my father annually administered. Health was only one of science's valuable deliverances, and naturalism was the metaphysics that fitted science best. John Dewey was the "Jesus" of naturalism and Alfred North Whitehead its "Saint Paul."

There could not have been a more fanatical zealot for naturalism than Huston Smith the graduate student at the University of Chicago, but shortly before I exited that university my naturalistic worldview collapsed like a house of cards. The collapse occurred in a single night through a book on mysticism that had fallen into my hands. Mysticism had scarcely figured in my education, but when it was placed before me undismissively, I saw it as true. There is a reality other than the one that science and common sense – a workable definition of naturalism – set before us, and it is more exciting by every standard that can be invoked. It is more ultimate, more powerful, more awesome, more significant, and more mysterious. My instructors had taught me that Plato's Allegory of the Cave was a brilliant piece of speculation, and yes, magnificent poetry too, but the book I was reading presented it as true. Or rather, as true as words can make out when (through the use of allegory and metaphor) they attempt to describe things that are beyond their reach. I can still feel the electricity of that discovery. Plato tells us that when he thought of the Sun that shines eternally outside the cave of everyday existence, "first a shudder runs through me, and then the old awe creeps over me." Move over, I found myself exclaiming. You have company.

I completed my dissertation perfunctorily, for I needed the union card to teach, but once my diploma was firmly in hand I turned to mysticism seriously. I soon discovered that its fundamental dichotomy – between this world and another world, *samsara* and *nirvana,* the profane and the sacred – needed to be elaborated, and that its consummate elaboration was the Great Chain of Being, the conception of the world as composed of an infinite number of links ranging in hierarchical order from the meagerest kind of existence through every possible grade up to the *ens perfectissimum,* the Perfect Being at its summit.

When the reader comes to the description of my first entheogen experience, the point of this biographical prelude will be glaringly evident. To the best of my ability, it establishes the "set" – to repeat, the psychological makeup through which my experiences of the entheogens, and my thoughts about them as well, have been filtered. There remains the matter of the "setting," which is to say, the circumstances in which I have encountered the substances, and I will devote the second half of this Introduction to that subject.

Setting

The trail goes back to the book that sprung me from my naturalism. In referring to that book, I deliberately withheld its title for fear that it would raise false expectations. One man's meaning is another man's mush, and this holds for the same individual at different stages of his life. Apparently, on the night that I read the book in question my mind had reached the state of a saturated solution which needed but the shock of the right contact for it to recrystallize in a form that startled me. The book I was reading applied that contact, but I have never gone back to it, feeling sure that (as my thoughts have changed) I would now find it disappointing. Still, there is no point in playing cat-and-mouse: the book was Gerald Heard's *Pain, Sex and Time.* There is another reason for mentioning its name, for its author turned out to figure importantly in the road that led me to the entheogens.

Today Gerald Heard is not widely remembered, but at mid-century he was known for having written a best-selling detective story, *A Taste of Honey,* and (to the cognoscenti) as the thinker who had moved his friend Aldous Huxley from the cynicism of his early, *Brave New World* period, to the mysticism of *The Perennial Philosophy.* Before immigrating with Huxley to the United States, Heard was science commentator for the British Broadcasting Company; H. G. Wells said at the time that he was the only man he bothered to listen to on the wireless. Part of the excitement of Heard's writings derived from the way he drew on his BBC background to cite scientific discoveries that seem to support the mystical worldview. After completing my doctorate I read everything he had written, whereupon I determined to meet the man. I wrote to him in care of his publisher and received a reply postmarked "Trabuco Canyon, California," which he explained was in the Los Angeles area.

I made the journey and spent a memorable twenty-four hours with Heard at what turned out to be a retreat center he had founded. As I was leaving he asked if I had met Aldous Huxley, adding that "he's interested in our kinds of things." The prospect of meeting Aldous Huxley at my young age staggered me, and I pursued the lead eagerly. A meeting was arranged by phone, and I returned to Denver having had a second unforgettable encounter, this one an afternoon with Huxley and his wife, Maria, in their hideaway cabin in the Mojave Desert.

An important consequence of my Los Angeles safari was my discovery of the importance that both Heard and Huxley attached to meditation. My Protestantism had taught that if I lived decently I would meet God beyond the grave, but the mystics claimed that God could be found in this life. In a general way I had encountered that belief in the mystical writings I had started to study, but somehow it didn't sink in until I found that Heard was meditating six hours a day to realize God, and that he had founded his retreat center, Trabuco College, for precisely that purpose; Huxley had spent six months there, dividing his time between meditating and writing *The Perennial Philosophy.* Vedanta (the philosophical expression of Hinduism as taught in America by monks of the Ramakrishna order) had sparked Heard's project, for he and Huxley had found those monks to be the most serious and knowledgeable mystics around. (Swami Prabavananda, the director of the Vedanta Society in Hollywood at the time, was the model for the enlightened swami in Somerset Maugham's novel *The Razor's Edge.*) When Heard learned that I was moving from the University of Denver to Washington University, he gave me the name of the swami who headed the Saint Louis Vedanta Society. That swami, Swami Satprakashananda, introduced me to meditation, and for twenty years – ten in Vedanta and ten in Zen training – it was my primary spiritual practice.

With disappointing results, I have to confess. I do not regret those years, and continue to meditate each day, but it does more to strengthen my life's trajectory and call me back to the here and now than it does to produce mystical visions and altered states of consciousness. As a direct experience of God was what I most wanted during those two decades – friends twitted me about whoring after the Absolute – when Huxley's *Doors of Perception* appeared, the mescaline it reported sounded like a Godsend – literally. Like me, Huxley was a poor visualizer – "even the pregnant words of poets do not evoke pictures in my mind; no hypnogogic visions greet me on the verge of sleep," he wrote – yet mescaline overcame that disability and introduced him to "the flow of beauty to heightened beauty, from deeper to ever deeper meaning." Perhaps it could do the same for me.

It took six years to find out. When those years were up, in the fall semester of 1960, Huxley joined the faculty of the Massachusetts Institute of Technology (where I was then teaching) for a semester as Distinguished Visiting Professor of the Humanities. It was the same semester in which Timothy Leary joined the faculty of Harvard University for a three-year appointment as Research Professor in its Center for Personality Research. On his way to that appointment, Leary had ingested seven mushrooms by the side of

a swimming pool in Cuernavaca, and, astounded by their effects, he chose as his first Harvard project researching the potentials of psychoactivating chemicals for correcting behavior disorders.

A mile and a half down the Charles River, Aldous Huxley was an obvious resource. When I told Huxley of my interest in the matter, he gave me Leary's phone number and we arranged to meet for lunch at the Harvard Faculty Club. Getting down to business, we pulled out our date books to schedule a session with mescaline. Several tries wouldn't work for one or the other of us, whereupon Leary flipped past Christmas and (with the faintest trace of a mischievous smile, as I remember the scene) asked, "What about New Year's Day?"

It proved to be a prophetic way to enter the "psychedelic sixties."

CHAPTER ONE

Empirical Metaphysics

As I noted in my Introduction, my initiation into the entheogens took place in 1961 under the auspices of the Center for Personality Research at Harvard University as part of a project directed by Professor Timothy Leary to determine if a certain class of virtually nonaddictive mind-altering chemicals – mescaline, psilocybin, and LSD – could facilitate behavior change in desirable directions. Such changes are not easy to gauge. Subjective reports are notoriously unreliable, but two populations do lend themselves to statistical measurement. Six months after an entheogen experience, is a paroled prisoner still on the streets or back behind bars, and is the recovered alcoholic still off the bottle? Such were the kinds of questions that the study hoped to answer, but it was necessary to start from scratch, for this was the first concerted effort to study these substances scientifically. (At one point Freud had hopes for cocaine, but he soon abandoned them, and besides, cocaine falls into a different class of drugs because it is addictive.) Accordingly, the first step was to get some idea of the range and kinds of experiences the drugs occasion when given in a supportive atmosphere. Volunteers were solicited to establish a data bank of phenomenological reports. Subjects were screened to rule out those with psychological problems, and precise doses of one of the three drugs being investigated were administered. A physician or psychiatrist was invariably present, with an antidote ready should it be needed – chapter 7 of this book reports the only case I witnessed when one was used. Every effort was made to keep the sessions unstressful. Flowers and music were encouraged, and subjects were invited to surround themselves with meaningful artifacts – family photos, candles, icons, incense – if they chose to do so. Often the "laboratory" was the subject's own living room, and family and friends were welcome to be present. A follow-up report was required in which the subject was asked to describe the experience and retrospective feelings about it.

What follows is the report I turned in. Ralph Metzner got wind of it and included it in the anthology he published, The Ecstatic Adventure.

New Year's Day, 1961. Eleanor (who now answers to the name Kendra) and I reached the home of Dr. Timothy Leary in Newton, Massachusetts, about 12:30 P.M. Present in addition to Leary were Dr. George Alexander, psychiatrist, and Frank Barron, on sabbatical from the department of psychology at the University of California, Santa Cruz.

After coffee and pleasantries, Tim sprinkled some capsules of mescaline onto the coffee table and invited us to be his guest. One, he said, was a mild dose, two an average dose, and three a large dose. I took one; Eleanor, more venturesome, took two. After about half an hour, when nothing seemed to be happening, I too took a second capsule.

After what I estimate to have been about an hour, I noticed mounting tension in my body that turned into tremors in my legs. I went into the large living room and lay down on its couch. The tremors turned into twitches, though they were seldom visible.

It would be impossible for me to fix the time when I passed into the visionary state, for the transition was imperceptible. From here on time becomes irrelevant. With great effort I might be able to reconstruct the order in which my thoughts, all heavily laden with feelings, occurred, but there seems to be no point in trying to do so.

The world into which I was ushered was strange, weird, uncanny, significant, and terrifying beyond belief. Two things struck me especially. First, the mescaline acted as a psychological prism. It was as if the layers of the mind, most of whose contents our conscious mind screens out to smelt the remainder down into a single band we can cope with, were now revealed in their completeness – spread out as if by spectroscope into about five distinguishable layers. And the odd thing was that I could to some degree be aware of them all simultaneously, and could move back and forth among them at will, shifting my attention to now this one, now another one. Thus, I could hear distinctly the quiet conversation of Tim and Dr. Alexander in the adjoining study, and follow their discussion and even participate in it imaginatively. But this leads to the second marked feature. Though the five bands of consciousness – I say five roughly; they were not sharply divided and I made no attempt to count them – were all real, they were not of equal importance. I was experiencing the metaphysical theory known as emanationism, in which, beginning with the clear, unbroken Light of the Void, that light then fractures into multiple forms and declines in intensity as it devolves through descending levels of reality. My friends in the study were present in one band of this spectrum, but it was far more restricted than higher bands that were in view. Bergson's notion of the brain as a reducing valve struck me as accurate.

Along with "psychological prism," another phrase occurred to me: empirical metaphysics. Plotinus's emanation theory, and its more detailed Vedantic counterpart, had hitherto been only conceptual theories for me. Now I was *seeing* them, with their descending bands spread out before me. I found myself amused, thinking how duped historians of philosophy had been in crediting the originators of such worldviews with being speculative geniuses. Had they had experiences such as mine (subsequent chapters of this book suggest that they *had* had such experiences) they need have been no more than hack reporters. But beyond accounting for the origin of these philosophies, my experience supported their truth. As in Plato's myth of the cave, what I was now seeing struck me with the force of the sun, in comparison with which everyday experience reveals only flickering shadows in a dim cavern.

How could these layers upon layers, these worlds within worlds, these paradoxes in which I could be both myself *and* my world and an episode could be both momentary *and* eternal – how could such things be put into words? I realized how utterly impossible it would be for me to describe such things tomorrow, or even right then to Tim or Eleanor. There came the clearest realization I have ever had as to what literary genius consists of: a near-miraculous talent for using words to transport readers from the everyday world to things analogous to what I was now experiencing.

It should not be assumed from what I have written that the experience was pleasurable. The accurate words are significance and terror.* In *The Idea of the Holy,* Rudolf Otto describes awe as a distinctive blend of fear and fascination, and I was experiencing at peak level that paradoxical mix. The experience was momentous because it showed me range upon range of reality that previously I had only believed existed and tried without much success to imagine. Whence, then, the terror? In part, from my sense of the utter freedom of the psyche and its dominion over the body. I was aware of my body, laid out on the couch as if on an undertaker's slab, cool and slightly moist. But I also had the sense that it would reactivate only if my spirit chose to reenter it. Should it so choose? There seemed to be no clear reason for it to do so. Moreover, could it reconnect if I willed it to? We have it on good authority that no man can see God and live – the sight would be too much for the body to withstand, like plugging a toaster into a power line. I thought of trying to get up and walk across the floor. I suspected that I could do so, but I didn't want to risk forcing this intensity of experience into my physical frame. It might shatter the frame.

Later, after the peak had passed and I had walked a few steps, I said to Tim, "I hope you know what you're playing around with here. I realize I'm still under the influence and that things probably look different from your side, but it looks to me like you're taking an awful chance in these experiments. Objective tests might reveal that my heart has been beating normally this afternoon, but there *is* such a thing as people being frightened to death. I feel like I'm in an operating room, having barely squeaked through an ordeal in which for two hours my life hung in the balance."

I have said nothing about the visual. Where it was important,

* Years later when I came upon Rilke's *Duino Elegies* I realized that he must have been sensing something like what I was experiencing when he wrote,

> *. . . Beauty is nothing*
> *but the beginning of terror, which we still are just able to endure,*
> *and we are so awed because it serenely disdains*
> *to annihilate us.*
>
> (Stephen Mitchell's translation)

it was abstract. Lights such as never were on land or sea. And space – not three or four dimensions but more like twelve. When I focused visually on my physical surroundings, I tended to be uninterested. Shapes and colors, however intensified, had little to contribute to the problem that obsessed me, which was what this experience implied for the understanding of life and reality. So I regarded the visual as largely an intrusive distraction and tended to keep my eyes closed. Only twice did physical forms command my attention. Once was when Dr. Alexander induced me to look at the pattern a lampshade was throwing on a taupe rug. That *was* extraordinary; the shapes stood out like three-dimensional blocks. They also undulated like writhing serpents. The other time was involuntary, when the Christmas tree, its lights unlit, suddenly jumped out at me. It had been in my visual field much of the afternoon, but this was transfiguration. Had I not been in the room throughout, I would have said that someone had re-trimmed the tree, increasing its tinsel tenfold. Where before there was a tree with decorations, now there were decorations with a clotheshorse of a tree to support them.

Interactions with Eleanor, who had dived inward and was reliving important phases of her childhood, form a happy but separate and essentially personal story. Around 10:30 P.M. we drove back to our incomparable, never-more-precious children who were sleeping as if the world was as it had always been, which it definitely was not for us. Neither of us fell asleep until about five, whereupon we slept until around nine. I was definitely into the cold that had been coming on, but my head was clear.

CHAPTER TWO

Do Drugs Have Religious Import?

For several years following my initiation, the entheogens were the center of my reflective and social life. Reflectively, to have become overnight a visionary – one who not merely believes in the existence of a more momentous world than this one but who has actually visited it – was no small matter. How could what felt like an epochal change in my life have been crowded into a few hours and occasioned by a chemical? I knew how my M.I.T. colleagues – Hans-Lukas Teuber, its renowned experimental psychologist, and its equally legendary professor of microbiology, Jerome Lettvin – would answer that question. The mescaline had scrambled the synapses in the nerve connections of my brain, creating irregular associations between its centers for vision, alarm, euphoria, and excitement, et cetera, et cetera – we get the idea. I was not persuaded. Still, if chemistry does not tell the whole story, what is that story? And what part do chemicals, replacing angels as divine intermediaries, play in it?

Questions like these assaulted me with an urgency that reconstructed my social life. Family and friends remained in place, but beyond those I sought out associates who shared my compulsion to talk about and understand our shared secret. This is the stuff of which churches are made, and within the Harvard Project an ad hoc *"church" emerged. Its glue was our resistance to epiphenomenal, reductionistic explanations of our revelations, and our certainty – equal to that of Huxley, Hofmann, Wasson, and William James, the giants in whose footsteps we thought of ourselves as walking – that it was impossible to close our accounts with reality without taking these revelations into consideration. What to make of the entheogens was the question, and we lived for the times when, like Socrates and his friends, we could hang out together to talk.*

The Harvard Project was hospitable. Open-ended, it wanted to explore the effects of psychoactive chemicals in all promising directions, so our "church" had its blessing and benefactions. Once every month or so we gathered to take our sacramental in a vaguely ritualistic context – incense, candles, favorite poems, passages from sacred texts, and spontaneous inputs in the style of Quaker meetings. In between those "services" we gathered to talk philosophically. We were but one satellite in Leary's project, which served as an umbrella under which those of its subjects who wanted to follow up on their experiences clustered and networked. An organization sprang up, the International Federation for Internal Freedom, which for ten years published a journal, The Psychedelic Review. *It attracted some notable contributors, among them Robert Graves, Aldous Huxley, Albert Hofmann, and Gordon Wasson.*

Lisa Bieberman and Peter John (who went on to become a Methodist minister) deserve to be mentioned for holding our "church" together; without pay or public recognition, they gave virtually their full time to it. Readers of this book may recognize the names of several other members of our group for the parts they played in the history of the entheogens. Walter Houston Clark, professor of the psychology of religion at Andover-Newton Theological Seminary, became the most ardent crusader for the substances, arguing that they were the only things in sight that could restore the experiential component to the mainstream churches, without which they would not survive. Walter Pahnke's Good Friday Experiment is the subject of chapter 7 in this book; it is difficult to imagine how the history of the entheogens might have been different had he not died in a scuba-diving accident, for he brought to his serious involvement with mysticism the scientific training of a medical doctor and his intention to devote his career to studying the resources of chemicals for religion. Paul Lee, who at the time was Paul Tillich's teaching assistant at Harvard, went on to teach philosophy at the University of California, Santa Cruz, before leaving academia to devote himself full-time to studying herbs.

The essay that follows sets forth the conclusions I reached about the entheogens in the course of the three years that I was involved with the Harvard Project. Titled "Do Drugs Have Religious Import?" it appeared in the 1 October 1964 issue of The Journal of Philosophy. *I am told that it has been anthologized more times than any other article in that journal, over twenty times to date. I have made a few minor alterations to clarify points that might otherwise be obscure.*

Until six months ago, if I picked up my phone in the Cambridge area and dialed KISS-BIG, a voice would answer, "If-if." These were coincidences: KISS-BIG happened to be the letter equivalents of an arbitrarily assigned telephone number, and I.F.I.F. represented the initials of an organization with the improbable name of the International Federation for Internal Freedom. But the coincidences were apposite to the point of being poetic. "Kiss big" caught the euphoric, manic, life-embracing attitude that characterized this most publicized of the organizations formed to explore the newly synthesized consciousness-changing substances; and the organization itself was surely one of the "iffyest" phenomena to appear on our social and intellectual scene in some time. It produced the first firings in Harvard's history – of professors Timothy Leary and his associate, Richard Alpert – and an ensuing ultimatum to Leary to get out of Mexico in five days; and also "the Miracle of Marsh Chapel," in which, during a two-and-a-half-hour Good Friday service, fifteen theological students and professors ingested psilocybin and were visited by what they generally reported to be the deepest religious experiences of their lives.

Despite the last of these phenomena and its numerous, if less dramatic, parallels, students of religion appear to be dismissing the psychedelic drugs that have sprung to our attention in the sixties as having little religious relevance. The position taken in one of the most forward-looking volumes of theological essays to have appeared in recent years – *Soundings,* edited by A. R. Vidler[1] – accepts R. C. Zaehner's *Mysticism, Sacred and Profane* as having "fully examined and refuted" the religious claims for mescaline which Aldous Huxley sketched in *The Doors of Perception.* This closing of the case strikes me as premature, for it looks as if the drugs have light to throw on the history of religion, the phenomenology of religion, the philosophy of religion, and religious life itself.

1. *Drugs and Religion Viewed Historically*

In their trial-and-error life explorations, peoples almost everywhere have stumbled upon connections between vegetables (eaten or brewed) and actions (yogic breathing exercises; fast, whirling-dervish dances; self-flagellations) that induce dramatic alterations in consciousness. From the psychopharmacological standpoint we now understand these states to be the products of changes in brain chemistry. From the sociological perspective we see that they tend to be connected in some way with religion. If we discount the wine used in Christian communion services, the instances closest to us in time and space are the Peyote sacrament of the Native American Church and Mexico's two-thousand-year-old tradition of using sacred mushrooms, the latter rendered in Aztec as "God's Flesh," a striking parallel to "the body of our Lord" in the Christian Eucharist. Further away are the *soma* of the Hindus, the *haoma* and hemp of the Zoroastrians, Dionysus the Greek who "everywhere taught men the culture of the vine and the mysteries of his worship and everywhere was accepted as a god,"[2] the *benzoin* of Southeast Asia, Zen's tea whose fifth cup purifies and whose sixth "calls to the realm of the immortals,"[3] the *pituri* of the Australian aborigines, and probably the mystic *kykeon* that was drunk at the climactic close of the sixth day of the Eleusinian Mysteries.[4] There is no need to extend the list, as a reasonably complete account is given in Philippe de Felice's comprehensive study of the subject, *Poisons Sacrés, Ivresses Divines.*

More interesting than the fact that consciousness-changing devices have been linked with religion is the possibility that they may actually have initiated many of the religious perspectives which, taking root in history, continued after their entheogenic origins were forgotten. Bergson saw the first movement of Hindus and Greeks toward "dynamic religion" as associated with the "divine rapture" found in intoxicating beverages,[5] and more recently Robert Graves, Gordon Wasson, and Alan Watts have suggested that most religions have arisen from such chemically induced theophanies. Mary Barnard is the most explicit proponent of this thesis. "Which was more likely to happen first," she asks; "the spontaneously generated idea of an afterlife in which the disembodied soul, liberated from the restrictions of time and space, experiences eternal bliss, or the accidental discovery of hallucinogenic plants that give a sense of euphoria, dislocate the center of consciousness, and distort time and space, making them balloon outward in greatly expanded vistas?"[6] Her own answer is that "the latter experience might have had an almost explosive effect on the largely dormant minds of men, causing them to think of things they had never thought of before. This, if you like, is direct revelation." Her use of the subjunctive "might" renders this formulation of her answer equivocal, but she concludes her essay on a note that is categorical: "Looking at the matter coldly, unintoxicated and unentranced, I am willing to prophesy that fifty theobotanists working for fifty years would make the current theories concerning the origins of much mythology and theology as out-of-date as pre-Copernican astronomy."

This is an important hypothesis – one which must surely engage the attention of historians of religion for some time to come. But as I am content here to forego prophecy and limit myself to the points where the drugs surface in serious religious study, I shall not pursue Ms. Barnard's thesis. Having tagged what appears to be the crux of the historical question, namely the extent to which drugs not merely duplicate or simulate theologically sponsored experiences but generate or shape theologies themselves, I turn to phenomenology.

2. *Drugs and Religion Viewed Phenomenologically*

Phenomenology attempts a careful description of human experience. The question that drugs pose for the phenomenology of religion, therefore, is whether the experiences they induce differ from religious experiences reached without them and, if so, how.

Even the Bible notes that substance-altered psychic states bear *some* resemblance to religious ones. Peter had to appeal to a circumstantial point – the early hour of the day – to defend those who were caught up in the Pentecostal experience against the charge that they were merely drunk: "These men are not drunk, as you suppose, since it is only the third hour of the day" (Acts 2:15); and Paul initiates the comparison when he admonishes the Ephesians not to "get drunk with wine, but to be filled with the spirit" (Ephesians 5:18). Are such comparisons, which have counterparts in virtually every religion, superficial? How far can they be pushed?

Not all the way, students of religion have thus far insisted. With respect to the new drugs, Professor R. C. Zaehner has drawn the line emphatically. "The importance of Huxley's *Doors of Perception,*" he writes, "is that in it the author clearly makes the claim that what he experienced under the influence of mescaline is closely comparable to a genuine mystical experience. If he is right, the conclusions are alarming."[7] Zaehner thinks that Huxley is not right, but I believe that it is Zaehner who is mistaken.

There are, of course, innumerable drug experiences that have no religious features; they can be sensual as readily as spiritual, trivial as readily as transforming, capricious as readily as sacramental. If there is one point about which every student of psychoactivating agents agrees, it is that there is no such thing as *the* drug experience per se – no experience that the drugs, as it were, secrete. Every experience is a mix of three ingredients: drug, set (the psychological makeup of the individual), and setting (the social and physical environment in which it is taken). But given the right set and setting, the drugs can induce religious experiences that are indistinguishable from such experiences that occur spontaneously. Nor need the sets and settings be exceptional. The way the statistics are currently running, it looks as if from one-fourth to one-third of the general population will have religious experiences if they take certain drugs under naturalistic conditions, meaning conditions in which the researcher supports the subject but does not interfere with the direction the experience takes. Among subjects who have strong religious proclivities, the proportion of those who will have religious experiences jumps to three-fourths. If such subjects take the drugs in religious settings, the percentage soars to nine out of ten.

How do we know that the experiences these people have really are religious? We can begin with the fact that they say they are. The "one-fourth to one-third of the general population" figure is drawn from two sources. Ten months after they had had their experiences, 24 percent of the 194 subjects in a study by the California psychiatrist Oscar Janiger characterized their experiences as having been religious.[8] Thirty-two percent of the seventy-four subjects in Ditman and Hayman's study reported, looking back on their LSD experience, that it looked as if it had been "very much" or "quite a bit" a religious experience; 42 percent checked as true the statement that they "were left with a greater awareness of God, or a higher power, or ultimate reality."[9] The statement that three-fourths of subjects having religious "sets" will have religious experiences comes from the reports of sixty-nine religious professionals who took the drugs while the Harvard project was in progress.[10]

In the absence of (a) a single definition of religious experience generally acceptable to psychologists of religion, and (b) foolproof ways of ascertaining whether actual experiences exemplify any definition, I am not sure there is a better way of telling whether the experiences of the 333 men and women involved in the above studies were religious than by their reports that they seemed so to them. But if more rigorous methods are preferred, they exist; they have been utilized, and they confirm the conviction of the man in the street that drug experiences can indeed be religious. In his doctoral study at Harvard University, Walter Pahnke worked out a typology of religious experience (in this instance, mystical experience) based on classic reports that Walter Stace included in his *Mysticism and Philosophy.* Pahnke then administered psilocybin to fifteen theology professors and students (half of the total population of thirty) in the setting of a Good Friday service. The drug was given in a "double-blind" experiment, meaning that neither Dr. Pahnke nor his subjects knew which fifteen were getting psilocybin and which fifteen received placebos to constitute a control group. Subsequently the subjects' reports of their experiences were rated independently by three former schoolteachers on the degree (strong, moderate, slight, or none) to which each report evinced the nine traits of mystical experience that Stace enumerates. The results showed that "those subjects who received psilocybin experienced phenomena which were indistinguishable from, if not identical with, the categories defined by our typology of mysticism."[11]

With the thought that the reader might like to test his or her own powers of discernment on the question being considered, I insert here a simple test I gave to members of the Woodrow Wilson Society at Princeton University. I presented them with two accounts of religious experiences, one drug-occasioned, the other not, and asked them to guess which was which.

I

> Suddenly I burst into a vast, new, indescribably wonderful universe. Although I am writing this over a year later, the thrill of the surprise and amazement, the awesomeness of the revelation, the engulfment in an overwhelming feeling-wave of gratitude and blessed wonderment, are as fresh, and the memory of the experience is as vivid, as if it had happened five minutes ago. And yet to concoct anything by way of description that would even hint at the magnitude, the sense of ultimate reality . . . this seems such an impossible task. The knowledge which has infused and affected every aspect of my life came instantaneously and with such complete force of certainty that it was impossible, then or since, to doubt its validity.

II

> All at once, without warning of any kind, I found myself wrapped in a flame-colored cloud. For an instant I thought of fire . . . the next, I knew that the fire was within myself. Directly afterward there came upon me a sense of exultation, of immense joyousness accompanied or immediately followed by an intellectual illumination impossible to describe. Among other things, I did not merely come to believe, but I saw that the universe is not composed of dead matter, but is, on the contrary, a living Presence; I became conscious in myself of

> eternal life. . . . I saw that all men are immortal: that the cosmic order is such that without any peradventure all things work together for the good of each and all; that the foundation principle of the world . . . is what we call love, and that the happiness of each and all is in the long run absolutely certain.

On the occasion referred to, twice as many students (46) answered incorrectly as answered correctly (23). I bury the correct answer in an endnote to enable the reader to test himself on the question if he wishes to do so.[12]

Why, in the face of this considerable evidence, does Zaehner hold that drug experiences cannot be authentically religious? There appear to be three reasons:

1. His own experience was "utterly trivial." This of course proves that not all drug experiences are religious, but not that none is.

2. He thinks the experiences of others that appear religious to them are not truly so. Zaehner distinguishes three kinds of mysticism: nature mysticism, in which the soul is united with the natural world; monistic mysticism, in which the soul merges with an impersonal absolute; and theistic mysticism, in which the soul confronts the living, personal God. He concedes that drugs can induce the first two species of mysticism, but not its supreme instance, the theistic. As proof, he analyzes Huxley's experience as recounted in *The Doors of Perception* to show that it produced at best a blend of nature and monistic mysticism. Even if we were to accept Zaehner's evaluation of the three forms of mysticism, Huxley's case (and indeed Zaehner's entire book) would prove only that not every mystical experience induced by the drugs is theistic. Insofar as Zaehner goes beyond this to imply that drugs do not and cannot induce theistic mysticism, he not only goes beyond the evidence but proceeds in the face of it. James Slotkin reports that the Peyote Indians "see visions, which may be of Christ Himself. Sometimes they hear the voice of the Great Spirit.

Sometimes they become aware of the presence of God and of those personal shortcomings which must be corrected if they are to do His will."[13] And G. M. Carstairs, reporting on the use of *bhang* in India, quotes a Brahmin as saying, "It gives good *bhakti,*" *bhakti* being precisely Hinduism's theistic way of relating to the divine.[14]

3. There is a third reason why Zaehner might doubt that drugs can induce genuinely mystical experiences. Zaehner is a Roman Catholic, and Roman Catholic doctrine teaches that mystical rapture is a gift of grace and as such can never be brought under human control. This may be true; certainly the empirical evidence cited does not preclude the possibility of a genuine ontological or theological difference between natural and drug-induced religious experiences. At this point, however, we are considering phenomenology rather than ontology, description rather than truth-claims, and on this level there is no difference. Descriptively, drug experiences cannot be distinguished from their natural religious counterparts. When the current philosophical authority on mysticism, W. T. Stace, was asked whether the drug experience is similar to the mystical experience, he answered, "It's not a matter of its being *similar to* mystical experience; it is mystical experience."

What we seem to be witnessing in Zaehner's *Mysticism, Sacred and Profane* is a reenactment of the age-old pattern in the conflict between science and religion. Whenever a new controversy arises, religion's first impulse is to deny the disturbing evidence science has produced. Seen in perspective, Zaehner's refusal to admit that drugs can induce experiences descriptively indistinguishable from spontaneous ones is a current counterpart of the seventeenth-century theologians' initial refusal to accept the evidence for the Copernican revolution. When the fact that drugs can trigger religious experiences becomes incontrovertible, discussion will move to the more difficult question of how this fact is to be interpreted. That question leads beyond phenomenology into philosophy.

3. *Drugs and Religion Viewed Philosophically*

Why do people reject evidence? Presumably because they find it threatening. Theologians are not the only professionals who can be defensive. In his *Personal Knowledge,*[15] Michael Polanyi recounts the way the medical profession ignored such palpable facts as the painless amputation of human limbs performed before their own eyes in hundreds of successive cases, concluding that the subjects were impostors who were either deluding their physicians or colluding with them. One physician, Esdaile, carried out about three hundred major operations painlessly under mesmeric trance in India, but neither in India nor in Great Britain could he get medical journals to print accounts of his work. Polanyi attributes this closed-mindedness to "lack of a conceptual framework in which their discoveries could be separated from specious and untenable admixtures."

The "untenable admixture" in the fact that psychotomimetic drugs can induce religious experiences is its apparent implication: that religious disclosures are no more veridical than psychotic ones. For religious skeptics, this conclusion is of course not untenable at all; it fits perfectly with their thesis that *all* religion is at heart an escape from reality. Psychotics avoid reality by retiring into make-believe dream worlds; what better evidence could there be that religious visionaries do the same than the fact that their visions can be chemically induced as well? Marx appears to have been more than metaphorically accurate in proposing that religion is the "opiate" of the people. And Freud too was more right than he realized. He "never doubted that religious phenomena are to be understood only on the model of the neurotic symptoms of the individual."[16] He should have said "psychotic symptoms."

So the religious skeptic is likely to reason. What about the religious believer? Convinced that religious experiences are basically veridical, can he admit that psychoactive drugs can occasion them? To do so he needs (to return to Polanyi's words) "a conceptual framework in which [the discoveries can] be separated from specious and untenable admixtures," the "untenable admixture" here being the conclusion that religious experiences are in general delusory.

One way to effect the separation would be to argue that, despite phenomenological similarities between natural and drug-induced religious experiences, they are separated by a crucial *ontological* difference. Such an argument would follow the pattern of theologians who argue for the "real presence" of Christ's body and blood in the bread and wine of the Eucharist despite their admission that chemical analysis, confined as it is to the level of "accidents" rather than "essences," would not disclose this presence. But this distinction will not appeal to many today, for it turns on an essence-accident metaphysics which is not widely accepted. Instead of fighting a rear-guard action by insisting that if drug and non-drug religious experiences cannot be distinguished empirically there must be some transempirical factor that distinguishes them and renders the drug experience profane, I wish to explore the possibility of validating drug-induced religious experiences on grounds that they come up with the same basic claims about reality that religions always do.

To begin with the weakest of all arguments, the argument from authority, William James (whom I class among the religious for his sensibilities) did not discount *his* insights that occurred while his brain chemistry was altered. The paragraph in which he retrospectively evaluates his nitrous oxide experiences has become classic, and I quote it here for its pertinence to the point under consideration.

> One conclusion was forced upon my mind at that time, and my impression of its truth has ever since remained unshaken. It is that our normal waking consciousness, rational consciousness as we call it, is but one special type of consciousness, whilst all about it, parted from it by the filmiest of screens, there lie potential forms of consciousness entirely different. We may go through life without suspecting their existence; but apply the requisite stimulus, and at a touch they are there in all their completeness, definite types of mentality which probably somewhere have their field of application and

> adaptation. No account of the universe in its totality can be final which leaves these other forms of consciousness quite disregarded. How to regard them is the question – for they are so discontinuous with ordinary consciousness. Yet they may determine attitudes though they cannot furnish formulas, and open a region though they fail to give a map. At any rate, they forbid a premature closing of our accounts with reality. Looking back on my own experiences, they all converge toward a kind of insight to which I cannot help ascribing some metaphysical significance.[17]

To this argument from authority, I add two arguments that try to provide something by ways of reasons. Drug experiences that assume a religious cast tend to have fearful and/or beatific features, and my hypotheses relate to these two features of the experience.

Beginning with the ominous, "fear of the Lord" awe-full feature, I have already registered in the frontispiece of this book Gordon Wasson's account, which (being short) I reenter here. "Ecstasy! In common parlance ecstasy is fun. But ecstasy is not fun. Your very soul is seized and shaken until it tingles. After all, who will choose to feel undiluted awe? The unknowing vulgar abuse the word; we must recapture its full and terrifying sense."[18]

Emotionally the drug experience can be like having forty-foot waves crash over you for several hours while you cling desperately to a life raft which may be swept from under you at any moment. It seems quite possible that such an ordeal, like any experience of a close call, could awaken fundamental sentiments respecting life, death, and destiny, and trigger the "no atheists in foxholes" effect. Similarly, as the subject emerges from the ordeal and realizes that he will not be permanently insane as he had feared, he may experience waves of overwhelming relief and gratitude like those that patients recovering from critical illnesses frequently report. Here is one such report.

> It happened on the day when my bed was pushed out of doors to the open gallery of the hospital. I cannot now recall whether the revelation came suddenly or gradually; I only remember

> finding myself in the very midst of those wonderful moments, beholding life for the first time in all its young intoxication of loveliness, in its unspeakable joy, beauty, and importance. I cannot say exactly what the mysterious change was. I saw no new thing, but I saw all the usual things in a miraculous new light – in what I believe is their true light. I saw for the first time how wildly beautiful and joyous, beyond any words of mine to describe, is the whole of life. Every human being moving across that porch, every sparrow that flew, every branch tossing in the wind, was caught in and was a part of the whole mad ecstasy of loveliness, of joy, of importance, of intoxication of life.[19]

If we do not discount religious intuitions because they are prompted by battlefields and physical crises; if we regard the latter as "calling us to our senses" more often than they seduce us into delusions, need comparable intuitions be discounted simply because the crises that trigger them are of an inner, psychic variety?

Turning from the fearful to the beatific aspects of the drug experience, some of the latter may be explainable by the hypothesis just stated; that is, they may be occasioned by the relief that attends the sense of escape from high danger. But this hypothesis cannot possibly account for all of the blissful episodes that chemicals occasion for the simple reason that the positive episodes often come first, or to persons who experience no negative episodes whatever. Dr. Sanford Unger of the National Institute of Mental Health reports that among his subjects "50 to 60% will not manifest any real disturbance worthy of discussion, yet around 75% will have at least one episode in which exaltation, rapture, and joy are the key descriptions."[20] How are we to account for the drug's capacity to induce peak experiences, such as the following, which are not preceded by fear?

> A feeling of great peace and contentment seemed to flow through my entire body. All sound ceased and I seemed to be floating in a great, very very still void or hemisphere. It is impossible to describe the overpowering feeling of peace, contentment, and being a part of goodness itself that I felt. I could feel

> my body dissolving and actually becoming a part of the goodness and peace that was all around me. Words can't describe this. I feel an awe and wonder that such a feeling could have occurred to me.[21]

Consider the following argument: Like every other form of life, human nature has become distinctive through specialization. Human beings have specialized in developing a cerebral cortex. The analytic powers of this instrument are a standing wonder, but the instrument seems less able to provide people with the sense that they are meaningfully related to their environment: to life, the world, and history in their wholeness. As Albert Camus describes the situation, "If I were a cat among animals, this life would have a meaning, or rather this problem would not arise, for I should belong to this world. I would *be* this world to which I am now opposed by my whole consciousness."[22] Note that it is Camus's consciousness that opposes him to his world. The drugs do not knock this consciousness out, but while they leave it operative they also activate areas of the brain that normally lie below its threshold of awareness. One of the clearest objective signs that the drugs are taking effect is the dilation they produce in the pupils of the eyes, and one of the most predictable subjective signs is the intensification of visual perception. Both of these responses are controlled by portions of the brain that lie deep, further to the rear than the mechanisms that govern consciousness. Meanwhile we know that the human organism is joined to its world in innumerable ways that our senses do not pick up – through gravitational fields, bodily respiration, and the like: the list could be multiplied until the human skin begins to look more like a traffic maze than a boundary. Perhaps the deeper regions of the brain which evolved earlier and are more like those of the lower animals – "If I were a cat I should belong to this world" – can sense this relatedness better than can the cerebral cortex which now demands our attention. If so, when the drugs rearrange the neurohumors that chemically transmit impulses across synapses between neurons, human consciousness and its submerged, intuitive, ecological awareness might for a spell become interlaced. This is, of course, no more than a hypothesis, but how else are we to account for the extraordinary incidence under the drugs of that kind of insight the keynote of which James described as "invariably a reconciliation"? "It is as if," he continues, "the opposites of the world, whose contradictoriness and conflict make all our difficulties and troubles, were melted into one and the same genus, but one of the species, the nobler and better one, is itself the genus, and so soaks up and absorbs its opposites into itself."[23]

4. *Drugs and Religion Viewed "Religiously"*

Suppose that drugs can induce experiences indistinguishable from religious experiences and that we can respect their reports. Do they shed any light, I now ask, not on life, but on the nature of the religious life?

One thing they may do is throw religious experience itself into perspective by clarifying its relation to the religious life as a whole. Drugs appear to be able to induce religious experiences; it is less evident that they can produce religious lives. It follows that religion is more than a string of experiences. This is hardly news, but it may be a useful reminder, especially to those who incline toward "the religion of religious experience"; which is to say toward lives bent on the acquisition of desired states of experience irrespective of their relation to life's other demands and components.

Despite the dangers of the "faculty psychology" that was in vogue in the first half of this century, it remains useful to regard human beings as having minds, wills, and feelings. One of the lessons of religious history is that to be adequate, a faith must activate all three components of human nature. Overly rationalistic religions grow arid, and moralistic ones grow leaden. Those that emphasize experience have their comparable pitfalls, as evidenced by Taoism's struggle (not always successful) to keep from degenerating into quietism, and the vehemence with which Zen Buddhism insists that once students have attained *satori*, they must be driven out of it and back into the world. The case of Zen is especially pertinent here, for it pivots on an enlightenment experience – *satori*, or *kensho* – which some (but not all) Zen Buddhists say resembles the LSD experience. Alike or different, the point is that Zen recognizes that unless the experience is joined to discipline, it will come to naught.

> Even the Buddha continued to sit. Without *joriki*, the particular power developed through *zazen* [seated meditation], the vision of oneness attained in enlightenment in time becomes clouded and eventually fades into a pleasant memory instead of remaining an omnipresent reality shaping our daily life. To be able to live in accordance with what the mind's eye has revealed through *satori* requires, like the purification of character and the development of personality, a ripening period of *zazen*.[24]

If the religion of religious experience is a snare and a delusion, it follows that no religion that fixes its faith primarily in substances that induce religious experiences can be expected to come to a good end. What promised to be a shortcut will prove to be a short circuit; what began as a religion will end as a religious surrogate. Whether chemical substances can be helpful adjuncts to faith is another question. The Peyote-using Native American Church seems to indicate that they can be; anthropologists give this church a good report, noting among other things that its members resist alcohol better than do nonmembers.[25] The conclusion to which the evidence seems currently to point is that it is indeed possible for chemicals to enhance the religious life, but only when they are set within the context of faith (conviction that what they disclose is true) and discipline (exercise of the will toward fulfilling what the disclosures ask of us).

Nowhere today in Western civilization are both of these conditions met. Faith has declined in churches and synagogues, and the counterculture lacks discipline. This might lead us to forget about the drugs, were it not for the fact that the distinctive religious emotion, and the one that drugs can unquestionably occasion – Otto's *mysterium tremendum, majestas, mysterium fascinans;* in a phrase, the phenomenon of religious awe – seems to be declining sharply. As

Paul Tillich said in an address to the Hillel Society at Harvard several years ago:

> The question our century puts before us is: Is it possible to regain the lost dimension, the encounter with the Holy, the dimension that cuts through the world of subjectivity and objectivity and goes down to that which is not world but is the mystery of the Ground of Being?

Tillich may be right; this may be the religious question of our century. For if (as I have argued) religion cannot be equated with religious experiences, neither can it long survive their absence.

CHAPTER THREE

Psychedelic Theophanies and the Religious Life

During the three years that followed the writing of the preceding essay, the hopes that had attended Timothy Leary's Harvard project began to fade. Albert Hofmann's LSD: My Problem Child *pretty much tells the story. Hofmann, the discoverer of LSD, had had high hopes for his compound (mainly as an adjunct to psychiatry), but he became alarmed when it proved to be no more containable than other drugs and was pounced on for recreational use. A drug culture emerged as a component of the counterculture that arose to protest the war in Vietnam, the backlash to the civil rights movement that led to the assassination of Martin Luther King, and disillusionment with a science that had created the atomic bomb, napalm, and environmental pollution. Timothy Leary was fired from Harvard University for transgressing the agreement that undergraduate students would not be involved in his program. Irish rebel that he was at heart, he joined the counterculture, became its idol, and coined its famous slogan, "Turn on, tune in, drop out."*

These developments called for a reassessment of my initial, rather optimistic appraisal of the promise entheogens hold for religion. Everything I had said in my Journal of Philosophy *essay still struck me as true, but I came to feel that the distinction between religious experiences and the religious life needed to be emphasized more than it was in that piece. This next essay, which appeared in* Christianity and Crisis *in 1967, supplies that missing emphasis. Written later in the sixties, it contains more social history – the history of that tumultuous decade – than do the other essays in this book. As with the other essays, I have made minor changes in the original text.*

Psychedelic experiences can be religious. Subjects often say that they are, and their reports can read like accounts of classic theophanies.

What concerns me here is their staying power. No theophany is certain to retain its force – backsliding, falling from grace, and the psalmist's lament, "restore unto me the joy of my salvation," are not discoveries of the psychedelic age. And psychedelic theophanies can have *some* staying power: among alcoholics (Saskatchewan), lawbreakers (Laguna Beach, California), severe neurotics (Spring Grove State Hospital, Maryland), and terminal cancer patients (Chicago and Spring Grove). Nevertheless, I suspect that psychedelic religious experiences are having, and for the foreseeable future will continue to have, less faith-filled carryover than those that occur spontaneously.

I say this less for inductive than for deductive reasons. With too little available data at this point for an inductive conclusion, my pessimism derives from the following syllogism:

Major premise: Religious history suggests that for theophanies to take hold, certain conditions must obtain.

Minor premise: Those conditions are lacking in the psychedelic movement.

Conclusion: Psychedelic theophanies are not likely at this juncture of history to have substantial staying power.

What are the conditions that are needed for theophanies to take hold? The most important one is conviction, carrying over into the non-drug state, that the insights that emerge in the theophany are true.

That a theophany's disclosures seem true while the theophany is in progress follows by definition: it would not be a theophany but some other kind of experience were it to be otherwise. As René Daumal says: "At that moment comes *certainty.* Speech must now be content to wheel in circles around that bare fact." The experience's content is certain because doubts could enter only from the perspective of this world, which world pales before (where it is not obliterated by) the world into which the see-er has stepped.

Except in the tragic case of psychotics, however, this world eventually reasserts itself and its claims press hard upon us, which claims in our culture challenge the validity of pharmacological theophanies. The two foremost Western models of the mind are those of hard science (artificial intelligence and its chemical counterpart) and soft science (depth psychology with Freudianism in the lead). Both of these stand ready to explain chemically assisted convictions in ways that explain them away.

Artificial Intelligence and the Freudian Model

Is one moved in the course of a psychedelic experience to the conviction that everything is perfect just as it is; that (as Hakuin put it in his "Song in Praise of Zazen") "this earth on which we stand / is the promised Lotus Land / and this very body is the body of the Buddha"? The cyberneticist will tell us why. What happens to produce that "realization" is that in the neurological reshuffling that LSD occasions, the conviction center of our brain flip-flops to yes/green/go while being wired to the euphoria center. Meanwhile, contravening impressions are short-circuited and shut out. This last is an important point, for it challenges the all-too-common, too vague, too uncritical claim that psychedelics expand consciousness. The balloon image that that phrase suggests is less apposite than that of a microscope. Does a microscope enlarge our vision? Yes, by enabling us to see deeper into nature; but concomitantly no, for it shuts out everything but the microscopic object at which we are looking. According to the artificial-intelligence account, everything seems wonderful because at the moment in question euphoria fills our horizon. The entire world seems wonderful because the world has been collapsed to include only the rose-tinted things we have in mind at the moment.

As for the noetic property of the experience – the conviction that what one witnesses are not subjective phantasms but realities that exist objectively in the world, William James listed this as one of the four marks of mystical experience, but artificial intelligence explains it away on grounds that novel experiences hit us with exceptional force because they are not glazed over by habituation, the loss in vividness that repetition breeds.

The Freudian proceeds by a different route, but he too can discredit the experiences of ego-transcendence, the dissolution of the subject/object dichotomy, and the peace that passeth understanding that entheogens occasion. They are variations of the "oceanic feeling" that results when subjects regress psychologically to their mothers' wombs.

I do not defend these contemporary models of the mind; I merely point out that they saturate contemporary thinking to the point that even those who consciously reject them are imprinted by them to a large degree. This makes it difficult to accept psychedelic theophanies at face value, but is not history studded with examples of sects that have managed to survive while holding beliefs that challenge reigning conventions?

The point is well taken, and it takes me to my second reason for doubting the staying power of psychedelic experiences. History shows that minority faiths are viable, but only when they are cradled in communities that are solid and structured enough to constitute what in effect are churches. To date, the psychedelic movement shows no signs of having within it the makings of such a church. Sporadic "happenings" in makeshift quarters, and periodic gestures toward institutionalization, do not challenge this assertion; they confirm it by their ineffectiveness.

I say "no signs of the makings of a church," but perhaps I should qualify that in one respect. The psychedelic movement does have a charismatic leader: a man of intelligence, culture, and charm who is completely self-assured and apparently absolutely fearless. When Arthur Kleps, head of a branch of the short-lived Neo-American Church, testified before the Special Senate Judiciary Subcommittee on Narcotics that "we regard Dr. Timothy Leary with the same special love and respect as was reserved by the early Christians for Jesus, by Muslims for Mohammed, and by Buddhists for Gautama," we sensed the presence of charisma, the magnetism of a person who is regarded by his followers as an embodiment of spiritual power.

A charismatic leader is a great asset to a movement, but his presence is not sufficient to insure its success. And the psychedelic movement possesses other features which, if religious history provides grounds for prediction, augur against its becoming a genuine church. It lacks a social philosophy. It is antinomian. And, ignoring the Taoist adage to know ten things but tell only nine, it draws no line between the exoteric (what can appropriately be made public) and the esoteric, which should be reserved for the initiated.

I begin with the first of these three lacks.

A flawed social program

The psychedelic movement lacks a blueprint for relating itself to society. That it rejects the claims of mainstream culture is glaringly apparent. "You must quit your attachments to American society," wrote Leary in the first installment of his column syndicated by the counterculture's *East Village Other* and carried by the *L.A. Free Press,* the *Fifth Estate,* and *The Paper.*

Many early Christians adopted a comparable stance. The author of The New Testament's First Letter of John admonishes his readers not to "love the world or the things in the world" (2:15). He condemns the world for "its sensuality, superficiality and pretentiousness, its materialism and its egoism" (C. H. Dodd's summary in *The Johannine Epistles*). I have not heard psychedelic spokesmen criticize today's establishment for its sensuality, but otherwise the parallel is exact, as are those between some of Leary's writings and Tertullian's directives to the Christian church as it confronted the Roman Empire.

Tertullian	*Leary*
Political life is to be eschewed.	It is impossible to live conventionally on this planet without joining the antilife social systems. So drop out.

Tertullian	*Leary*
Trade is scarcely fitting for a servant of God, for apart from covetousness there is no real motive for engaging in it.	American social institutions lust for material things, so quit your job. For good.
Academicians, typified by the philosophers, have nothing in common with the disciples of heaven. They corrupt the truth. They seek their own fame. They are talkers rather than doers.	Present educational methods are neurologically crippling and antagonistic to your cellular wisdom. Quit school. For good.

Equally striking is the parallel between the reason Clement gives for why Christians should sit loose to society – they represent a new race of people – and the claims of the psychedelic prophets who regard persons under thirty as a new breed, a mutation in human evolution.

These similarities are impressive, but there is an important difference. Early Christians were apocalyptic; they expected the imminent end of history through divine intervention. This gave them a philosophy of history that underwrote their opposition to a social order. The psychedelic movement has no such philosophy.

Obviously, apocalypticism is not the only alternative to a prevailing social system; there is the possibility of improvements that people themselves effect. If the psychedelic leaders had a social philosophy that pointed in this direction, they would be (like Muhammad) not just rebels but revolutionaries. They have no such philosophy.

There is a third possibility. Neither apocalyptic nor revolutionary, the psychedelic movement might be utopian in seeking to create humane enclaves within a society that as a whole is considered to be beyond redemption. The utopian tradition has a respectable history in the West; the nineteenth century witnessed more than two hundred utopian ventures in the United States alone. But to date the psychedelic movement has failed to create a viable utopian community. Several attempts have been made, but they have been short-lived.

If the psychedelic movement were apocalyptic, revolutionary, or utopian, it would present an alternative to the status quo. Being none of these, its social message comes down to "Quit school. Quit your job. Drop out." The slogan is too negative to command respect.

Antinomianism

The psychedelic movement is antinomian. Derived from the Greek word meaning "law," antinomianism is the belief that it is possible to advance in virtue to a point where one stands above the law and is entitled to lay aside its commands in the name of a higher morality.

An historical example that sheds light on the psychedelic movement's antinomian tendencies and the problems that attend that stance is the Oneida Community, a New York product of the religious revival that swept America in the 1830s. Its founder, John Humphrey Noyes, became convinced as a result of his religious awakening and subsequent studies that the second coming of Christ did not lie in the future but had occurred at the close of the Apostolic Age. He found the corollaries of this conclusion to be far-reaching. First, original sin having been effectively eradicated by Christ's unrecognized return, nothing remained in human nature to keep people from living perfectly right now. Second, once (through the exercise of understanding and resolve) human beings do begin to live perfectly, no external guides for living can rival their own consciences and intelligence.

Noyes was absolutely convinced that salvation without the law presented the central idea of Christ's gospel. But, as his biographer, G. W. Noyes, points out, he found the doctrine "exceedingly liable to be perverted," and the perversions produced three disruptive forces in his community: sexual irresponsibility, anarchy, and lethargy. Noyes's descriptions of these problems read like a warning to the psychedelic movement. According to the Oneida Community's founder,

> right action had two essential components, right intent and intelligence. Since these internal monitors might conflict with external law, right action without freedom from external law was a contradiction in terms. Hence Noyes and his followers, though brought up in the strictest school of New England morality, declared themselves free from law. But here a new danger appeared. In escaping from the law, many of the Perfectionists, like the medieval mystics, fell into antinomianism. Antinomianism takes different forms according to the temperamental susceptibilities of its subjects. In those inclined to sensuality it takes the form of lasciviousness; in those whose leading trait is self-esteem, it takes the form of anti-organization; in those of an indolent disposition it takes the form of lethargy. During the prevalence of the antinomian aberration in 1835–1836 it seemed as though the Oneida Community would be completely given over to anarchy and imbecility. The Perfectionists did not abandon the principle of freedom from law, but they were brought gradually to the conviction that even the spiritually-minded in the present stage of human development needed to be restrained by moral forces which, though consistent with personal freedom, were nevertheless in effect equivalent to the law.

I commend to the psychedelic movement the example of John Humphrey Noyes. Here was a man who, in the name of religious convictions, advocated practices in comparison with which drug-taking seems tame. In the interests of spiritual eugenics he advocated what the law defined then and continues to define as adultery and bastardy. Yet – and here is the genius of the man – on this socially scandalous platform and in the face of enormous public opposition, he founded a community of several hundred persons that prospered financially (we still use its silverware) and spanned three generations, as against the average three-year life-span for utopian communities. An important factor in his success was his facing up to the tension between freedom and antinomianism in a manner more substantial than I find in his contemporary counterparts.

Comparable lessons come from Asia, the source of much of the psychedelic movement's inspiration. The West picks up Zen's endorsements of freedom and spontaneity enthusiastically, and Zen does indeed celebrate these qualities extravagantly. But careful reading of the record shows that it presents them as the crown of years of arduous discipline. Kenneth Ch'en's *Buddhism in China* points out that advancement in the hierarchy of Ch'an Buddhism was contingent on a dozen or so years of intensive study of, and discipline in, the *vinaya,* the first and moral disciplines basket of the three-basket *Tipitaka,* a work so detailed in its moral injunctions that it would fill about a dozen good-sized volumes in a western library. D. T. Suzuki picks up the thread and adds that even with this requirement in place, Ch'an and Zen still had to exercise great vigilance to keep from degenerating into stultifying passivity, the heresy known as quietism. To guard against this dry rot, Ch'an leaders introduced the requirement that monks work. The first rule of Ch'an monastic life was "A day of no work is a day of no eating."

This prompts me to ask if the corollary of "turn on" and "tune in" has to be "drop out"? To appeal to Asian traditions in support of antinomianism – be the traditions Ch'an, Zen, Tibetan Buddhism, or even Tantra in its Hindu and Buddhist versions – is to take their names in vain.

The esoteric/exoteric divide

To argue that there are things in religion that are best kept secret cuts against our democratic grain, yet tested religions do so argue.* There are pearls which, cast before swine, will be damaged themselves (by trampling) or damage the swine (should the swine eat them). In its early centuries, Christianity reserved a number of

* As do tested philosophies at times. A major Platonic scholar, J. N. Findlay, says that the two following sentences from Plato's Seventh Letter should be put at the head of every translation of Plato's works: "The publicizing of those secrets [with which that letter is concerned] I do not deem a boon for men, excepting for those initiates who are able to discover them with no more than a given hint. For the others it would produce a stupid derision or else a self-glorification in a mistaken idea that they have eaten wisdom with spoons."

its dogmas for those who had undergone probationary instruction and been baptized. The promise that was exacted of them, "I will not speak of thy mysteries to thine enemies," still appears in orthodox Christian liturgies.

Asia followed the same route. India developed the guru system in which disclosures to disciples are calibrated to the disciples' capacities to comprehend them. In the *Bhagavad Gita,* Krishna forbids imparting higher knowledge to those who are not ready for it, and in the *Katha Upanisad,* Yama tests Nachiketa's fitness in various ways before consenting to impart his highest teachings. In the *Taittiriya Upanisad,* Varuna puts off his son Bhrigu four times (with instructions to kindle the fire of his soul by aspiration, self-discipline, and meditation) before imparting to him the knowledge of Brahman. In raja yoga, *yama* and *niyama,* preliminary rules of discipline and purification are held to be indispensable prerequisites. India honors higher states of consciousness fully as much as today's psychedelic proponents do, but insists that if they are accessed by persons who are unprepared for them, one of two things will happen. Either (as I have said) the subject will be damaged, or the significance of the experience will be missed and the encounter trivialized. Thus either the subject is damaged or the *dharma* is damaged, usually both. The psychedelic movement pays lip service to these dangers by advising screening and preparing subjects, but on the whole it honors the esoteric/exoteric distinction only perfunctorily.

Inability to integrate the psychedelic experience with daily life is not without precedent. In the Ch'an/Zen tradition, early texts (as John Blofeld has pointed out) tend to cite *satori* as the goal of training. Later texts do not. The reason seems clear. When *satori* first arrives, its momentousness is likely to make it seem ultimate, the be-all and end-all of existence. As life goes on, however, one recognizes that this is not the case. Routine reasserts itself, and one discovers that even those who have had powerful *satoris* sometimes misbehave. At this point there enters the realization that comes to be stressed increasingly in later texts. In those texts *satori* is not the goal; it is the first major hurdle in the unending endeavor to work the *satori* experience into the fabric of one's daily life until one's entire life takes on a *satoric* quality. "Drawing water, hewing wood – this is the supernatural power; this the marvelous activity."[1]

Despite the fact that I do not see within the psychedelic movement the makings of a viable church, I hope that (as legal use of the entheogens seems destined for the immediate future to be restricted to research) "religious research" will not be considered a contradiction in terms. If a sincere group wishing to use the entheogens for genuinely religious purposes were permitted to do so while qualified observers kept close check on what happens to the group and in the individual lives of its members, the results would at least be interesting, and might be instructive.

That is my conclusion. I append a short postscript.

Strange things seem to be happening to human religiousness in our time, especially among the young. On the one hand, students are making a left end run around the prophetic, this-worldly wing of institutional religion to tackle directly such issues as Vietnam, racial justice, and the problems of poverty. This has been evident for several years.

What is new is that they are now making a right end run around the priestly, other-worldly wing of institutional religion to link up with Zen, Tibetan, and Asian gurus of wide variety, Native American spirituality, channeling, and the New Age cornucopia generally, including pharmacological mysticism. Theological supernaturalism is being replaced by psychological supernaturalism, defined as belief in the existence of paranormal powers and their deliverances.

Whether the current chapter of human religiousness is being written more in churches and ecumenical councils or on college campuses and in experimental communes is a question whose answer is blowing in the wind.

CHAPTER FOUR

Historical Evidence: India's Sacred Soma

This essay steps back from the current scene to see what history can tell us about the entheogens. Preceding chapters have touched on that question, but this one zeroes in on an important chapter in religious history – Hinduism in its formative period – to treat it as a case study.

The contextual facts are these: Among the gods of the Vedic pantheon, Soma appears to have been the most revered. His home was a plant, and in the holiest of rituals priests ingested the god by drinking a brew that was made from this plant. Somewhere along the way the plant's identity was lost, and any Indologist who retrieved it was assured of a permanent place in the annals of his discipline. It came as a surprise, therefore, when the prize went to an amateur – a retired banker named R. Gordon Wasson.

I considered the subject sufficiently important to give a summer to researching it, and this essay reports my findings. Its copious footnotes show that (like the second essay in this book) it was written for an academic journal; titled "Wasson's SOMA: A Review Article," it appeared in the December 1972 issue of the Journal of the American Academy of Religion.[1] *I pause for a moment to indulge myself. Because my best-known work,* The World's Religions, *is an undergraduate text that appeared early in my career, I have had to struggle against the fear – self-imposed perhaps, but real nonetheless – of being written off by my colleagues as a popularizer. It has, therefore, encouraged me no end that the foremost linguist of my time, Roman Jakobson, called this essay "a magnificent survey," and that one of the two leading historians of religion in my generation, Wilfred Cantwell Smith (the other was Mircea Eliade) credited it with being "a model of a piece: superbly organized, marvelously informative, engagingly written, and altogether exactly right."*

If I were writing it today, I would have to temper my claim that Gordon Wasson solved the soma enigma conclusively. The quarter-

century that has elapsed has brought new criticisms of his arguments, and rival candidates for the soma plant have been proposed. I continue to think that Wasson's arguments for his candidate are the strongest in the field, but the debate continues.

To get an immediate sense of the relevance of Wasson's work for this book, I suggest that the reader begin by reading the long footnote that appears on page 51. It is as compelling an account of the entheogenic experience as I know.

I have spent so much time recently reviewing the work of others that I am growing impatient to get on with my own pursuits, but the thesis here considered is important enough to warrant another detour. Moreover, the excursion is bound to prove interesting, for it leads through one of the most colorful intellectual exploits of our century.

Having mentioned both importance and interest, let me begin with the former. Alfred North Whitehead is reported to have remarked that Vedanta is the most impressive metaphysics the human mind has conceived.[2] The extent to which it may have influenced our own western outlook after Alexander's invasion of India does not concern us here; what is at issue is its origins.

Etymologically and otherwise, Vedanta is "the culmination of the Vedas," and the Vedas derive, more than from any other single identifiable source, from Soma.* Would it not be useful, then, to

* As this statement may seem excessively categorical, I give my reasons for it. Soma enjoys a special place in the Vedic pantheon. I will indicate the specifics of that place shortly, but let me acknowledge that its position warrants my allegation only when supported by recognition of the extent to which the Upanisadic metaphysics could have been facilitated by the entheogen that Soma was, and in the Vedas was exclusively. My arguments supporting this recognition fall into three categories: personal experience, the role of the entheogens in engendering religious perspectives generally, and the distinctive character of the Soma experience in Vedic religion.

(a) Personal experience. I quote from the account of my own first ingestion of an entheogen, mescaline. "Another phrase came to me: 'empirical metaphysics.' The emanation theory and elaborately delineated layers of Indian cosmology and psychology had hitherto been concepts and inferences. Now they were objects of direct, immediate perception. I saw that theories such as these were required by the experience I was having. I found myself amused, thinking how duped historians of philosophy had been in crediting those who formulated such worldviews with being speculative geniuses. Had they had experiences such as mine they need have been no know what Soma was? Not particularly, India herself seems to have answered, judging from her scholars' lack of interest in identifying

more than hack reporters. Beyond accounting for the origin of these philosophies, my experience supported their truth. As in Plato's myth of the cave, what I was now seeing struck me with the force of the sun in comparison with which normal experience was flickering shadows on the wall" ("Empirical Metaphysics," in Ralph Metzner, ed., *The Ecstatic Adventure* [New York: The Macmillan Company, 1968], p. 73).

(b) On the role of entheogens in occasioning religious purviews generally, I quote again, as I did in an earlier essay, Mary Barnard who asks, "Which was more likely to happen first: the spontaneously generated idea of an afterlife in which the disembodied soul, liberated from the restrictions of time and space, experiences eternal bliss, or the accidental discovery of hallucinogenic plants that give a sense of euphoria, dislocate the center of consciousness, and distort time and space, making them balloon outward in greatly expanded vistas? The [latter] experience might have had an almost explosive effect on the largely dormant minds of men, causing them to think of things they had never thought of before. [I interrupt to note that in reading for the present review I came across a pointed support of Ms. Barnard's conjecture, specifically the part connecting the concept of an afterlife to hallucinogens. Concerning certain Algonquin Indians in the region of Quebec, Father Charles Lallemand wrote in 1626, "They believe in the immortality of the Soul; and in troth they so assert that after death they go to Heaven, where they do eat Mushrooms" (21).] Looking at the matter coldly, unintoxicated and unentranced, I am willing to prophesy that fifty theo-botanists working for fifty years would make the current theories concerning the origins of much mythology and theology as out-of-date as pre-Copernician astronomy," *The Mythmakers* (Athens, Ohio University Press, 1966), pp. 21–22, 24. On the same theme, by the author of the book under review: "As man emerged from his brutish past there was a stage in the evolution of his awareness when the discovery of [an indole] with miraculous properties was a revelation to him, a veritable detonator to his soul, arousing in him sentiments of awe, reverence, gentleness and love, to the highest pitch of which mankind is capable, all those sentiments and virtues that mankind his ever since regarded as the highest attribute of his kind. It made him see what this perishing mortal eye cannot see. What today is resolved into a mere drug was for him a prodigious miracle, inspiring in him poetry and philosophy and religion" (1:162). (Numbers preceding colons refer to numbered items in the bibliography; those following colons to page numbers therein.)

(c) Finally, on the specific place of the entheogen experience in Vedic religion, these words by Daniel Ingalls, Wales Professor of Sanskrit at Harvard University, written to register a perception that came to him on reading through Book IX, the Soma Book, of the Rig-Veda after reading Wasson's book here under review. "Soma and Agni represent the two great roads between this world and the other world. They are the great channels of communication between the human and the divine." But, Ingalls goes on to note, there is a difference. "The Agni hymns seek for a harmony between this world and the sacred, but are always aware of the distinction. The Soma hymns, on the other hand, concentrate on an immediate experience. There is no myth, no past, no need, for harmony. It is all here, all alive and one. The Soma experience was always an extraordinary event, exciting, immediate, transcending the logic of space and time."

the lost plant – that characteristic Indian casualness toward history again. Western scholars, by contrast, have been curious from the first. In the two centuries since Indology became an academic discipline in Europe, forty-three candidates for Soma were proposed in the nineteenth century, and in the twentieth the number rose to a total of over one hundred. Any Indologist who settled the issue would have been assured of a permanent place in the annals, not only of Indian and religious scholarship, but of historical scholarship generally. Most ranking scholars had abandoned the quest as hopeless.

This is where the story picks up, for when the answer arrives – and it will be the burden of my review that it has arrived – it comes not from a Sanskritist, Indologist, or academician of any stripe. It comes from outside the world of professional scholarship altogether, from an amateur – a retired banker, and a high-school dropout at that. But more. Let the master clue be one of the most improbable lines in all Sacred Writ: "Fullbellied the priests piss the sacred Soma"; a line which, verging on scatology, had regularly thrown the pundits into confusion and leveled the exegetes. Let the discovery surface in a bibliophile's dream that is a story in itself – printed in limited edition on handmade paper, the book became a collector's item overnight. Finally, let the subject fall squarely in taboo domain – the chaotic, puzzling, passion-filled world of the "psychedelics," with all that word has come to mean to America in the last fifteen years – and the reader can see why I felt that my own work could wait. The immediate occasion for my review is the appearance of *SOMA* in a popular edition, but it is also time for a general stock-taking, because the three years since the book's initial publication have allowed time for reviews to appear in the major critical journals.

I. Where Things Stood

In the pantheon the Aryans brought with them when they swept into Afghanistan and the Indus Valley in the second millenium B.C.E., Soma occupied a unique position. Indra with his thunderbolt was more commanding, and Agni evoked the awe that fire so readily inspired before the invention of matches made it commonplace. But Soma was special, partly (we may assume) because one could become Soma through ingestion, but also because of what one then became: "We have drunk Soma and become immortal." The Soma hymns are vibrant with ecstasy. It appears to be virtually the only plant man has deified; the Mexican Indians regard mushrooms, *peyotl,* and morning glories as "god's flesh" or in other ways mediators of the divine, but the plants do not figure in their pantheons. The crucial Mandala IX consisting of 114 of the Rig-Veda's 1,028 hymns is dedicated exclusively to Soma, as are six other hymns, but his significance extends far beyond these hymns in which he is invoked in isolation. "Soma saturates the Rig-Veda" (7:169); the entire corpus is "shot through with Soma." "The Soma sacrifice was the focal point of the Vedic religion," writes W. D. O'Flaherty, adding,

> Indeed, if one accepts the point of view that the whole of Indian mystical practice from the *Upanisads* through the more mechanical methods of yoga is merely an attempt to recapture the vision granted by the Soma plant, then the nature of that vision – and of that plant – underlies the whole of Indian religion, and everything of a mystical nature within that religion is pertinent to the identity of the plant. (4:95)

Louis Renou once said that the whole of the Rig-Veda is encapsulated in the themes Soma presents.

In the course of the Soma sacrifice dried plants were steeped in water and their juice pounded out with stones and wooden boards covered with bull hides. This juice was then forced through wooden filters and blended with milk, curds, barley water, ghee, and occasionally honey. To the priests who drank the holy brew it is said to have given strength, magnitude, and brilliance. "One has only to read the Soma hymns," Daniel Ingalls observes, "to grant some truth to the claim" (15:15).

Then, even as the last parts of the Vedas were being composed, Soma disappears. The Brämanas, codified around 800 B.C.E., contain no mention of it. Reverence for the god persisted; his sacrifice continues to be performed right down to today. But surrogates replaced the original plant. For nearly three thousand years, Soma retreats to the mountain fastnesses from whence it came. Like a yogi in training, deliberately isolated so his austerities won't be interrupted, Soma drops out of history – to the historians' dismay, as I earlier remarked.

II. Enter Gordon Wasson

In certain respects Gordon Wasson was an unlikely candidate for the discoverer. He knew no Sanskrit, had no special interest in India, and his years were against him; born in 1898, he was already in his sixties and had retired from his banking career when he turned to Soma. But it goes without saying that he didn't just stumble on his find. He was equipped for the search – ideally so, we can say with wisdom of hindsight. To begin with, he was intelligent. His career bears this out from beginning to end. Without having completed high school he was appointed to teach English at Columbia University. Turning from that to journalism, he served as financial reporter for the *New York Herald Tribune* until his uncanny sense of the business world caused J. P. Morgan and Company to take him on and advance him, in time, to a vice-presidency. And atop this basic intelligence Wasson had erected a specialist's repertoire. Though he was neither scholar nor scientist by profession, there was a field in which he was a master, and it was the one that proved to be decisive: ethnomycology. Assisted by his wife, Valentina Pavlovna, a pediatrician who died in 1959, his work in this area had led to (a) rediscovery of *teonanactl,* the sacred mushroom of Mexico[3] and the worldwide attention it subsequently received; (b) publication in 1957 of a monumental two-volume treatise, *Mushrooms, Russia and History (3)*, which argued the possibility of the mushroom cult being man's oldest surviving religious institution; (c) reputation as founder of a science of "ethnomycology," a name analogous to "ethnobotany"; (d) appointment as Research Fellow (later Honorary Research Fellow) of the Botanical Museum of Harvard University; and (e) Honorary Research Fellow of the New York Botanical Garden and Life Member of the Garden's Board of Managers.

These talents alone might have sufficed, but the longer one ponders the Soma discovery, the more facets of Gordon Wasson appear relevant until one has to remind oneself that it wasn't the preordained purpose for which he was born. Though advanced in years when he hit the Soma trail, his health and zest for research, including fieldwork, had held up; ten years later he continues to sleep in a sleeping bag on a screened porch the year 'round in Connecticut temperatures that can dip to fifteen degrees below zero. His depth-exploration of the Mexican mushroom – for years he and his wife spent their annual vacation in joint expeditions with the great French mycologist, Roger Heim – had made him directly, experientially knowledgeable about entheogens and the way they can function in a religious setting.* Even the careers Wasson pursued on his

* I do not consider this an incidental resource. I find it not only aetiologically natural but metaphysically apposite that Soma's identity should have been discovered by an initiate – not, to be sure, in the Soma cult itself, but in a western counterpart. We both search and find according to our sensibilities, a point which (if I may be pardoned a personal reference) has been borne in on me by the one empirical discovery of my career. Had I not possessed, first, a musical ear which alerted me immediately to the fact that in the Gyüto (Tibetan) chanting I was in the presence of something subtly astonishing; and second, a musical temperament which required that I get to the bottom of what had so moved me, the "important landmark in the study of music," which *Ethnomusicology* (January 1972) credited the find as being, would not have been forthcoming. Something comparable, I am certain, was at work in Wasson's discovery of Soma. To indicate what it was, I quote at length from Wasson's response to the sacred mushroom of Meso-America which he came upon twenty years earlier.

"When we first went down to Mexico, we felt certain, my wife and I, that we were on the trail of an ancient and holy mystery, and we went as pilgrims seeking the Grail. To this attitude of ours I attribute such success as we have had. A simple layman, I am profoundly grateful to my Indian friends for having initiated me into the tremendous Mystery of the mushroom.

"In the uplands of southern Mexico the rites take place now, in scattered dwellings, humble, thatched, without windows, far from the beaten track, high in the mountains of Mexico, in the stillness of the night, broken only by the distant barking of a dog or the braying of an ass. Or, since we are in the rainy season, perhaps the Mystery is accompanied by torrential rains and punctuated by terrifying thunderbolts.

"Then, indeed, as you lie there bemushroomed, listening to the music and seeing the visions, you know a soul-shattering experience. The orthodox Christian must accept by faith the miracle of Transubstantiation. By contrast, the mushroom of the way to Soma were only seeming detours. English and journalism gave him a feel for language which was to grace his report when it

Aztecs carries its own conviction; every communicant will testify to the miracle that he has experienced. 'He who does not imagine in stronger and better lineaments, and in stronger and better light than his perishing eye can see, does not imagine at all,' Blake writes. The mushroom puts many (if not everyone) within reach of this state. It permits you to see, more clearly than our perishing mortal eye can see, vistas beyond the horizons of this life, to travel backwards and forwards in time, to enter other planes of existence, even to know God. It is hardly surprising that your emotions are profoundly affected, and you feel that an indissoluble bond unites you with the others who have shared with you in the sacred *agape.* All that you see during this night has a pristine quality: the landscape, the edifices, the carvings, the animals – they look as though they had come straight from the Maker's workshop. This newness of everything – it is as though the world had just dawned – overwhelms you and melts you with its beauty. Not unnaturally, what is happening to you seems to you freighted with significance, beside which the humdrum events of the everyday are trivial. All these things you see with an immediacy of vision that leads you to say to yourself, 'Now I am seeing for the first time, seeing direct, without the intervention of mortal eyes.'

"And all the time that you are seeing these things, the priestess sings, not loud, but with authority. You are lying on a *petate* or mat; perhaps, if you have been wise, on an air mattress and in a sleeping bag. It is dark, for all lights have been extinguished save a few embers among the stones on the floor and the incense in a sherd. It is still, for the thatched hut is apt to be some distance away from the village. In the darkness and stillness, that voice hovers through the hut, coming now from beyond your feet, now at your very ear, now distant, now actually underneath you, with strange, ventriloquistic effect. Your body lies in the darkness, heavy as lead, but your spirit seems to soar and leave the hut, and with the speed of thought to travel where it listeth, in time and space, accompanied by the shaman's singing. You are poised in space, a disembodied eye, invisible, incorporeal, seeing but not seen. In truth, you are the five senses disembodied, all of them keyed to the height of sensitivity and awareness, all of them blending into one another most strangely, until the person, utterly passive, becomes a pure receptor, infinitely delicate, of sensations. As your body lies there in its sleeping bag, your soul is free, loses all sense of time, alert as it never was before, living an eternity in a night, seeing infinity in a grain of sand. What you have seen and heard is cut as with a burin in your memory, never to be effaced. At last you know what the ineffable is, and what ecstasy means. Ecstasy! For the Greeks *ekstasis* meant the flight of the soul from the body. Can you find a better word than that to describe the bemushroomed state? In common parlance ecstasy is fun. But ecstasy is not fun. Your very soul is seized and shaken until it tingles. Who will choose to feel undiluted awe, or to float through that door yonder into the Divine Presence?

"A few hours later, the next morning, you are fit to go to work. But how unimportant work seems to you, by comparison with the portentous happenings of that night! If you can, you prefer to stay close to the house, and, with those who lived through the night, compare notes, and utter ejaculations of amazement." (Condensed and slightly transposed from 1:149–62.)

appeared,* and banking, being lucrative, enabled him to travel when fieldwork beckoned and to consult the authorities whose diverse areas of expertise – Sanskrit, history, philology, comparative mythology, folklore, art, poetry, literature, ecology, ethnobotany, phytochemistry, and pharmacology – he was to fit with his own mycological knowledge to craft the solution. Also, when it became apparent that the Vedic references would be crucial, he could employ a talented Sanskritist, Wendy Doniger O'Flaherty of the School of Oriental and African Studies of the University of London, to translate the relevant passages. Wasson's comfortable circumstances bear, too, on *SOMA* as a *de luxe* publication, to which a later section of this review will be devoted. Its author is an aristocrat; every dimension of his life has style.

Finally, it was in Wasson's favor that he was not an academic. We need not go as far as Robert Graves and credit his innocence of a university education with preserving his genius. It is enough to share Professor Ingalls's suspicion, voiced at a testimonial dinner at the Harvard Faculty Club on the occasion of the publication of the book under review, that the specialists, each burrowing deeper and deeper down the narrowing shaft of his own specific competence, would never have discovered Soma's secret. The problem called for an amateur, a man who could approach it with innocence and love and across disciplinary boundaries.

The *Concise Oxford Dictionary* defines "amateur" as "one who is fond of; one who cultivates a thing as a pastime." The French is stronger; my dictionary renders it "lover, virtuoso." Wasson was an amateur mycologist in the French sense. His love and consequent

* I content myself with a single example: "Often have I penetrated into a forest in the fall of the year as night gathered and seen the whiteness of the white mushrooms, as they seemed to take to themselves the last rays of the setting sun and hold them fast as all else faded into the darkness. When fragments of the white veil of the fly-agaric still cling to, the cap, though night has taken over all else, from afar you may still see Soma, silver white; resting in his well-appointed birth-place close by some birch or pine tree. Here is *how* three thousand years ago a priest-poet of the Indo-Aryans gave voice to this impression: "By day he appears the color of fire, by night, silver white (IX 97[9d])." Soma's scarlet coat dominates by day; by night the redness sinks out of sight, and the white patches, silvery by moon and starlight, take over" (4:41–42).

virtuosity respecting the mushrooms rooted back into nothing less decisive than his love for his wife. In August 1927, newly wed and enjoying a vacation in the Catskills, they chanced on a forest floor that was covered with wild fungi. Their responses were exact opposites: he was indifferent, even distrustful, while she was seized by wild glee. Some couples might have left the difference at that, but the Wassons were of an inquiring bent. Examining their difference, they found it to be rooted in a difference between entire peoples. Dr. Wasson, a White Russian who practiced pediatrics in New York, had absorbed almost *cum lacte* (with her mother's milk) a solid body of empirical knowledge about mushrooms and a passionate regard for them; even "worthless" varieties were arranged with moss and stones into attractive centerpieces. By contrast, Gordon, of Anglo-Saxon heritage, had been shielded from the plants. Given to pejoratives like "toadstool" and exaggerated rumors of their toxicity, his people had been as mycophobic as hers had been mycophilic. In Russian literature mushrooms figure in love scenes and pastoral idylls; in English they are emblems of death. For over thirty years the Wassons devoted much of their leisure to dissecting, defining, and tracing this difference until it led to the thesis – supported by comparative philology, mythology, legends, fairy tales, epochs, ballads, historical episodes, poetry, novels, and scabrous vocabularies that are off-limits to proper lexicographers – that at some point in the past, perhaps five thousand years ago, our European ancestors had worshiped a psychoactive mushroom, and that their descendants had divided according to whether the *facinans* (fascination) or the *tremendum* (fear) of its holy power predominated.

Two years after the publication of *Mushrooms, Russia and History* in 1957, Mrs. Wasson died, and Gordon, forced into life changes and with a pension sufficient for his needs, retired from banking and promoted ethnomycology from his hobby into a second career. Soma was not on his docket. He wanted to look into the "mushroom madness" of New Guinea (still unsolved) and why the Maoris of New Zealand share the Eurasian association of mushrooms with lightning. Somewhere down the line he intended to examine India's largely negative attitude toward mushrooms, and this led him to spend some weeks in 1964 at the American Institute of Indian Studies at Poona where he began reading Renou's translation of the Vedas. It proved to be the turning point. During the days that followed, on a freighter to Japan, a number of disjointed things he had learned during forty years of research fell into place. The hypothesis that Soma was a mushroom, specifically the *Amanita muscaria* or fly-agaric,[4] came to view. From that point on it was a matter of corroborating his hypothesis.

III. The Evidence

To enter all the evidence Wasson uncovered in his five ensuing years of concentrated work in the libraries and botanical centers of the United States and Europe, and in the field in Asia, would be to duplicate his book. Instead I shall summarize the evidence he marshals for his conclusion under six points.

1. The references to Soma contain no mention of the leaves, flowers, fruit, seeds, and roots that pertain to chlorophyll-bearing plants. They refer repeatedly to stems and caps.

2. All the color references fit the *Amanita muscaria*. There is no mention of its being green, black, gray, dark, or blue (the colors of vegetation), while the colors that are mentioned conform without exception to the mushroom's cap (bright red), the membrane, unique to the *A. muscaria*, that protects it in its early stages (brilliant white), or its pressed – *sauma* means "to press" – juice (golden or tawny yellow). Wasson makes the latter point by using quotations from the Rig-Veda to caption a series of stunning photographs of the fly-agaric. The color-epithet that is invoked most often is *hari*, which in Sanskrit "seems to have run from red to light yellow" (4:37), always accenting its dazzling and resplendent character which the photographs that Wasson himself took capture brilliantly. "The hide is of a bull [red bulls are favored in India], the dress of sheep" (IX 70[7]). This "dress of sheep," the white membrane, is invoked by a variety of analogies: "He makes from milk

his robe of state" (IX 71^{2}), and "with unfading vesture, brilliant, newly clothed, the immortal [Soma] wraps himself all around. He has taken to clothe himself in a spread-cloth like to a cloud" (IX 69^{5}). The mushroom's rupture of its embryonic envelope, too, is noted. "He sloughs off the Asurian colour that is his. He abandons his envelope" (IX 72^{2}). "Like a serpent he creeps out of his old skin" (IX 86^{44e}). The flecks of the veil that cling to the mushroom's crown after the veil bursts give meaning to "he lets his color sweat when he abandons his envelope" (IX 71^{2}).

3. References to shape are equally apposite. The mushroom's head, peering through the undergrowth while still in its white skin, is "the single eye" (IX 9^{4}). When its cap is fully formed, it mirrors the vault of heaven and is "the mainstay of the sky." Or again, its curved cap can look like an udder – "the swollen stalks were milked like cows with [full] udders" (VIII 9^{19ab}) – and its puffy foot like a teat: "The priests milk this shoot like the auroral milch cow" (I 137^{ab}).

4. Soma altered consciousness but was not alcohol; it was an entheogen. The Aryans knew alcohol in the form of *sura,* a beer, but the time allotted for Soma's preparation in the sacrifices precludes fermentation. Moreover, whereas the Vedas generally disapprove of *sura,* noting the muddleheadedness and other bad effects it produces, Soma is not only *aducchuna,* without evil effects; it leads to godliness:

> We have drunk the Soma, we have become immortal, we have arrived at the light, we have found the gods.
>
> What now can the enemy do to harm us, and what malice can mortals entertain? Amplify, O Soma, our lives for the purpose of living.
>
> These splendid waters, granting much, protecting.
>
> Like fire produced by friction, may the waters inflame us! May they cause us to see afar and to have increasing welfare (Rig-Veda, VIII 48).

5. Geography fits. *Amanita muscaria* requires, for host, the north temperate birch forest, and the Indus Valley is bordered by lofty mountains whose altitude compensates for its southern latitude. South of the Oxus River, *A. muscaria* grows only at altitudes of eight thousand feet or more, and this fits with the fact that Soma was confined to mountains. Parts of Afghanistan, where the Aryans resided before continuing their southeastward push, and the Hindu Kush through which they entered the Indian subcontinent, are *A. muscaria* country.

6. Finally, there is the line of the Rig-Veda that I quoted at the beginning of this essay which has priests urinating diluted Soma. The *Amanita* is an entheogen whose vision-producing properties are known to survive metabolic processing. Ritualistic urine-drinking forms a part of a number of fly-agaric ceremonies that have survived to the present in Siberia and elsewhere. As translated by most Indologists, a verse in the Rig-Veda, (IX 74[4]) reads, "The swollen men urinate the on-flowing Soma." There is, in addition, the fact that the Vedas mention a "third filter" for Soma while describing only two; Wasson thinks this third filter could have been the human organism which, there is reason to believe, reduces the nauseous properties of the fly-agaric while retaining for as many as five ingestions the chemical, musicimol, which in the dried mushroom is the entheogenic agent.

IV. Critical Response

Wasson's *SOMA* appeared in 1969; this review is being written three years later. The interval has allowed time for authorities to review the use Wasson makes of their respective fields, and I categorize the most significant of their verdicts.

A. NONCOMMITTAL

F. B. J. Kuiper, Vedist, University of Leiden: "Wasson may be perfectly right in assuming that the original Soma plant was the

Amanita muscaria, but the problem cannot be solved beyond doubt" (18:284).

Winthrop Sargent, critic: "Wasson has given us the most persuasive hypothesis that has yet appeared, but nobody really can say what Soma was" (25).

B. CONFIRMING

Sanskritists and Indologists

Daniel Ingalls, Harvard University: The "basic facts about the Soma plant as described in the Rig-Veda cannot well be accounted for by any of the previous identifications. . . . They are all perfectly accounted for by the identification with the mushroom *Amanita muscaria* or fly agaric. Not all the epithets remarked on by Wasson need be taken just as he takes them, but enough still remains to be convincing. Wasson's identification is a valuable discovery" (14:188).

Stella Kramrisch, Institute of Fine Arts, New York University: "Wasson proves beyond doubt that Soma was prepared from *Amanita muscaria.* He has set right almost three thousand years of ignorance about the 'plant of immortality'" (17).

Wendy Doniger O'Flaherty, University of London: "For long she [O'Flaherty] was skeptical about my thesis, but she now authorizes me [Wasson] to say here today [at the International Congress of Orientalists, Canberra, January 1971] that she is a full-fledged convert" (7:169).

Ulrich Schneider, University of Freiburg: In his book *Der Somaraub des Manu,* 1971, he concludes that Soma is *Amanita muscaria.*

Botanists and mycologists

Albert Pilat in the Swiss bulletin of mycology: "In this interesting and magnificently produced work, the noted American ethnomycologist, R. Gordon Wasson, proves that the religious drug known under the name of 'Soma' is *Amanita muscaria*" (24:11).

Richard Evans Schultes, Botanical Museum Harvard University: "The data fit together as tightly as pieces of an intricate jig-saw puzzle. Wasson provides, so far as I am concerned, incontrovertible proof of the strongest kind that Soma must have been *Amanita muscaria.* Once and for all he has provided the identification" (27:101–5).

Anthropologists

Claude Lévi-Strauss, College de France: "Mr. Wasson's work establishes convincingly that, among all the possible candidates for Soma, *Amanita muscaria* is far and away the most plausible" (20).

Weston La Barre, Duke University: "The closure of linguistic, botanical, ethnographic, and ecological evidence is exhilarating. The identification of soma with *Amanita muscaria* is definite and the Sanskrit puzzle of two millennia, from the *Brahmanas* to this day, can now be regarded as finally solved" (19:371).

Linguists

Calvert Watkins, Harvard University: "I accept Wasson's identification of Soma with *A. muscaria.* I am myself by way of being an amateur mycologist, and in my review article (in preparation for Wolfgang Meid [ed.], *Gedenkschrift für Hermann Güntert* [Innsbruck, 1973]), I hope to show that there is considerably more evidence for his hypothesis in the Rig-Veda, and also in the Iranian, Avestan, data, with which he was not concerned" (from a letter, 19 June 1972, to the author of this review).

Generalists

Robert Graves, poet, mythologist, savant: "Wasson has identified Soma, without any possibility of scientific or scholarly doubt, as the *Amanita muscaria,* or 'fly-agaric'. The argument is as lucid as it is unanswerable. His book satisfies me completely. I congratulate him on his feat" (12:109, 113).

C. REJECTING

John Brough, Professor of Sanskrit, Cambridge University: "It is with regret that I find myself unable to accept that Wasson has proved his theory that the original Vedic Soma was *Amanita muscaria*" (10:362).

D. ROMAN JAKOBSON

As Professor Emeritus at Harvard and M.I.T., Jakobson merits a category to himself, not only because he is the world's greatest living linguist (which he is), but by virtue of his special relationship to the book. The fact that the *de luxe* edition is dedicated to him removes him from controversies over it, and it is unlikely that he will write about it. He permits me to report, however, that although, not being a Vedist, he feels unqualified to pronounce on Wasson's conclusion, he has been impressed from the first with the caliber of his search. Wasson is free of stereotypes and prejudices that have impeded the Soma quest, his standards of scholarship are of the highest, and he has consistently checked his findings with ranking authorities in every field he has entered.

[*I omit the next, long section of my original review, titled "Disputed Points," because it deals with technicalities that are likely to be of interest only to professional Indologists. The issues the experts debate there are: (a) whether evidence outside of India and Iran is relevant for identifying the* soma *plant; (b) whether the Vedic tropes and epithets for* soma *refer primarily to the* soma *plant or to its indwelling god; and (c)* soma's *relation to urine – whether the startling line in question says that priests piss* soma *or, metaphorically, that the god Indra does that.*]

V. The Book

There remains the book as a physical object, lying open on my desk, inviting comment in its own right as an exhibit in bookmaking.

Wasson's first book, his two-volume opus written with his wife titled *Mushrooms, Russia and History,* appeared in a limited edition of 512 numbered copies. I recall that it rated a multi-paged spread in *Life* magazine, which may help account for the fact that, announced at $125, its sales became so brisk that the surprised publishers started raising the price, and its last copies retailed at twice the original figure. Of *SOMA,* twelve years later, 680 copies were printed of which 250 were allotted to the United States. Being a single volume, its price was kept to $200 (inflation must be kept in mind), and again the stock was exhausted within months.

Is it known what is the most expensive book that has ever been published? Regardless, *SOMA* is by all accounts a sumptuous production. Wasson lavished on the bookmaking dimension of his work the same meticulous attention he devoted to the Soma search itself. The volume is in blue half-leather with a dark blue spine, stamped in gold and slip-covered in fine blue linen cloth. The book was designed by Giovanni Mardersteig and set in Dante type; the text and illustrations were printed by the Stamperia Valdonega, Verona. I have already spoken of the stunning photographs: thirteen color tip-ins of the fly-agaric in its natural habitat. The paper was handmade by Fratelli Magnani, Pescia; pages are of International Size A–4. In all, it is a book lover's dream, and in the three years that have elapsed since its publication it has been hard to come by. In the face of the declining quality in the format of botanical publications in the 1930s, Professor Oakes of Harvard argued that "the results of a scientist's research are jewels worthy of a proper setting." Wasson's book would have satisfied him.

As I was telling the SOMA story in class last fall, noting that to get at the book itself students would have to get the key to the Houghton Rare Book Room at Harvard University, one of them raised his hand to say that he had seen the book in the Tech Coop, M.I.T.'s bookstore, on his way to class. I told him he must have been mistaken, for I felt certain that Wasson's aristocratic tastes precluded a popular edition in principle. Happily, it was I who was mistaken. Popular editions have appeared in both cloth ($15) and paperback ($7.50). They lack the jacket watercolor and generous margins of the original and their paper is not handmade, but in other respects they are faithful to the *de luxe* edition.

VI. Conclusion

Soma seems to have been rediscovered, but why was its identity lost in the first place? Wasson believes that its importance, coupled with the famed mnemonic capacities of the Vedic priests, rules out its having simply been forgotten; it must have been deliberately suppressed. In *SOMA* he proposes, as the reason for suppressing it, distribution problems. As the Aryans moved down the Gangetic plains, this high-altitude mushroom became increasingly more difficult to procure. Inconsistency – now the fly-agaric, now a substitute – proved ecclesiastically unworkable; a patron discovering that rhubarb was used in the sacrifice while his neighbor got the genuine article could be difficult. A crisis developed and the governing Brahmins decided that the originals had to be eliminated completely.

Recently Wasson has been inclining toward a different reason: that the substance may have started to get out of hand. Quality declines in the last *Soma* hymns, and some border on irreverence. Three thousand years in advance of our times, India may have found herself on the brink of a psychedelic mess like the one America created in the 1960s. She wasn't able to close the door on it completely – plenty of *bhang* smoking *sadhus* (wandering ascetics) in whom it is impossible to determine whether *sattva* (illumination) or *tamas* (sloth) predominates, can be found in India right down to the present. But at a critical moment, Wasson hypothesized, the Brahmins did everything they could to prevent such abuse. They would rather have the botanical home of their god forgotten than let him be subjected to profanation. If the hypothesis is correct, it would help to explain why the Buddha felt strongly enough about drugs to list them with murder, theft, lying, and adultery as one of the Five Forbidden Things. It could also throw light on Zarathustra's angry excoriation of those who use inebriating urine in their sacrifices: "When wilt thou do away with the urine of drunkenness with which the priests delude the people" (Avesta, Yasna 48:10).

I will myself stretch this line of thought to its conclusion. Even among those who are religiously responsible, entheogens appear to have (in the parlance of atomic decay) a half-life; their revelations decline. They are also capricious. Opening the gates of heaven at the start, there comes a time – I can attest to this myself – when they begin to open either onto less and less or onto the demonic. It is precisely apposite that the book that introduced the entheogens to the contemporary West, Aldous Huxley's *Doors of Perception,* was followed quickly by his *Heaven and Hell.* It seems that if God can manifest himself through anything, it is equally the case that nothing can commandeer him and guarantee his arrival. It is compatible with the notion that the Absolute entered India by way of a mushroom to hold that sometime later it stopped doing so.

CHAPTER FIVE

The Sacred Unconscious

Having reached midpoint in this collection of essays, it is time to take stock.

The problem, or rather mystery, that stalks our understanding of the entheogens is how Ultimate Reality or God can disclose him/her/ itself – the pronouns never work – to us through changes in brain chemistry. This chapter addresses that issue head-on. It builds on my essay, "The Sacred Unconscious," which was written for Roger Walsh and Deane Shapiro's book, Beyond Health and Normality: Explorations of Exceptional Psychological Well-Being, *but I have reworked that essay substantially to bring it closer to the concerns of this book.*

The point of including it reduces to this. In the view of the mind that evolutionary biology, the cognitive sciences, and clinical psychology have assembled, there is no way that entheogenic certainties can be accepted at face value, for the only explanations that that model can offer are ones that explain the certainties away. (I am speaking of course of certainties that retain their force after the conventional world re-forms. Some things that seem certain at the time are easily recognized as nonsense the moment the chemicals wear off. "The entire universe is pervaded by a strong odor of turpentine" is a frequently cited example.) This leaves those who accept those certainties as authentic theophanies needing a different model of the mind to work with.

The model here presented is so at odds with the current model that – fearing that mine will be dismissed out of hand – I will take my cue from football coaches who hold that there are times when the best defense is a good offense. Accordingly, I shall attack.

When it comes to improbabilities, advocates of the current model of the mind are in no position to throw stones. According to this model, the mind is the culmination of a twenty-billion-year history in which – from a mysterious substratum not rightly described as "matter" that the Big

Bang left in its wake – matter evolved from its most elementary constituents into ones that became progressively more complex until organisms appeared. These, in the course of their three-billion-year history on this planet, have developed increasingly ingenious strategies for relating to their environments, with human intelligence the most sophisticated of these.

In its way this is an impressive scenario, but it leaves four things unexplained:

First, how the universe originated in the first place. One of the world's foremost astronomers, Allan Sandage of the Observatories of the Carnegie Institution in Pasadena, California, recently proposed that the Big Bang could only be understood as "a miracle," in which some higher force must have played a role. A fair number of cosmologists are now saying the same.

Second, how, following the Big Bang, matter derived from nonmatter. That it did so derive seems now to be accepted, for (as a theoretical physicist at the University of California put it to me recently) if you begin with matter as simply given, you're lost. Yet how quark containment with its rest mass – as serviceable a definition of matter as we have – gets into the picture is something physicists don't like to think about.

Third, how qualities derive from quantities. Even people who recognize the limitations of science assume that it can handle the corporeal world, but it cannot – not in that world's fullness. The world comes to us clothed in sounds and colors and fragrances, which in textbook science have no right to be there, for the electromagnetic waves that underlie those qualities are as close to them as science can get.

Finally, how thoughts and feelings – in short, mind – derive from brain. This fourth difficulty takes us directly to the psychoactive drugs, so it deserves more extended treatment.

To characterize the relation between consciousness and brain activity as "the hard problem" in cognitive science is an understatement, for efforts to understand that connection have failed to a degree unparalleled in any other scientific endeavor. In 1992 a symposium on Experimental and Theoretical Studies of Consciousness was convened by the Ciba

Foundation in London. After three days of deliberations, its participants (all of whom were leading philosophers, neurobiologists, and cognitive scientists) were forced to conclude that the mind-body problem not only remains unsolved; there was not even a consensus on how it might be solved. As one of the participants, Thomas Nagel, put the point, "Unless conscious points of view can be subjected to outright physical reduction [which Nagel doubted they can be], it will not be possible to understand how they necessarily arise in certain kinds of physical systems, described only in terms of contemporary physics and chemistry." Since the Ciba conference, this position has gained strength. Under the ungainly name "mysterianism," it has garnered the support of a number of leading cognitive scientists, including Steven Pinker, head of the Center for Cognitive Science at M.I.T. (See his book How the Mind Works *and Colin McGinn's* The Mysterious Flame.*)*

This impasse that the reigning model of the mind seems to have run into can be taken as an invitation to consider alternatives. That the inspiration for the model that I present here comes from India should not surprise us, given the fact that historically India has been the world's introspective psychologist, having poured roughly as much energy into exploring inner space as the West has devoted to probing the external world.

Dropping several of the opening paragraphs in "The Sacred Unconscious" essay that I wrote for the Walsh/Shapiro volume, I pick up with its discussion of depth psychology.

That a great deal of what goes on in our minds is out of sight is beyond question. Much of our knowledge gets programmed into neural pathways in ways that we neither perceive nor understand in detail. Thus, I rightly say that I know how to ride a bicycle and to type using the touch system, but I am not aware of the workings of this "tacit knowledge," as Michael Polanyi christened it. Atop this physiologically impounded knowledge is the storehouse of memories that can be recalled but are otherwise not in full view. Fragments from this storehouse often arrive uninvited through free associations. Psychologists study the pathways these associations lay down in our early years, and they help us understand ways in which those pathways influence our responses for the rest of our lives. The totality of these pathways constitutes the "individual unconscious" each of us possesses, individual because no childhood duplicates another. To that individual unconscious Marx added a "social unconscious," for our position in the social hierarchy also influences the way we see things. Reading this paragraph backward, we can say that the West has identified three quite well-delineated unconsciouses: a social unconscious (constructed by our place in society), an individual unconscious (which our individual childhood experiences have produced), and a physiological unconscious (which impounds our tacit knowledge).

Learning about these three layers of our unconscious minds has proved useful, but there is a problem. Set within the scientific worldview which sees the mind as having evolved from what lacks mind, they do not provide us with a very inspiring image of ourselves. Artists tend to see this most clearly, and I will quote one of the most discerning among them, Saul Bellow. Speaking of the "contractual daylight" view of ourselves that psychologists, sociologists, historians, and journalists have moved into place, Bellow, in his 1976 Nobel Prize Address, continues:

> The images that come to us in this contractual daylight, so boring to us all, originate in the contemporary worldview. We put into our books the consumer, civil servant, football fan, lover, television viewer. And in the contractual daylight version their life is a kind of death. There is another life, coming from an insistent sense of what we are, that denies these daylight formulations and the false life – the death in life – they make for us. For it is false, and we know it, and our secret and incoherent resistance to it cannot stop, for that resistance arises from persistent intuitions. Perhaps humankind cannot bear too much reality, but neither can it bear too much unreality, too much abuse of truth.

In this context, Roger Walsh and Deane Shapiro's *Beyond Health and Normality: Explorations of Exceptional Psychological Well-Being* comes like a burst of fresh air. As psychiatrists, Walsh and Shapiro are aware of how much Freud contributed to the "boring, contractual daylight" model of the self that Saul Bellow recounts, but they don't stop there. Noting that the complete edition of the collected works of Sigmund Freud contains over four hundred entries for neurosis and none for health, they go on to point out that Freud was a physician and that physicians deal with sick patients. Freud dealt with sick unconsciouses. Walsh and Shapiro ask us in effect to turn this on its head by tying to imagine what a model of the human self might look like if its database numbered saints rather than neurotics.

I shall use the second half of this essay to address that question directly, but because saints constitute a small sample of the human population, I shall begin more neutrally by proposing a model of the human mind that holds for human beings generally and work up from there to the supreme actualization it allows for.

Among other virtues in the paragraph by Saul Bellow that I quoted is its recognition that self-images are shaped by the worldviews that prompt them, so I will say in a sentence what it is in the traditional and modern worldviews that causes them to generate opposing anthropological models. Modernity sees humanity as having ascended from what is inferior to it – life begins in slime and ends in intelligence – whereas traditional cultures see it as descended from its superiors. As the anthropologist Marshall Sahlins puts the matter: "We are the only people who assume that we have ascended from apes. Everybody else takes it for granted that they are descended from gods."

I shall be tapping here into the Indian account of this divine lineage which traditional peoples universally assumed, and a vivid memory takes me directly to the heart of that account.

In 1970, while conducting thirty students around the world for an academic year to study cultures on location, I availed myself of my professional friendship with a distinguished philosopher at the University of Madras, T. M. P. Mahadevan, to ask him to speak to my students. I felt awkward about the invitation for I assigned him an impossible topic, to explain to neophytes in one short morning how Indian philosophy differs from Western philosophy. I needn't have been concerned, for he rose to the occasion effortlessly. Beginning with a sentence that I remember verbatim for the scope it covered, he said, matter-of-factly: "Indian philosophy differs from Western in that Western philosophers philosophize from a single state of consciousness, the waking state, whereas India philosophizes from them all." From that arresting beginning, he went on to explain that India sees waking conscious as one state among four, the other three being the dream state, the state of dreamless sleep, and a final state that is so far from our waking consciousness that it is referred to simply as "the fourth."

I pass over the fact that it is only in the last fifty years that the West has taken serious notice of the difference between dream and dreamless sleep, which difference yogis have worked with for millennia. What is important is not the time scale, but the different ways the two civilizations characterize dreamless sleep. The West does not assume that it includes awareness, whereas India holds that we are then more intensely aware than we are when we are awake or dreaming.

I battled my Vedanta teacher for seven years over that issue, I claiming that I was not aware while sleeping dreamlessly and he insisting that I was aware. When I pointed out that I certainly wasn't *aware* that I was aware, he dismissed my riposte as sophomoric. Think of how rapidly even your dreams evaporate, he said – most of them don't survive until morning. That much I had to grant him, and he went on to press his advantage. Dreamless sleep transpires in a far deeper stratum of the mind than dreams occupy, he said, so it stands to reason that years of (yogic) attention are required to bring its content back to wakeful memory. And what the yogis report is that dreamless sleep is a state of bliss, bliss so intense in fact that it is exceeded only by "the fourth" state of consciousness wherein *atman* (the self's foundation) merges with *Brahman*. Were it not for the fact that we recharge our batteries every twenty-four hours by experiencing this bliss of dreamless sleep, my teacher concluded, the trials and disappointments of daily life would wear us down and we would give up on life.

My swami and I ran out his clock in this deadlock, but several years after he died something happened that brought me around to his position. To have my wisdom teeth extracted I was administered total anesthesia, and in the improbable circumstances of a cramped recovery room with a nurse shaking me to return me to wakefulness, I heard myself exclaiming, "It's so beautiful!" Even as I pronounced those words, my grip on the experience had slipped to the point that I could no longer remember *what* it was that was beautiful, but *that* it was beautiful, staggeringly so, I can remember so vividly that to this day it continues to send chills up my spine. When I reported the incident to one of our daughters, she said that the same thing had happened to her, only her first words were "I love you," professed to a total stranger. "I'm so happy" was the variation a third party reported, while a fourth acquaintance reports that "this is so neat!" was the best his high school vocabulary could manage in describing the experience when it occurred to him.

The question this raises is, what prompted those ecstatic (I use the word advisedly) utterances in their respective modes of beauty, love, and bliss – three faces of God or the Good? (I omit "neat" because of its adolescent vagueness.) Clearly, nothing that was going on in the empirical world. My own conclusion is that all three were reports from the state of "dreamless sleep" that the anesthetics had transported their patients to and whose importance my swami had been trying to persuade me of.

India's model of the mind incorporates these three states of consciousness in a way that positions them (figuratively speaking) as Chinese boxes. *Atman* – the pure, effulgent, undiluted consciousness that is our essence – is encased in what the Indians call the "causal body" which we experience in deep sleep. This in turn nestles in the "subtle body," which fabricates dreams, and that in turn is housed in the "gross body," which generates wakeful experience. As the Pure Light of the Void works its way through these "boxes" – which in Sanskrit are called "sheaths," and which in many traditions are referred to as veils – it grows progressively dimmer and we are brought back to the fundamental difference between the traditional and modern models of the mind that I previously mentioned. Modernity sees the mind as having arisen from what is inferior to it, whereas traditional people see it as deriving from what is greater than itself and carrying within itself traces of its noble ancestry.

Even today we hear echoes of this traditional, less-from-more model of the mind in voices like those of Wordsworth ("trailing clouds of glory we come from heaven, our home"), Eddington (who concluded that the world is more like a mind than a machine), Schrödinger (who likened it to *Brahman:* infinite being, infinite awareness, and infinite bliss), and Bergson (who considered the mind a reducing valve), but I must move on to the second half of this essay. Having called attention to seemingly insuperable problems that plague the modern model of the mind, and following that by dubbing in (in Vedantic idiom) the basics of the generic traditional model, I turn now to the direct object of Walsh and Shapiro's book, which is to explore possibilities of extreme psychological well-being. What kind of person does the traditional model of selfhood allow for at its best?

It is easier to describe flawed instances of the model than perfected ones, for those are the only instances we actually encounter – even Christ asked rhetorically, "Why callest thou me good?" Paradoxically, these tarnished instances are our best resource for trying to imagine a perfected one, as the "tragic flaw" theory of art attests. No playwright would dream of trying to create a perfect hero, for matched with real life he would appear as bogus as a cardboard cutout. But endow the hero with a fatal weakness – Hamlet's indecision is the classic example – and the reader will supply the missing virtue instinctively. Master psychologist that he was, the Buddha anticipated this tragic-flaw device by twenty-five centuries. He never played to the galleries with previews of coming attractions, tantalizing descriptions of what *nirvana* would feel like. What he hammered at was the three poisons that stand in its way: greed, hatred, and delusion. Human egos that result from the veilings of the three sheaths are shot through with these poisons. Interests and thoughts normally extend outward, but our egos work like magnets to bend them back in U-turns onto ourselves. This creates a hot spot of being that hugs us tightly; things that concern us directly we view with feverish intensity, while regarding those at a distance with cold indifference. Our glasses are prescription-ground to cause us to see things our way. It's as the Tibetans say: when a pickpocket meets a saint, what he sees is his pockets. More poignantly, when poor children are asked to draw a penny they draw it larger than do children for whom pennies are commonplace; it looms larger in their minds' eyes. What we take to be objective facts are largely psychological constructs, as the Latin, *factum,* "that which is made," reminds us. The conclusion is inescapable. Our normal self is little more than an amalgam of desires and aversions that we have wound around it as tightly as the elastic of a golf ball.

This tight, constricted, golf-ball self is in for hard knocks, but what concerns me here is that on average it doesn't feel very good. Anxiety hovers over it. It can feel victimized and grow embittered. It is easily disappointed and can become unstrung. To others it often appears no prettier than it feels to itself: petty, self-centered, drab, and bored.

I am deliberately putting down this golf-ball self – hurling it to the ground, as it were, to see how high our total self can bounce: how far toward heaven it can ascend. In order to ascend, it must break out of the hard rubber strings of our attachments that we have stretched so tightly around it. If we change our image from rubber to glass and picture the Three Poisons as a lens that refracts light waves in the direction of our private, importunate demands, freedom from these egocentric distortions will come by, progressively decreasing our lens's curve – reducing its bulge. The logical terminus of this would be clear glass. Through this glass we would be able to see things objectively, as they are in their own right.

This clear glass, which for purposes of vision is equivalent to no glass at all, is our sacred unconscious. It is helpful to think of it as an absence because (like window glass) it functions best when it calls no attention to itself, and it is precisely this absence that makes the world available to us – "the less there is of self, the more there is of Self," as Meister Eckhart said. From clear glass we have moved to no glass – the removal of everything that might separate subject from object, self from world. Zen Buddhists use the image of a Great Round Mirror. When the obscuring deposits of the Three Poisons are removed from it, it reflects the world just as it is.

To claim that human consciousness can move permanently into this condition is (as I have said) probably to go too far, but advances along the asymptotic curve that slopes in its direction are clearly perceptible. When our aversion lens is bloated, humping toward a semicircle, we like very little that comes our way. The same holds, of course, for our desire lens which is only the convex side of aversion's concave arc – the more it distorts our evaluations toward our own self-interests, the less energy remains to appreciate things in their own right. Blake's formulation of the alternative to this self-centered outlook (which I draw upon for the title of this book) has become classic. "If the doors of perception were cleansed, everything would appear to man as it is, infinite."

The fully realized human being is one whose doors of perception have been cleansed – I have myself referred to these doors as windows and envisioned them as successive layers of our unconscious minds.* Those that are near the surface vary from per-

* Daniel Brown has uncovered something here that is interesting and perhaps important. Writing in the *International Journal of Clinical and Experimental Hypnosis* (XXV, 4, October 1977), he notes that the steps in Tantric Buddhist meditation reverse the stages of perceptive and cognitive development as these have been discovered by the Constructivist school in child psychology: Piaget, Gesell, Kagan, Lois Murphy, Brunner, *et al.* Paralleling in reverse the processes by which the infant successively acquires (constructs), first a sense of self around which to organize his experience and then structures for organizing his perceptions and after that his thoughts, Tantric meditation dismantles those constructs. After an initial stage that trains the lama to introspect intently, a second state disrupts his thought-structures, regressing him to the world of pure perception. Step three takes over from there and disrupts the perception-patterning processes that moved into place in infancy. The fourth and final stage breaks through the organizing mechanisms that constructed the infant's sense of ego and enables the lama to experience a world in which there is no obstructing sense of self. In the vocabulary of the present essay, such meditation peels back intermediate layers of our unconscious minds and allows us to be in direct touch with our sacred unconscious.

I am indebted to Kendra Smith for the substance of the preceding paragraph and to Jeffrey Becker (a former student of mine who is now a medical student) for calling my attention to this sequel. It appears that direct pharmacological intervention into the brain can produce experiences that may be roughly parallel to those the Tibetans access by meditating, particularly the phenomenon of self-annihilation.

son to person, for they are deposited by our idiosyncratic childhood experiences. At some level, though, we encounter the Three Poisons (once again: desire, aversion, and ignorance) that are common to humankind and to some degree may be necessary for us to function as individuals. But the deepest layer, I have suggested, is really a no-layer, for – being a glass door ajar, or a mirror that discloses other things rather than itself – it effectively isn't there. Even if it were there, in what sense could we call it ours? For when we look toward it we see simply – world.

This opening out onto the world's infinity is one good reason for calling this deepest stratum of the mind sacred, for surely holiness has something to do with the whole. But the concreteness of Blake's formulation is instructive. He doesn't tell us that a cleansed perception discloses the Infinite per se. It finds it in the things at hand, in keeping with the Buddhist teaching that the most sacred scriptures are in unwritten pages – an old pine tree gnarled by wind and weather, or a skein of geese traversing the autumn sky.

Thus far I have defined a realized human being, what the Indians call a *jivamukta*. It remains to describe one. What would it feel like to be such a person, and how would one appear to others?

Basically she lives in the unvarying presence of the numinous. This does not mean that she is excited or "hyped"; her condition has nothing to do with adrenaline flow or manic states that call for depressive ones to balance the emotional account. It's more like what Kipling had in mind when he said of one of his characters, "He believed that all things were one big miracle, and when a man knows that much he knows something to go upon." The opposite of the sense of the sacred is not serenity or sobriety. It is drabness; taken-for-grantedness. Lack of interest. The humdrum and prosaic. The deadly sin of acedia.

The activity of a certain neuronal receptor – the N-methyl-D-aspartate (NMDA) receptor – is particularly essential to the emergence of many higher human processes, including memory and one's sense of self. Blocking the activity of that receptor with nitrous oxide or ketamine can help bring about ego dissolution. Awareness remains, but with ego boundaries lowered or even leveled, the awareness is now experienced as boundless.

All other attributes of a realized being must be relativized against this one absolute: an acute sense of the astonishing mystery of everything. Everything else we say of him must have a yes/no quality. Is he always happy? Well, yes and no. On one level he emphatically is not; if he were, he couldn't "weep with those who mourn" – he would be an unfeeling monster, a callous brute. If anything, a realized soul is *more* in touch with the grief and sorrow that is part and parcel of the human condition, knowing that it too needs to be accepted and lived as all life needs to be lived. To reject the shadow side of life and pass it by with averted eyes, refusing our share of common sorrow while expecting our share of common joy, would cause the unlived, closed-off shadows in us to deepen into fear, including the fear of death.

A story that is told of the recent Zen master Shaku Soen points up the dialectical stance of the realized soul toward the happiness I am noting. In the evening he liked to take a stroll through the outskirts of his village. One evening he heard wailing in a house he was passing, and, on entering quietly, he found that the householder had died and his family and neighbors were crying. Immediately he sat down and started weeping with them. An elderly gentleman, shaken by this display of emotion by an accomplished master, berated him. "I thought you were beyond this kind of thing," he said indignantly. Through his sobs the master managed to falter, "It is this that puts me beyond it."[1]

The master's tears we can understand; the sense in which he was beyond them is more difficult, it being the peace that passeth understanding. The peace that comes when a man is hungry and finds food, is sick and recovers, or is lonely and finds a friend – peace of this sort is comprehensible. But the peace that passeth understanding comes when the pain of life is not relieved. It shimmers on the crest of a wave of pain; it is the spear of frustration transformed into a shaft of light. The master's sobs were real, yet paradoxically they did not erode the yes-experience of the East's "it is as it should be" and the West's "Thy will be done."

In our efforts to conceive the human best, everything turns on an affirmation that steers between cynicism on the one hand and sentimentality on the other. A realized self isn't incessantly and oppressively cheerful – oppressively, not only because we suspect some dissembling in his unvarying smile, but because it underscores our moodiness by contrast. Not every room a *jivamukta* enters floods with light; he can flash anger, and upset money changers' tables. Not invariance but appropriateness is his hallmark, an appropriateness that has the whole repertoire of emotions at its command. The Catholic church is right in linking radiance with sanctity, but the paradoxical, "in spite of" character of this radiance must again be stressed. Along with being a gift to be received, life is a task to be performed. The adept performs it. Whatever her hand finds to do, she does it with a will. Even if it proves her lot to walk stretches of life as a desert waste, she walks it rather than pining for its alternative. Happiness enters as by-product. What matters focally, as the Zen master Dogen never tired of noting, is resolve.

If a *jivamukta* isn't forever radiating sweetness and light, neither does he constantly emit blasts of energy. He can be forceful when need be; we find it restoring rather than draining to be around him, and he has reserves to draw on, as when Socrates stood all night in trance and outpaced the militia with bare feet on ice. In general, though, we sense him as composed rather than charged – the model of the dynamic and magnetic personality tends to have a lot of ego in it for needing to be noticed. Remember: everything save the adept's access to inner vistas, the realms of gold I am calling the sacred unconscious, must be relativized. If leadership is called for, he steps forward; if not, he is pleased to follow. He isn't debarred from being a guru, but he doesn't need disciples to prop up his ego. Focus or periphery, limelight or shadow, it doesn't really matter. Both have their opportunities, both the demands they exact.

All these relativities that I have mentioned – happiness, energy, prominence, impact – pertain to the *jivamukta's* finite self which she progressively pushes aside as she makes her way toward her sacred unconscious. As her goal is an impersonal, impartial one, her identification with it involves a dying to her finite selfhood. Part of her being is engaged in a perpetual vanishing act, as Coomaraswamy suggested when he wrote,

"Blessed is the man on whose tomb can be written, '*Hic jacet nemo,*' here lies no one."[2]

But having insisted that there is only one absolute or constant in the journey toward this self-naughting; namely, the sense of the sacred, that luminous mystery in which all things are bathed, I must now concede that there is one other: the realization of how far we all are from the goal that beckons, how many precipices must yet be climbed. As human beings we are created to surpass ourselves and are truly ourselves only when transcending ourselves. Only the slightest of barriers separates us from our sacred unconscious; it is infinitely close to us. But we are infinitely far from it, so for us the barrier looms as a mountain that we must dig through with bare hands. We scrape away at the earth, but in vain; the mountain remains. Still we go on digging at the mountain, in the name of God or whatever.

Of the final truth we for the most part only hear; very rarely do we experience it. The mountain isn't there. It never was there.

This is the one essay in this collection in which entheogens are not mentioned. The reader probably understands why I have included it, but at risk of belaboring the obvious I will be explicit. In ways (and for reasons that no one understands) entheogens hold the possibility of opening the doors of perception to the sacred unconscious.

CHAPTER SIX

Contemporary Evidence: Psychiatry and the Work of Stanislav Grof

In 1976 I published a book, Forgotten Truth: The Primordial Tradition, *which sketched in broad outlines the worldview, and correlative model of the human self, to which all traditional societies subscribed.*

Shortly before completing that book, I came upon the groundbreaking work of a Czech psychiatrist, Dr. Stanislav Grof, who had researched the entheogens more systematically than any other scientist. In Czechoslovakia he was assigned to determine whether LSD is a potential cure for schizophrenia. Child of his time and in the Soviet orbit – the Iron Curtain was solidly in place then – he had been schooled on the mechanistic, Pavlovian model of the mind which his psychiatric professors claimed was scientific. His own psychoanalysis, together with what came to light in the seventeen years (and more than twelve thousand clinical hours) he spent researching LSD with his patients, convinced him of the opposite.

I found his findings so in keeping with the traditional concept of the self that I had outlined in my book that I added an Appendix to my book to summarize his work. This chapter reprints that Appendix. It relates to the present book as follows: Whereas the preceding chapter proposed a model of the human self that allows room for authentic entheogenic revelations, this one doubles back on that chapter to use the entheogens to validate that model itself.

To tie those two essays together, I have included here the Sanskrit terms that I employed in the preceding chapter but which are not in the original version of this essay. In the interests of smoother reading I have deleted the footnotes that show where the statements by Grof that I quote verbatim *appear in his corpus. Those who want that information can find it in the Appendix of* Forgotten Truth *(HarperSanFrancisco, 1976, 1994).*

Know ten things, tell nine, the Taoists say – one wonders whether it is wise even to mention the entheogens in connection with God and the Infinite. For though a connection exists, as is the case with sex in Tantra, it is next to impossible to speak of it in the West today without being misunderstood. Such potential misunderstanding may be the reason that the identity of the Eleusinian sacrament is one of history's best-kept secrets, and why Brahmins came eventually to conceal (and then deliberately forget) the identity of Soma.

If the only thing to say about the entheogens were that they seem on occasion to disclose higher planes of consciousness and perhaps the Infinite itself, I would hold my peace. For though such experiences may be veridical in ways, the goal (it cannot be stressed too often) is not religious experiences, but the religious life. And with respect to the latter, chemically occasioned "theophanies" can abort a quest as readily as they can further it.

It is not, therefore, the isolated mystical experiences that entheogens can occasion that leads me to add this Appendix to *Forgotten Truth,* but rather evidence of a different order. Long-term, professionally garnered, and carefully weighed, this second kind of evidence deserves to be called (if anything in this murky area merits the attribution) scientific. I report that evidence here because of the ways in which (and extent to which) it seems to corroborate the primordial anthropology that *Forgotten Truth* presents and my preceding chapter summarizes. In contradistinction to writings on the entheogens that are occupied with experiences the mind can *have,* my concern here is with evidence they afford as to what the mind *is*.

The evidence in question is not widely known, for it has been reported only in a few relatively obscure journals and a single book. Still, judged both by the caliber of the data encompassed and the explanatory power of the hypotheses that make sense of the data, it is the most impressive evidence that the entheogens have thus far produced. It came together through the work of Stanislav Grof.

Grof's work began in Czechoslovakia, where for four years he worked in an interdisciplinary complex of research institutes in

Prague, and for another seven in the Psychiatric Research Institute in the same city. On coming to the United States in 1967 he continued his investigations at the Research Unit of Spring Grove State Hospital in Baltimore. Two covering facts about his work are worth noting before I turn to its content. First, in the use of LSD for therapy and personality assessment, his experience is by far the vastest that any single individual has amassed, covering as it does over twenty-five hundred sessions in each of which he spent a minimum of five hours with its subject. In addition, his studies cover another eight hundred cases that his colleagues in Prague and Baltimore conducted. Second, in spanning the Atlantic his work straddles the two dominant approaches to LSD therapy that have been developed: psycholytic therapy (used at Prague and favored in Europe generally), which involves numerous administrations of low-to-medium doses of LSD or its variants over a long therapeutic program, and high-dose therapy (confined to America), which involves one or several high-dose sessions in a short period of treatment.

The first thing Grof and his associates discovered was that there is no specific pharmacological effect that LSD invariably produces: "I have not been able to find a single phenomenon that could be considered an invariant product of the chemical action of the drug in any of the areas studied – perceptual, emotional, ideational, and physical." Not even mydriasis (prolonged dilation of the pupils), one of the most common symptoms, occurs invariably. Psychological effects vary even more than do physiological, but the range of the latter – mydriasis, nausea and vomiting, enhanced intestinal movements, diarrhea, constipation, frequent urination, acceleration as well as retardation of pulse, cardiac distress and pain, palpitations, suffocation and dyspnea, excessive sweating and hypersalivation, dry mouth, reddening of the skin, hot flushes and chills, instability and vertigo, inner trembling, fine muscle tremors – exceeds that of any other drug that affects the autonomic nervous system. These somatic symptoms are practically independent of dosage and occur in all possible combinations. Variability between subjects is equaled by variation in the symptoms a single subject will experience under different circumstances; particularly important from the clinical point of view are the differences that appear at different stages in the therapeutic process. All of this led Grof to conclude that LSD is not a specific causal agent, but rather a catalyst. It is an unspecific amplifier of neural and mental processes. By exteriorizing for the therapist, and raising to consciousness for the patient himself, material that is otherwise buried, and by enlarging this material to the point of caricature so that it appears as if under a magnifying glass, the LSD-like drugs are (Grof became convinced) unrivaled instruments: first, for identifying causes in psychopathology (the problem that is causing the difficulty); second, for personality diagnosis (determining the character type of the subject in question); and third, for understanding the human mind generally. "It does not seem inappropriate to compare their potential significance for psychiatry and psychology to that of the microscope for medicine or of the telescope in astronomy. Freud called dreams the 'royal way to the unconscious.' The statement is valid to a greater extent for LSD experiences."

Of the drug's three potentials, it is the third – its resources for enlarging our understanding of the human mind and self – that concerns us in this book. The nature of man is so central to our study that even flickers of light from Grof's work would make it interesting. That the light proves to be remarkably clear and steady makes it important.

Let me move at once to the point. The traditional view of human beings presents them as multilayered creatures, and Grof's work points to that same conclusion. As long as the matter is put thus generally it signals nothing novel, for standard depth psychology – psychiatry and psychoanalysis – concurs: the adjective "depth" implies as much, and metaphors of archaeology and excavation lace the writings of Freud, Jung, and their ilk. The novelty of Grof's work lies in the precision with which the levels of the mind that it uncovers correspond with the levels of selfhood the primordial tradition postulates.

In chemo-excavation the levels come to view sequentially. In this respect, too, images of archaeology apply: surface levels must be uncovered to get at ones that lie deeper. In high-dose therapy the deeper levels tend to appear later in the course of a single session; in psycholytic (low-dose) therapy they usually surface later in the sequence of therapeutic sessions. The sequences are the same, but since the levels first came to Grof's attention during his psycholytic work in Prague, and since that earlier work was the more extensive – it spanned eleven of the seventeen years he worked with the drugs – I shall confine myself to it in reporting his experimental work.

The basic study at Prague included fifty-two psychiatric patients. All major clinical categories were represented, from depressive disorders through psychoneuroses, psychosomatic diseases, and character disorders to borderline and clear-cut psychoses in the schizophrenic group. Patients with above-average intelligence were favored to obtain high-quality introspective reports; otherwise, cases with dim prognosis in each category were chosen. Grof himself worked with twenty-two of the subjects, his two colleagues with the remainder. The number of psycholytic sessions ranged from fifteen to one hundred per patient with a total of over twenty-five hundred sessions being conducted. Each patient's treatment began with several weeks of drug-free psychotherapy. Thereafter the therapy was punctuated with doses of 100 to 250 micrograms of LSD administered at seven- to fourteen-day intervals.

The basic finding was that "when material from consecutive LSD sessions of the same person was compared it became evident that there was a definite continuity between these sessions. Rather than being unrelated and random, the material seemed to represent a successive unfolding of deeper and deeper levels of the unconscious with a very definite trend."

The trend regularly led through three successive stages preceded by another which, being less important psychologically, Grof calls a preliminary phase. In this opening phase the chemical works primarily on the subject's body. In this respect it resembles what earlier researchers had called the vegetative phase, but the two are not identical. Proponents of a vegetative phase assumed that LSD directly causes the manifold somatic responses that patients typically experience in the early stages of the sessions. I have already mentioned that Grof's more extensive evidence counters this view. Vegetative symptoms are real enough, but they vary so much between subjects (and for a single subject under varying circumstances) that it seems probable that they are occasioned more by anxieties and resistances than by the chemical's direct action. There is also the fact that they are far from confined to early phases of the LSD sequence. These considerations led Grof to doubt that there is a vegetative phase per se. The most he is prepared to admit is that the drug has a tendency at the start to affect one specific part of the body: its perceptual and particularly its optical apparatus. Colors become exceptionally bright and beautiful, objects and persons are geometrized, things vibrate and undulate, one hears music as if one were somehow inside it, and so on. This is as close as the drug comes to producing a direct somatic effect, but that effect suffices to warrant speaking of an introductory phase which Grof calls "aesthetic."

With this preliminary phase behind him, the subject begins his psycholytic journey proper. Its first stage is occupied with material that is psychodynamic in the standard sense: Grof calls it the psychodynamic or Freudian stage. Experiences here are of a distinctly personal character. They involve regression into childhood and the reliving of traumatic infantile experiences in which Oedipal and Electra conflicts and ones relating to various libidinal zones are conspicuous; first and last, pretty much the full Freudian topography is traversed. The amount of unfinished business this layer of the self contains varies enormously; as would be expected, in disturbed subjects there is more than in normal ones. But the layer itself is present in everyone and must be worked through before the next stratum can be reached. "Worked through" again means essentially what psychiatry stipulates: a reliving not only in memory but also in emotion of the traumatic episodes that have unconsciously crippled the patient's responses. Freud and Breuer's hypothesis – that insufficient emotional and motor abreaction during early traumatic episodes produces a "jamming" of affect that provides energy for ensuing neurotic symptoms – is corroborated, for when patients in the course of a number of sessions enter into a problem area to the point of reliving it completely and integrating it into consciousness, the symptoms related to that area "never reappear" and the patient is freed to work on other symptoms.

This much was in keeping with Grof's psychiatric orientation; it came as "laboratory proof of the basic premises of psychoanalysis." But there that model gave out. For the experiences that followed, "no adequate explanation can be found within the framework of classical Freudian psychoanalysis."

Negatively, the new stage was characterized by an absence of the individual, biographically determined material that had dominated the sessions theretofore. As a result, the experiential content of this second stage was more uniform for the population than was the content of the first. I have already cited Grof's contention that LSD is not so much an agent that produces specific effects as it is an amplifier of material that is already present, and in the first stage the enlarging process worked to magnify individual differences: "The sessions of patients belonging to various diagnostic categories were characterized by an unusual inter- and also intra-individual variability." In the second stage the results reversed. With the magnifying glass still in place, variations receded: "The content seemed to be strikingly similar in all of the subjects." This is already important, for the emergent similarity suggests that the subjects were entering a region of the mind which they shared in common, a region that underlay the differing scrawls their separate biographies had incised upon it. As to content, "the central focus and basic characteristics of the experience on this level are the problems related to physical pain and agony, dying and death, biological birth, aging, disease and decrepitude" – the Buddha's First Noble Truth, Grof somewhere observes, and the three of the Four Passing Sights that informed it. Inevitably, he continues,

> the shattering encounter with these critical aspects of human existence and the deep realization of the frailty and impermanence of man as a biological creature is accompanied by an agonizing existential crisis. The individual comes to realize through these experiences that no matter what he does in his

> life, he cannot escape the inevitable: he will have to leave this world bereft of everything that he has accumulated, achieved and has been emotionally attached to.

Among the phenomena of this second stage, the theme of death and rebirth recurred so frequently that it sent Grof to a book he had heard of in his psychiatric training but had not read, it having been written by a psychoanalytic renegade, Otto Rank. It bore the title *The Trauma of Birth,* and, to use Grof's own word, he was "flabbergasted" to find how closely second-stage psycholytic experiences conformed to it. He and his colleagues fell to calling the second stage perinatal or Rankian.

During the weeks through which this stage extends, the patient's clinical condition often worsens. The stage climaxes in a session in which the patient experiences the agony of dying and appears – to himself – actually to die.

> The subjects can spend hours in agonizing pain, with facial contortions, gasping for breath and discharging enormous amounts of muscular tension in various tremors, twitching, violent shaking and complex twisting movements. The color of the face can be dark purple or dead pale, and the pulse rate considerably accelerated. The body temperature usually oscillates in a wide range, sweating can be profuse, and nausea with projectile vomiting is a frequent occurrence.

This death experience tends to be followed immediately by rebirth, an explosive ecstasy in which joy, freedom, and the promise of life of a new order are the dominant motifs.

Outside the LSD sequence, the new life showed itself in the patients' marked clinical improvement. Within the sequence it introduced a third experiential landscape. When Grof's eyes became acclimated to it, it appeared at first to be Jungian, Jung being the only major psychologist to have dealt seriously and relatively unreductively with the visions that appear. Later it seemed better to refer to the stage as "transpersonal."

Two features defined this third and final stage. First, its "most typical characteristics . . . were profound religious and mystical experiences."

> Everyone who experientially reached these levels developed convincing insights into the utmost relevance of spiritual and religious dimensions in the universal scheme of things. Even the most hardcore materialists, positivistically oriented scientists, skeptics and cynics, uncompromising atheists and anti-religious crusaders such as the Marxist philosophers, became suddenly interested in spiritual search after they confronted these levels in themselves.

Grof speaks of levels in the plural here, for the "agonizing existential crisis" of the second stage is already religious in its way – death and rebirth are ultimates or there are none. The distinguishing feature of the third stage is not, strictly speaking, that it is religious, but that it is (as Grof's works indicate) mystically religious: religious in a mode in which (a) the whole subsumes its parts, and (b) evil is thereby rescinded. This connects with the stage's other feature, its transpersonal aspect, which was so pronounced as to present itself in the end as the best name for that stage. A trend toward transpersonal experiences, in which one is occupied with things other than oneself, had already shown itself in stage two. Suffering, for example, which in the first stage presented itself in the form of recollected autobiographical traumas, had in the second stage taken the form of identifying with the suffering of others, usually groups of others: famine victims, prisoners in concentration camps, or humankind as a whole with its suffering as symbolized by Christ on his cross, Tantalus exposed to eternal tortures in Hades, Sisyphus sentenced to roll a boulder uphill incessantly, Ixion fixed on the wheel, or Prometheus chained to his rock. Likewise with death. Already by stage two "the subjects felt that they were operating in a framework which was 'beyond individual death.'" The third stage continues this outbound, transpersonal momentum. Now the phenomena with which the subject identifies are not restricted to humankind or even to living forms. They are cosmic, having to do with the elements and forces from which life proceeds. And the subject is less conscious of himself as separate from what he perceives. To a large extent the subject-object dichotomy itself disappears.

So much for description of the three stages. Now to interpretation and explanation.

Grof was and is a psychiatrist. Psychiatry is the study and practice of ontogenetic explanations: it accounts for present syndromes in terms of its patients' antecedent experiences. Freud had mined these experiences as they occur in infancy and childhood, but Grof's work had led to regions that Freud's map did not reveal. As a psychiatrist, Grof had nowhere to turn for explanations save further in the same direction – further back. His methodology forced him to take seriously the possibility that experiences attending birth and even gestation could affect ensuing life trajectories.

Taking his cues from *The Trauma of Birth* while amending it in important respects, Grof worked out a typology in which second- and third-stage LSD experiences are correlated with four distinct stages in the birth process: (a) a comfortable, intrauterine stage before the onset of labor; (b) an oppressive stage at labor's start when the fetus suffers the womb's contractions and has "no exit" because the cervix has not opened; (c) the traumatic ensuing stage of labor during which the fetus is violently ejected through the birth canal; and (d) the freedom and release of birth itself. Stages (b) and (c) seemed to Grof to vector the second or Rankian stage in the LSD sequence. In the reliving of (b), the oppressiveness of the womb is generalized, causing the entire world and existence itself to seem oppressive; and when (c) – the agony of labor and forced expulsion through the birth canal – is relived, this produces the experience of dying: traumatic ejection from the only life-giving context the fetus has known. The rebirth experience (in which the Rankian stage climaxes) derives from reliving the experience of physical birth (d) and paves the way for the ensuing transpersonal stage. The sense of unshadowed bliss that dominates this final stage plugs into the earliest memories of all, those that precede the time when the womb was congested, back when the fetus was blended with its mother in mystic embrace (a). Some cognitive scientists reject this aspect of Grof's work on grounds that the brain structures known to support memory do not develop until months after birth, but Grof replies that this assumes a specificity regarding the neural underpinnings of memory in its entirety that has not been proven. When subjects in their Rankian stage report first suffocation and then a violent, projective explosion in which not only blood but urine and feces are everywhere, one is persuaded that revived memories of the birth process play some part in triggering, shaping, and energizing later-stage LSD experiences. The question is: Are they the only causes at work? As I have noted, in the psychiatric model of humankind, once the Freudian domain has been exhausted there is nowhere to look for causes save where Rank did, and where Grof followed him in looking. Driven back to earlier and yet earlier libido positions, the ego finally reenters the uterus.

In the anthropological model that *Forgotten Truth* and the preceding chapter of the present book describe, things are different. Therein, the social and biological histories of the organism are not the sole resources for explanation. "The soul that rises with us, / Hath had elsewhere its setting, / And cometh from afar: / Not in entire forgetfulness . . . / But trailing clouds of glory do we come." If we ask from whence we come, Wordsworth answers, "from God" and traditional people in principle agree. More proximately, however, it is from the psychic plane – our subtle bodies or souls – that our gross bodies derive. In the psychiatric perspective, bodies are basic, and explanations for mental occurrences are sought in the body's endowments and history, whereas in the traditional model physical bodies represent a kind of shaking out of mental phenomena that precede and are more real than their physical printouts.

Thus, to Grof's finding that later stages in the LSD sequence conform sufficiently to the stages of the birth process to warrant our saying that they are influenced by those stages, tradition adds: "influenced by" only, not caused by. The experiences of those stages put the subject in direct touch with the psychic and archetypal forces of which his life is a distillation and product. For birth and death are not physical only. This much everyone knows, but it is less recognized that physical birth and death are relatively minor manifestations of forces that are cosmic in blanketing the manifest world. Buddhism's *pratitya-samutpada* (Formulation of

Dependent Origination) speaks profoundly to this point, but all I shall say is that when a psychic quantum, the germ of an ego, decides – out of ignorance, the Buddhists insert immediately – that it would be interesting to go it alone and have an independent career, in separating itself from the whole (and in ways setting itself against the whole) the ego shoulders consequences. Because it is finite, things will not always go its way: hence suffering. And the temporal side of the self's finitude ordains that it will die – incrementally from the start as cells and early ambitions die, and eventually in its entirety. Energy is indestructible, however, so in some form there is rebirth. Confrontation of these principal truths in their transpersonal and trans-species generality is the stuff of the later-stage LSD experience. Biological memory enters, but conceivably with little more than a "me too": I too (that memory insists) know the sequence from the time when I was formed and delivered.

Spelled out in greater detail, the primordial explanation of the sequence runs as follows. Accepting LSD as a "tool for the study of the structure of human personality; of its various facets and levels," we see it uncovering the successively deeper layers of the self which Grof's study brings to light. Grof's psychiatric explanation for why it does so is that "defense systems are considerably loosened, resistances decrease, and memory recall is facilitated to a great degree. Deep unconscious material emerges into consciousness and is experienced in a complex symbolic way." The traditional explanation shifts the emphasis. Only in the first stage are the loosened defense systems ones that the individual ego has built to screen out painful memories. For the rest, what is loosened are structures that condition the human mode of existence and separate it from modes that are higher: its corporeality and compliance with the spatio-temporal strictures the gross body must conform to. The same holds for the memory-recall that LSD facilitates. In the first stage it is indeed recollections of experiences that are activated as the subject relives, directly or in symbolic guise, the things the subject individually experienced that are brought to light. But as the archeology delves deeper, what the psychiatrist continues to class as individual recollections – an even earlier, intrauterine memory – the ontologist (short of invoking reincarnation) sees as the discovery of layers of selfhood that are present from conception but are normally obscured from view. Likewise with the "peculiar double orientation and double role of the subject" that Grof describes. "On the one hand," he writes, the subject "experiences full and complex age regression into the traumatic situations of childhood; on the other hand, he can assume alternately or even simultaneously the position corresponding to his real age." This oscillation characterizes the entire sequence, but only in the first stage is its not-immediate referent the past. In the later sessions, that which is not immediate is removed not in time but in space – psychological space, of course. It lies below the surface of the exterior self that is normally in view.

The paradigm of the self that is sketched in chapter 4 of *Forgotten Truth* and the preceding chapter in the book in hand shows it to have four components: body, mind, soul, and Spirit; in the nomenclature of this book, gross body, subtle body, causal body, and "the Fourth." Working with spatial imagery, we can liken LSD to an MRI scan that sweeps progressively toward the core of the subject's being. In the early sessions of the LSD sequence it moves through the subject's body in two steps. The first of these triggers peripheral somatic responses (most regularly ones relating to perception) to produce the aesthetic phase. The second moves into memory regions of the brain where traces of past experiences are stored. That the events that were most important in the subject's formation are the ones that rush forward for attention, stands to reason. We are into the first of the three main stages of the psycholytic sequence, the psychodynamic or Freudian stage.

Passage from the Freudian to the Rankian stage occurs when the chemicals enter the region of the mind that outdistances the brain and swims in the psychic plane which is the subtle body's medium. The phenomenological consequences could almost have been predicted:

1. Biographical data – which imprinted themselves on the subject's body, specifically the memory region of his brain – recede.

2. Their place is taken by "existentials," the conditioning structures of human existence in general. The grim affect of this stage could be due in part to memories of the ordeals of gestation and birth; but the torment – the anguish arising from the realization of the amount of suffering that is endemic to life – derives mainly from the fact that the larger purview of the intermediate plane renders the inherent contractions of the terrestrial plane (*dukkha*) more visible than when the subject is individually immersed in them.

3. In the death-and-rebirth experience that climaxes this phase, Rankian considerations could play a part while again not preempting the show. The self had entered the psychic plane through the causal body's assumption of – compression into – the subtle body's sheath. Now, in reversing this sequence, the subtle body must die in order to return to its more august causal body.

Descriptions of this return appear regularly in subjects' reports of the transpersonal stage in their treatment. Here are some of its features:

1. Whereas in the Rankian stage "there was a very distinct polarity between very positive and very negative experience," experience is now predominantly beatific, with "melted ecstasy" perhaps its most-reported theme. Subjects "speak about mystic union, the fusion of the subjective with the objective world, identification with the universe, cosmic consciousness, the intuitive insight into the essence of being, the Buddhist's *nirvana*, the Veda's *samadhi*, the harmony of worlds and spheres, the approximation to God, etc."

2. Experience is more abstract. At its peak it "is usually contentless and accompanied by visions of blinding light or beautiful colors (heavenly blue, gold, the rainbow spectrum, peacock feathers, etc.)" or is associated with space or sound. When its accoutrements are more concrete they tend to be archetypal, with the archetypes seeming to be limitless in number.

3. The God who is almost invariably encountered is single and so far removed from anthropomorphism as to elicit, often, the

pronoun "it." This is in contrast to the gods of the Rankian stage who tend to be multiple, Olympian, and essentially human beings writ large.

4. Beyond the causal body (soul) lies only "the Fourth," (Spirit), an essence so ineffable that when the seeing eye perceives it, virtually all that can be reported is that it is "beyond" and "more than" all that had been encountered theretofore.

The correlations between traditional anthropology and the LSD sequence can be diagrammed as follows.

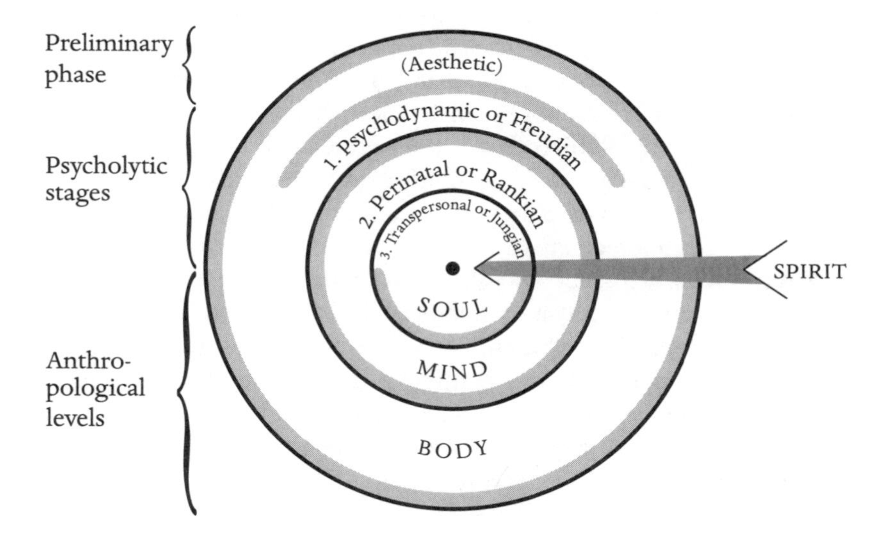

Up to this point I have summarized Grof's empirical findings and pointed to how they can be explained by the traditional model of the human self as readily as by his psychiatric model. It remains to note how the findings of his seventeen years of research affected his own thinking.

Engaged as he was in "the first mapping of completely unknown territories," he could not have foreseen where his inquiry would lead. What he found was that in "the most fascinating intellectual and spiritual adventure of my life, it opened up new fantastic areas and forced me to break with the old systems and frameworks."

The first change in his thinking has already been noted: the psycholytic sequences showed the birth trauma to have more dynamic consequences than Grof and his strictly Freudian associates had supposed. This change psychoanalysis could accommodate, but not the one that followed. "I started my LSD research in 1956 as a convinced and dedicated psychoanalyst," he writes. "In the light of everyday clinical observations in LSD sessions, I found this conception untenable." Basically, what proved to be untenable was "the present gloomy image of man, which is to a great extent influenced by psychoanalysis."[1] This picture of man,

> that of a social animal basically governed by blind and irrational instinctual forces, contradicts the experiences from the LSD sessions, or at least appears in their light superficial and limited. Most of the instinctual tendencies described by psychoanalysis (incestuous and murderous wishes, cannibalistic impulses, sadomasochistic inclinations, coprophilia, etc.) are very striking in the early LSD sessions; they are so common that they could almost be considered experimental evidence for some of the basic assumptions of psychoanalysis. Most of them, however, appear in the sessions for only a limited period of time. This whole area can be transcended, whereupon we are confronted with an image of man that is diametrically opposed to the previous one. Man in his innermost nature appears then as a being that is fundamentally in harmony with his environment and is governed by intrinsically high and universal values.

This change in anthropology has been the solid effect of entheogenic evidence on Grof's thinking. In psychoanalytic terms, Freud discovered the importance of infantile experience on ontogenetic development, Rank the importance of the experience of birth itself, and Grof's discoveries carry this search for ever earlier etiologies – in psychoanalytic theory earlier means stronger – to its logical limit: his optimistic view of man derives from discovering the influence and latent power of early-gestation memories of the way things were when the womb was still uncongested and all was well.

Beyond this revised anthropology, however, Grof has toyed with a changed ontology as well. Endowments that supplement his psychiatric competences have helped him here: he has a "musical ear," so to speak, for metaphysics, and an abiding interest in the subject. These caused him to listen attentively from the start to his subjects' reports on the nature of reality, and in one of his recent papers, "LSD and the Cosmic Game: Outline of Psychedelic Cosmology and Ontology," he gives these reports full rein. Laying aside for the interval his role as research psychiatrist, which required his seeing his patients' experiences as shaped by (if not projected from) early formative experiences, in this paper Grof turns phenomenologist and allows their reports to stand in their own right. The view of reality that results is so uncannily like the traditional one that *Forgotten Truth* and the preceding chapter of this book outline that, interweaving direct quotations from Grof's article (without bothering with quotation marks) with ones from those two sources, the picture that emerges looks like this:

> The ultimate source of existence is the Void, the supracosmic Silence, the uncreated and absolutely ineffable Supreme.
>
> The first possible formulation of this source is Universal Mind. Here, too, words fail, for Universal Mind transcends the dichotomies, polarities, and paradoxes that harass the relative world and our finite minds' comprehension of it. Insofar as description is attempted, the Vedantic ternary – Infinite Existence, Infinite Intelligence, Infinite Bliss – is as serviceable as any.
>
> God is not limited to his foregoing, "abstract" modes. He can be encountered concretely, as the God of the Old and New Testaments, Buddha, Shiva, or in other modes. These modes do not, however, wear the mantle of ultimacy or provide final answers.
>
> The phenomenal worlds owe their existence to Universal Mind, which Mind does not itself become implicated in their categories. Man, together with the three-dimensional world he experiences, is but one of innumerable modes through which Mind experiences itself. The heavy physicality and seemingly objective finality of man's material world, its space-time grid and the laws of nature that offer themselves as if they were the *sine qua nons* of existence itself – all these are in fact highly provisional and relative. Under exceptional circumstances, people can rise to a level of consciousness at which they see that taken

together they constitute but one of innumerable sets of limiting constructs that Universal Mind assumes. To saddle that Mind itself with those constructs would be as ridiculous as trying to understand the human mind through the rules of chess.

Created entities tend progressively to lose contact with their original source and the awareness of their pristine identity with it. In the initial stage of this falling away, those entities maintain contact with their source, and the separation is playful, relative, and obviously tentative. An image that illustrates this stage is that of a wave of the ocean. From a certain point of view the wave is a distinct entity – we can speak of it as large, fast-moving, green, and foamy. But its individuation doesn't keep it from belonging to the ocean proper.

At the next stage, created entities assume a partial independence and we can observe the beginnings of cosmic screenwork, the Absolute's assumption of veils that are gossamer-like in the beginning but grow increasingly opaque. Here unity with the source can be temporarily forgotten in the way an actor can forget his own identity as he identifies with the character he depicts.

Eventually the veiling process reaches a point where individuation looks like the normal state of things and the original wholeness is perceived only intuitively and sporadically. This can be likened to the relationship between cells of a body and the body as a whole. Cells are separate entities but function as their body's parts. Individuation and participation are dialectically combined. Complex biochemical interactions bridge provisional boundaries to ensure the functioning of the organism as a whole.

In the final stage, the separation is practically complete. Liaison with the source is lost sight of and the original identity forgotten. The screen is now all but impermeable, and a radical change of consciousness is required to break through it. A snowflake can serve as a symbol. In outward appearance it doesn't look like water; to understand that nevertheless it is water we have to get down to H_2O.

Human beings who manage to effect the requisite breakthrough find thereafter that life's polarities paradoxically do and do not exist. This holds for such contraries as matter and spirit, good and evil, permanence and change, heaven and hell, beauty

> and ugliness, and agony and ecstasy. In the end, there is no difference between subject and object, observer and observed, experiencer and experienced, creator and creation.

In the early years of psychoanalysis when hostility was shown to its theories on account of their astonishing novelty and they were dismissed as products of their authors' perverted imaginations, Freud used to hold up against this objection the argument that no human brain could have invented such facts and connections had they not been persistently forced on it by a series of converging and interlocking observations. Grof might argue in the same way: to wit, that the cosmology and ontology that his patients came up with is as uninventable as Freud's own system. Actually, however, he does not do so. In the manner of a good phenomenologist, he lets the evidence speak for itself, neither undermining it by referring it back to causes which (in purporting to explain it) would explain it away nor arguing that it is true. As phenomenologists themselves would say, he "brackets" his own judgment regarding the truth question and contents himself with summarizing what his patients said.

> The idea that the "three-dimensional world" is only one of many experiential worlds created by the Universal Mind appeared to them much more logical than the opposite alternative that is so frequently taken for granted, namely, that the material world has objective reality of its own and that the human consciousness and the concept of God are merely products of highly organized matter, the human brain. When closely analyzed the latter concept presents at least as many incongruities, paradoxes and absurdities as the concept of the Universal Mind. Problems such as the finitude versus infinity of time and space; the enigma of the origin of matter, energy and space; and the mystery of the prime impulse appear to be so overwhelming and defeating that one seriously questions why this approach should be given priority in our thinking.

CHAPTER SEVEN

The Good Friday Experiment

Mention has already been made of the Good Friday Experiment, in which – to test the power of entheogens to occasion mystical experiences in a religious setting – Walter Pahnke conducted a study in which theological professors and students were, in 1962, given psilocybin preceding the traditional Good Friday service at Boston University. The project was the research topic for the doctoral degree he received from Harvard University.

In addition to the detailed report of the experiment that appeared in Pahnke's doctoral dissertation, there have been a number of journalistic accounts of it and a study of the long-term consequences of that day for its subjects (see references at the back of this book). However, one significant incident that occurred during the experiment had not appeared in the reports, and as I was party to it, Thomas Roberts, Professor of Educational Psychology at Northern Illinois University, interviewed me about it to read it into the record. Robert Jesse of the Council on Spiritual Practices was in attendance.

THOMAS ROBERTS

This is October First, 1996. Huston Smith will be telling us about an event that happened in the Good Friday Experiment in 1962. Huston, do you want to tell us about the student who ran out on the experiment?

HUSTON SMITH

Just keep me on course.

ROBERTS

O.K.

SMITH

The basic facts of the experiment have been recorded elsewhere and are fairly well known, but I will summarize them briefly. In the early sixties, Walter Pahnke, a medical doctor with strong interests in mysticism, wanted to augment his medical knowledge with a doctorate in religion. He had heard that the entheogens often occasion mystical experiences, so he decided to make that issue the subject of his research. He obtained the support of Howard Thurman, Dean of Marsh Chapel at Boston University, for his project, and also that of Walter Houston Clark who taught psychology of religion at Andover Newton Theological Seminary and shared Wally's interest in the entheogens.

Clark procured twenty volunteer subjects, mostly students from his seminary. Ten more volunteers, of whom I was one, were recruited as guides. Howard Thurman's two-and-a-half-hour 1962 Good Friday service at Boston University would be piped down to a small chapel in the basement of the building where the volunteers would participate in it. Fifteen of us would receive, double blind, a dose of psilocybin, and the remaining fifteen a placebo: nicotinic acid, which produces a tingling sensation that could make its recipients think they had gotten the real thing. The day after the experiment we would write reports of our experiences, and Pahnke would have them scored by independent raters on a scale of from zero to three for the degree to which each subject's experience included the seven traits of mystical experience that W. T. Stace lists in his classic study, *Mysticism and Philosophy*. There was one borderline case, but apart from that, the experiences of those who received psilocybin were dramatically more mystical than those in the control group. I was one of those who received the psilocybin, and I will say a word about my experience before I proceed to the student that you asked about.

The experiment was powerful for me, and it left a permanent mark on my experienced worldview. (I say "experienced worldview" to distinguish it from what I think and believe the world is like.) For as long as I can remember I have believed in God, and I have experienced his presence both within the world and when the world was transcendentally eclipsed. But until the Good Friday Experiment, I had had no direct personal encounter with God of the sort that *bhakti yogis,* Pentecostals, and born-again Christians describe. The Good Friday Experiment changed that, presumably because the service focused on God as incarnate in Christ.

For me, the climax of the service came during a solo that was sung by a soprano whose voice (as it came to me through the prism of psilocybin) I can only describe as angelic. What she sang was no more than a simple hymn, but it entered my soul so deeply that its opening and closing verses have stayed with me ever since.

My times are in Thy hands, my God, I wish them there;
My life, my friends, my soul, I leave entirely in Thy care. . . .

My times are in Thy hands, I'll always trust in Thee;
And after death at Thy right hand I shall forever be.

In broad daylight those lines are not at all remarkable, but in the context of the experiment they said everything. The last three measures of each stanza ascended to a dominant seventh which the concluding tonic chord then resolved. This is as trite a way to end a melody as exists, but the context changed that totally. My mother was a music teacher, and she instilled in me an acute sensitivity to harmonic resonances. When that acquisition and my Christian nurturance converged on the Good Friday story under psilocybin, the gestalt transformed a routine musical progression into the most powerful cosmic homecoming I have ever experienced.

Having indicated how I experienced the service I can turn now to the incident that is the main point of this interview.

As the psilocybin began to take its effect, I became aware of a mounting disorder in the chapel. After all, half of our number were in a condition where social decorum meant nothing, and the other half were more interested in the spectacle that was unfolding before them than in the service proper. In any case, from out of this bizarre mix, one of our number emerged. He arose from his pew, walked up the aisle, and with uncertain steps mounted the chapel's modest pulpit. Thumbing through its Bible for a few moments, he proceeded to mumble a brief and incoherent homily, blessed the congregation with the sign of the cross, and started back down the aisle. But instead of returning to his pew, he marched to the rear entrance of the chapel and through its door.

Now, before the experiment began, we had been arranged in groups of four subjects plus two guides. We knew that two subjects and one guide in each group would receive psilocybin, and the guides were instructed to look after the others as needed. "John" (I withhold his actual name) was not my charge, but as no one got up to follow him, I did so.

This introduces an interesting parenthetical point. More than once I have been struck by the widely corroborated fact that however deep one may be into the chemical experience, short of the dissociation that John was experiencing, one can snap back to normal if need arises to so do. So it was in this case. When John's guide didn't respond to his leaving the chapel, I sprang to my feet and followed him out.

He had made a right turn and was striding down the hall, but that didn't worry me because we had been told that the entire basement had been sealed off for the experiment. But when he reached the door at the end of the corridor and jammed down its latch bar, it swung open. Something had misfired in the instructions to the janitor, and my charge, totally transported by his altered state, was loose on Commonwealth Avenue. I ran after him, but my remonstrances to return to the chapel fell on deaf ears, and he shook off my grip as if it were cobwebs.

What to do? I was afraid to leave him lest I lose track of him, but alone I was powerless to dissuade him from what appeared to be an appointed mission.

Providentially, help arrived from an unexpected quarter. Eva Pahnke, Wally's wife, was having a picnic on the grass with their child, and I shouted to her to keep track of John while I rushed back to the chapel for help. The strategy worked. When I returned on the double with Wally and another helper, John was still visible a block and a half ahead. Before we reached him we saw him enter what turned out to be 745 Commonwealth Avenue, the building

that houses Boston University's School of Theology and parts of its College of Liberal Arts. We caught up with him on the stairs to the third floor, but Wally's remonstrances cut no more ice than mine had. Together, however, the three of us were able to block his further ascent.

Things were at a standstill when a postman rounded the corner of the steps from below. He was carrying a brown envelope copiously plastered with special-delivery stickers, and as he was passing us John's arm shot out and snatched it from him. I was too occupied to notice the expression on the postman's face, for two of us had all we could do pinning John's arms while finger by finger, Wally pried the crumpled envelope from John's viselike grip and returned it to its stunned carrier. I have often wondered how that postman explained the mangled condition of his packet to its recipient, and how he described the incident to his wife at dinner that evening.

The rest is simply told. Realizing that he was overpowered – barely, for under the influence his strength was like Samson's – John, tightly flanked, submitted to being walked back to the chapel where Wally injected thorazine, an antidote. That returned him to his right mind, but with no recollection of what had happened. It took twenty-four hours for all the pieces of the episode to come back to him and fall into place, whereupon this was his story.

God, it emerged, had chosen him to announce to the world the dawning of the Messianic Age, a millennium of universal peace. (As often happens in such cases, the actual wording of the message made little sense to normal ears.) In his homily in the chapel, John broke the good news to our congregation, but he needed to get it to the world at large, which was what caused him to leave the chapel. When, walking down Commonwealth Avenue, he saw the plaque announcing "Dean of the College of Liberal Arts" by the entrance to 745 Commonwealth Avenue, it occurred to him that deans have influence, so if he could get to him, the dean would call a press conference that would complete John's mission. The postman's packet, he assumed, was for the dean, so if he attached himself to it, it would carry him to the dean himself.

John's long-term feelings about the experiment were heavily negative. He was the only one of its subjects who refused to participate in Rick Doblin's twenty-fifth-year retrospective study of the long-term effects of the experiment on its subjects, and he threatened to sue if his name was included.

ROBERTS

Was there something in Howard Thurman's sermon that prompted him to his messianic mission?

SMITH

Not to my notice. It was a typical Good Friday service with meditations on each of the seven last words of Christ. Howard Thurman was a remarkable man, both spiritually and in his ability to inspire people, but I don't remember the content of his words, only their moving impact. Did you ever meet him?

ROBERTS

No. I wish I had.

SMITH

Just relating this story I feel my spine tingling.

ROBERTS

Are there other memories of the afternoon that come to mind?

SMITH

I come back to the hubbub that erupted at times in our chapel. I was too deep into my own experience to be distracted by it, but I was peripherally aware of it and realized in retrospect that an observer would have found us a pretty unruly bunch. Half of us were enraptured, while the other half (as I learned from several of them the next day) felt left out and were not above acting out their resentment in derisive laughter and incredulous hoots over the way the rest of us were behaving.

I also recall a short exchange with one of our number in the foyer to the chapel just before the service began. I was already feeling my psilocybin, and sensing – wrongly, it turned out – that he was as well, I said to him from the depths of my being, "It's true, isn't it?" By "it," I meant the religious outlook, God and all that follows from God's reality. He didn't respond and told me when we next met that he had gotten only the placebo and hadn't a clue as to what I was talking about. So I was dead wrong in inferring from our eye contact that our minds were in sync.

ROBERTS

Anything else?

SMITH

Only the gratitude I feel toward Wally for having mounted the experiment – as you know, it's a poignant gratitude for he died nine years later in a tragic scuba diving accident. I have explained how it enlarged my understanding of God by affording me the only powerful experience I have had of his personal nature. I had known and firmly believed that God is love and that none of love's nuances could be absent from his infinite nature; but that God loves *me,* and I *him,* in the concrete way that human beings love individuals, each most wanting from the other what the other most wants to give and with everything that might distract from that holy relationship excluded from view – *that* relation with God I had never before had. It's the theistic mode that doesn't come naturally to me, but I have to say for it that its carryover topped those of my other entheogenic epiphanies. From somewhere between six weeks and three months (I should judge) I really *was* a better person – even at this remove, I remain confident of that. I slowed down a bit and was somewhat more considerate. I was able to some extent to prolong the realization that life really is a miracle, every moment of it, and that the only appropriate way to respond to the gift that we have been given is to be mindful of that gift at every moment and to be caring toward everyone we meet. To carry those sentiments with one onto the campus of the Massachusetts Institute of Technology requires empowerment.

ROBERTS

Thank you, Huston.

SMITH

You are welcome.

CHAPTER EIGHT

The Case of Cardinal John Henry Newman

Twenty years had elapsed since I had last written about the entheogens, and I might have kept my peace had not the editors of the journal ReVISION *dedicated one of its 1988 issues to the topic and requested an entry from me. My initial impulse was to decline, thinking that I had said all I had to say on the subject, but then I remembered that I had come upon a significant item in the interval that I had not reported. It relates to a towering nineteenth-century intellect, Cardinal John Henry Newman, and I used my entry for* ReVISION *to tell the story.*

For their relevance to the Newman story, I have (in this printing of it) added Dostoevsky's conclusions about his epilepsy. I urge the reader not miss those conclusions, as he might be likely to do because they appear in a footnote (I have placed them in that category only to keep them from interrupting the flow of the central, Newman narrative). Far from being incidental, Dostoevsky's perplexity and "paradoxical conclusion" (which he camouflages as fiction in The Idiot*) are so close to the ones that have kept me circling the entheogens for forty years that I could easily have paraphrased them to provide a completely satisfying rationale for my publishing this book.*

I thought that I had published everything I have to say about the entheogens, but *ReVISION*'s invitation to contribute to this issue of the journal reminds me that there is one item that has come my way which I have not seen reported. I begin with the circumstances that brought it to light, and I shall dwell on them somewhat, for the improbabilities involved are enough to remind one once again of life's strangeness. If this had not occurred, and that – and that, and that, and that, each in its seemingly appointed time and place – a pertinent fact about

John Henry Newman's conversion might never have come to light.

The story centers in the three-week Salzburg Seminar of 1972. Endowed by an Austrian immigrant to the United States who used the fortune he made to throw a bridge from his native homeland to the country that had treated him well, these seminars, which take place annually, are designed to inform influential Europeans about current aspects of American life, and the turbulent sixties drew me into their orbit. Taken together, the convulsive events of that decade – civil rights demonstrations (Martin Luther King), protests against the war in Vietnam (Daniel Ellsberg), musical breakthroughs (Bob Dylan and the Beatles), and Woodstock with its flower children and summer of love – led the public to reify the decade, making it a freestanding thing in itself. And because psychedelics (Timothy Leary and Haight-Ashbury) were a flamboyant part of that decade and I had paid more attention to them than had most academics, I was invited to be one of the four leaders of the 1972 seminar. Martin Marty, Robert McAfee Brown, and Michael Novak dealt with the civil rights revolution and Vietnam, while I did what I could to interpret the rise of the counterculture in America with its psychedelics included. One of the participants in the seminar, Hillary Jenkins, had just completed a book on John Henry Newman, *Newman's Mediterranean Voyage,* and when his knowledge of Newman's life intersected with what I was telling the seminar about the entheogens, some disparate pieces coalesced in a hypothesis that startled us all.

I begin with Jenkins's part of the story. In researching Newman's life, Jenkins learned that in his early twenties he had been severely depressed. He had no idea what to do with his life and was plagued by a crippling anxiety in which he oscillated between fear of failure on the one hand and an ambitious but daunting desire to advance himself on the other. In the hope of relieving his depression, his parents sent him on a Mediterranean vacation, and their plan succeeded in a way no one could have anticipated. In the course of his vacation he had a religious experience that was so powerful that not only did it pull him out

of his depression, it caused him quickly to become a public figure and made him in time an intellectual giant of his century. His *Idea of the University* is still regarded as one of the best books on education ever written. "Lead, Kindly Light" is one of Christendom's best-loved hymns. And Newman Centers – his tangible memorial – grace every major college and university campus in America.

The interests of this book enter when we learn that the experience that literally turned Newman into a "new man" occurred in the course of a near-fatal bout with a disease now judged to have been typhoid fever, and that Newman himself was aware in retrospect that (in his own words) "at the time that I had a most consoling and overpowering thought of God's electing love and seemed to feel I was His, all my feelings, painful and pleasant, were, I believe, heightened somewhat by delirium."

Here my own studies entered the picture, specifically something I picked up at a conference on psychoactive substances that the R. M. Bucke Society convened in Canada in the late 1960s. In the paper that Raymond Prince – a medical anthropologist in the Section of Transcultural Psychiatric Studies at McGill University – presented at that conference, he argued that infectious diseases, and other afflictions such as starvation and exhaustion on long hunting expeditions, probably figured more prominently than hallucinogenic plants in opening early man to the supernatural. These afflictions, too, affect brain chemistry in visionary directions and probably beset people quite regularly.* The bacteria

* Prince does not mention the brain states that immediately precede epileptic fits, but it seems reasonable to include them in his list. Dostoevsky's generalized description of the experience that accompanies those states – which he knew at first hand and which he puts in the mouth of Myshkin in *The Idiot* – has become classic: "[Myshkin] remembered among other things that he always had one minute just before the epileptic fit (if it came on while he was awake), when suddenly in the midst of sadness, spiritual darkness and oppression, there came at moments a flash of light in his brain, and with extraordinary impetus all his vital forces suddenly began working at their highest tension. The sense of life, the consciousness of self, were multiplied ten times at these moments which passed like a flash of lightning. His mind and his heart were flooded with extraordinary light; all his uneasiness, all his doubts, all his anxieties were relieved at once; they were all merged in a lofty calm, full of serene, harmonious joy and hope. But these moments, these flashes,

and other micro-organisms that cause infectious diseases are themselves plants and fungi, or closely akin to them. Moreover, it is they (not the fever their attacks provoke) that account for the visions that occur, for (a) deliriums may precede temperature rise and follow its return to normal, (b) fevers experimentally produced without infectious organisms do not alter consciousness, and (c) some febrile illnesses occasion deliriums, whereas

were only the prelude of that final second (it was never more than a second) with which the fit began. That second was, of course, unendurable. Thinking of that moment later, when he was all right again, he often said to himself that all these gleams and flashes of the highest sensation of life and self-consciousness, and therefore also of the highest form of existence, were nothing but disease, the interruption of the normal conditions; and if so, it was not at all the highest form of being, but on the contrary must be reckoned the lowest. And yet he came at last to an extremely paradoxical conclusion. 'What if it is a disease?' he decided at last. 'What does it matter that it is an abnormal intensity, if the result, if the minute of sensation, remembered and analyzed afterwards in health, turns out to be the acme of harmony and beauty, and gives a feeling, unknown and undivined till then, of completeness, of proportion, of reconciliation and of ecstatic devotional merging in the highest synthesis of life?' These vague expressions seemed to him very comprehensible, though too weak. That it really was 'beauty and workship,' that it really was the 'highest synthesis of life' he could not doubt, and could not admit the possibility of doubt. It was not as though he saw abnormal and unreal visions of some sort at that moment, as from hashish, opium, or wine, destroying the reason and distorting the soul. He was quite capable of judging of that when the attack was over. These moments were only an extraordinary quickening of self-consciousness – if the condition was to be expressed in one word – and at the same time of the direct sensation of existence in the most intense degree. Since at that second, that is at the very last conscious moment before the fit, he had time to say to himself clearly and consciously, 'Yes, for this moment one might give one's whole life!' then without doubt that moment was really worth the whole of life. He did not insist on the dialectical part of his argument, however. Stupefaction, spiritual darkness, idiocy stood before him conspicuously as the consequence of these 'higher moments'; seriously, of course, he could not have disputed it. There was undoubtedly a mistake in his conclusion – that is in his estimate of that minute, but the reality of the sensation somewhat perplexed him. What was he to make of that reality? For the very thing had happened; he actually had said to himself at that second, that, for the infinite happiness he had felt in it, that second really might be worth the whole of life. 'At that moment,' as he told Rogozhin one day in Moscow at the time when they used to meet there, 'at that moment I seem somehow to understand the extraordinary saying that *there shall be no more time*. Probably,' he added smiling, 'this is the very second which was not long enough for the water to be spilt out of Mahomet's pitcher, though the epileptic prophet had time to gaze at all the habitations of Allah.'"

others do not – smallpox, typhoid, and pneumonia fall into the first category; diphtheria, tetanus, and cholera into the second.

From this important paragraph, two relevant points emerge. The first – that austerities, including starvation and exhaustion, can cause somatic changes which at minimum accompany dramatic theophanies – has been known for some time and has been well covered; one thinks of Moses neither eating nor drinking for forty days on Mount Horeb before he saw the mountain in flames, and Christ's forty days in the wilderness where he too fasted for forty days and forty nights before Satan appeared visibly to him with his three temptations. What is new here is the prospect that an infectious disease may have opened John Henry Newman to his life-transforming visitation.

My brief entry for *ReVISION* concluded with that paragraph, but in the context of this book I will append this coda.

Some will find disturbing the conclusion that the facts here point toward; namely, that the religious experience that produced a giant of the Roman Catholic Church seems to have occurred while Newman was undergoing an entheogenic disease. To some extent I am in their company, and if it turns out that the facts fuel "the hermeneutics of suspicion" more than they serve the cause of truth, I shall regret not having kept them to myself. But in a dramatically incisive way, those facts highlight the basic object of this book, which is to ask if it is possible to honor the noetic deliverances of entheogenic theophanies without contradicting what we know about brain chemistry.

Newman himself faced that question at its incipient stage, and I respect the way he answered it. After acknowledging (as above reported) that his delirium may have heightened his feelings when he sensed that God had him under his care and had marked him for leadership, he went on to add, "but they still are from God in the way of Providence."

CHAPTER NINE

Entheogenic Religions: The Eleusinian Mysteries and the Native American Church

In the effort to see what the entheogens have to teach us about what we human beings are (human nature), about the inclusive context in which we live our lives (the world), and about the connection between the two (religion), the essays in this book approach those interlocking issues from various angles. This chapter touches on two instances where they produced full-fledged religions. Chapter 4 anticipated this topic, for Soma figured prominently in Hinduism's formation, but the concern in that chapter was restricted to the plant's botanical identity.

Beginning with the Eleusinian Mysteries, the first part of this chapter reprints the Preface to the revised edition of Wasson, Ruck, and Hofmann's groundbreaking book, The Road to Eleusis *(Los Angeles: Hermes Press, 1998). I was especially pleased to have been asked to write that Preface, for I am a Platonist at heart, and it seems likely that Plato's basic outlook derived from his Eleusinian initiation.*

My involvement with the Native American Church is a longer story. A former student of mine works for the Native American Rights Fund, and when in 1990 the United States Supreme Court ruled that the First Amendment does not guarantee a right to use peyote, he asked me if I wanted to become involved. I said that I did, and spent the next two years helping a remarkable Native American leader, Reuben Snake, compile a book titled One Nation Under God: The Triumph of the Native American Church *(Santa Fe: Clear Light Publishers, 1996). It originated with a political aim – to help win the right in all fifty states for the sacramental use of Peyote – but things moved faster than we had expected, and before the book went to press the Church had established its right through Congress. This caused us to turn the book into a celebration of the Native Americans' victory over the highest court of the land, and an account (for the historical record) of the impressive but little-publicized movement the Native Americans mounted to realize their aim.*

My chief contribution to that book was to interview members of the Native American Church in a number of its tribal branches to discover what the Church meant to them. The second half of this chapter presents samples of their testimonials, included here to show how the faith that entheogens generate can reach proportions that virtually require its institutionalization, even if only in the decentralized, congregational fashion of the Native American Church.

First, my Preface to the Eleusis book.

Two allegories frame Western civilization like majestic bookends – Plato's Allegory of the Cave at its start, and for now, Nietzsche's madman charging through the streets announcing that God is dead. We should not be sanguine about the direction in which those two landmarks point, for the notion of what it means to be human is far lower in the second allegory than in the first. "We do not think well of ourselves," Saul Bellow told his audience in accepting his Nobel Laureate prize, and he was right. Plato tells us that when his vision of reality swept him up and he sensed its life-giving implications, "First a shudder runs through me, and then the old awe creeps over me." Today we have the eminent physicist Steven Weinberg's report, "The more intelligible we find the universe, the more meaningless it seems."

Athens, of course, is only one of the sources of western civilization, its companion being Jerusalem whose assessment of things is as impressive as Athens'. We need only think of Moses trembling with awe on seeing Mount Sinai in flames; of Isaiah beholding the Lord high and lifted up, filling the whole earth with his glory; and of Jesus to whom the heavens opened at his baptism. To contrast this again with modernity, only recently has it come to light (through Frau Overbeck's report in *Conversations with Nietzsche*) that Nietzsche was himself deeply troubled by the loss that his madman reported. He was not sure that humanity could survive godlessness.

The bearing of all this on the book in hand is quite direct. Some theophanies seem to occur spontaneously, while others are facilitated by ways that seekers have discovered – one thinks of the place of fasting in the vision quest, the nightlong dancing of the Kalahari bushmen, prolonged intonings of sacred mantras, and the way peyote figures in the all-night vigils of the Native American Church. We do not know if (on the human side) it was anything more than absolute faith that joined earth to heaven on Mount Sinai and when three of Jesus' disciples saw him transfigured on Mount Hermon, his face shining like the sun and his clothes dazzling white. The Greeks, though, created a holy institution, the Eleusinian Mysteries, which seems regularly to have opened a space in the human psyche for God to enter.

The content of those Mysteries is, together with the identity of India's sacred Soma plant, one of the two best-kept secrets in history, and this book is the most successful attempt I know to unlock it. Triangulating the resources of an eminent classics scholar, Carl A. P. Ruck; the most creative mycologist of our time, R. Gordon Wasson; and the discoverer of LSD, Albert Hofmann, it is a historical tour de force while being more than that. For by direct implication it raises contemporary questions which our cultural establishment has thus far deemed too hot to handle.

The first of these is the already-cited question Nietzsche raised: Can humanity survive godlessness, which is to say, the lack of ennobling vision – a convincing, inspiring view of the nature of things and life's place in it?

Second, have modern secularism, scientism, materialism, and consumerism conspired to form a carapace that Transcendence now has difficulty piercing? If the answer to that second question is affirmative, a third one follows hard on its heels.

Is there need, perhaps an urgent need, to devise something like the Eleusinian Mysteries to get us out of Plato's cave and into the light of day?

Finally, can a way be found to legitimize, as the Greeks did, the constructive, life-giving use of entheogenic heaven-and-hell drugs without aggravating our serious drug problem?

The Road to Eleusis does not answer (or even directly address) these important, possibly fateful, questions. What it does do is to raise them by clear implication, elegantly and responsibly.

I turn now to the testimonials of members of the Native American Church, preceding them with the two opening paragraphs of the Preface that I wrote for the book from which the testimonials are drawn.

In a place apart, closer to nature than to the human scene, a tepee throws its outline against the night sky, a sacred silhouette. Inside, thirty or so Indians, men and women, sit on blankets and mats around a fire. Several children are sleeping in their parents' laps or on the ground by their sides. A seven-stone water drum pounds loudly and rapidly – the fetal heartbeat raised to cosmic proportions. Songs are sung with piercing intensity, interspersed with prayers and confessions. Tears flow, and a sacrament is ingested. It is a congregation of the Native American Church in one of its appointed meetings.

The Native American Church is the spiritual bulwark of an estimated quarter-million of the original inhabitants of this continent. Its roots extend into the twilight zone of prehistory, before the rise of Christianity or any of the historical religions.

[The testimonials that now unfold begin with one by Reuben Snake, who co-edited the book from which they are drawn.]

> I am a Hochunk or Winnebago, and the Native American Church has been the center of my adult life. Everything we do in our Church is to honor the Creator and find our place in his creation. We try to respect and honor our families and friends; we try to have compassion for our fellow men, for that's what our Creator tells us to do. This attitude comes to us through a sacred herb, one that is sacred because it is, in fact, divine. We call it Peyote, but more often, because of what it does for us, we call it our Medicine. It is the most powerful of all the plants, because God endowed it with his love and compassion. He put those qualities into this lowly herb so that when we eat it we can feel that the love that God is – I emphasize the love that God *is,* not that God *has* – is physically inside us. From there it overflows in compassion for human beings and all other kinds of creatures. It enables us to treat one another tenderly, and with joy, love, and respect.

[I continue with the remaining testimonials I have selected, arranged according to the aspect of peyote that each addresses most directly.]

The Medicine

We can't explain our religion. To understand it you have to eat its Medicine.

—Anonymous

Our favorite term for Peyote is Medicine. To us it is a portion of the body of Christ, even as the communion bread is believed to be a portion of Christ's body for Christians.

In the Bible, Christ spoke of a comforter who was to come. Sent by God, this comforter came to the Indians in the form of this holy Medicine. We know whereof we speak. We have tasted of God and our eyes have been opened.

It is utter folly for scientists to attempt to analyze this Medicine. Can science analyze God's body? It is a part of God's body, and God's Holy Spirit envelops it. It cures us of our temporal ills, as well as ills of a spiritual nature. It takes away the desire for strong drink. I myself have been cured of a loathsome disease too horrible to mention. So have hundreds of others. Hundreds of confirmed drunkards have been rescued from their downward ways.

—Albert Hensley, Winnebago

I'll be ninety-four on my next birthday, so if there's an example of someone who's been using Peyote all his life, I guess I am he. I feel that I would die for this Medicine, it has meant so much to my life. My people use it and find spiritual guidance. When I sit in the tepee and partake of the Medicine, I concentrate and think. I think of how I want to be blessed and who I want to pray for. The outside world disappears. I feel humble, and the good thoughts that come to me help me.

—Truman Dailey, Otoe-Missouri

I have never seen colors or experienced delusions of any sort while taking Peyote. What it feels like is that I am sitting right by God the Creator. I communicate with him. Of course he isn't there physically, but spiritually I sense that he is near me. And whatever I pray for, I feel that he hears me. Other than that special sense of closeness, the experience isn't remarkable.

Part of the experience of being close to God is that the Medicine gets bad stuff out of you. If you have evil thoughts or are in a poor frame of mind generally, you are going to see all that clearly. In this way, if you're not living your life well, Peyote purifies you. It helps clean your spirit. I have heard many testimonials of gratitude to the Creator and prayers for forgiveness.

—Patricia Mousetail Russell,
Southern Cheyenne

The Medicine is the main thing of all. It's our life. Nothing else can accomplish much without the herb the Creator gave us.

—A Washoe

Peyote is power. A tremendous power pervades the tepee during a meeting. It will take all your lifetime to know only a small part of the power that is there.

—Similar statements by a Navajo and
Crow, here interwoven

In the first creation God himself used to talk to people and tell them what to do. Long after, Christ came among the white people and told them what to do. Then God gave us Indians Peyote. That's how we found God.

—A Kiowa

Peyote goes all over my body – I feel it, its workings. My mind is clear. Before I didn't think much about what's right, but with Peyote I know it's God working, the God who gave it to us. I feel good because God is going to take care of me. I have nothing to worry about, nothing to be afraid of because the Almighty is at work.

—Dewey Neconish, Menomini

Reverence, Humility, Awe, and Love

There are certain times in a meeting when you can feel a presence. It's a holy feeling, the presence of the spirit of God that's in the midst of these people. It makes you want to pray deep in your heart.

—A Menomini man in his forties

Last week we admitted a newspaperwoman who wanted to find out about the Native American Church. When she came out of the tepee in the morning she told us how humble she felt. That's true of us too. Our Holy Sacrament, Peyote, teaches us humility.

—Paris Williams, Ponca

That night I realized that in all the years that I had lived on earth I had never known anything holy. Now, for the first time, I knew something holy.

—John Rave, Winnebago, after his first Peyote meeting

I'm glad I joined the Native American Church and used the Medicine, since it makes me think about the Almighty and how far away I have gotten from him.

—Beatrice Weasel Bear

Sometimes I am sitting here at home all alone, but I am not alone. I have my drum, water, and a gourd. I sing the songs of my church and my worries are gone. I feel good again, and refreshed. Peyote is not like a narcotic. When you eat it, your mind turns to the Great Spirit.

—Bernard Ice, a blind Oglala Lakota

Peyote to me is my Bible. I know what I should be doing and shouldn't be doing. When I take that Peyote, I feel humble and respectful all the time.

—Larry Etsitty, Navajo

This Medicine's got hope in it. It's got faith in it. It's got love in it. And it's got charity in it. So fill up all the fireplaces. Fill them up with those four words.

—Willie Riggs, Sr., Navajo Roadman

Moral Impact

This Peyote has done me a world of good. It put me on the right road. It has caused me to put aside all intoxicating liquor. I now have no desire for whiskey, beer, or any strong drink. I have no desire for tobacco. If I keep on using this Peyote, I'm going to be an upright man toward God.

—A Menomini

When I started eating that Medicine it told me something. I found out I was a sinner. Then I commenced to think why I was like that. I ate some more, and I found out that Peyote teaches me what is right. From then on I've tried to behave myself.

—A Menomini

Chief Peyote tells us that our meetings are to make Indians good, to make them friends, and to make them stop fighting. When we eat Peyote we feel towards others a warm glow in our hearts as if they were our brothers.

—Ralph Kochampanaskin

This is all that I know. When I started eating this Medicine I began to see everything. I no longer quarreled with anyone. I no longer was angry with anyone. That's it. When I started to eat this Medicine I began to think of the Great Spirit always, every day.

—A Menomini woman in her seventies

Petition, Prayer, Thanksgiving, and Guidance

Well Great Spirit, the time has now come. I am going to pray to you where I am standing. Please let everything be abundant, so that we may exist well on this earth where we live. We thank you, Great Spirit above, for allowing us to live up to the present. We pray you to give strength to every one of us. Hold the hands of each one of my children. Give them strength. Give them that which is good in the future, and make them stand erect here on earth where we live.

—An anonymous Menomini

Great Spirit who is all, I am going to give you thanks now. And now I am telling you in advance that we have all come here to this house, which you gave us, to pray to be purified. Now we are going to enter to pray to you all night. These, my brothers and sisters, will pray to you. Please keep carefully in mind whatever they will ask of you. Also, my relatives have come to visit us. They are going to pray to you for whatever is in their thoughts. So in advance we are telling you, here, that we are going to enter this house which you have given us to pray to you.

—Thomas Wayka, Menomini

Healing

I am ninety-six years old now, and when people ask me how long I have been taking this Medicine I say since before I was born. I say that because my mother took some Peyote the night before she delivered me. First she offered a prayer, and then she swallowed the Medicine. I was born about noon the following day, and as soon as she could get up and around, the first thing she did was make some Medicine tea and give me some of it. To understand all this you need to know that I wasn't her first child. There had been two boys before me, but neither of them survived. My mother wanted to be sure that I would make it, and that's why she took the Medicine at the time of my delivery and gave me some as soon as she was able.

To help you understand her great faith in our Medicine I have to go back four or five years. About three years before she gave birth to me she got so she couldn't walk, and to explain what happened then I need to go back still farther and tell you about her upbringing.

Her mother – my grandmother – was an orphan who was brought up in a Christian way and raised my mother that way. But when she was of marriageable age she went to visit the nearby Otoe tribe. One of the boys there liked her and they got married. The family she married into was one hundred percent Indian, but her upbringing had been Christian. Then a strange malady beset her. The doctors never did understand what it was, but she became crippled. She couldn't walk. With her Christian upbringing she prayed as hard as she could to Jesus, but she kept getting worse.

Then something happened that turned out to be important. Her family was on its way to a tribal meeting of some sort and the house where they stopped to spend the night had a tepee behind it – there was going to be a Peyote meeting that night. That was the first Peyote meeting my mother had ever attended. Around midnight a bucket of water was passed around and she drank some. That water hadn't touched the ground because it had been collected in pans that people had put out to catch rain. After they drank the water they stared to sing, and then they offered my mother the Medicine. They told her that if she took the Medicine and prayed with all the faith she could manage, it might help her. She consumed it, prayed, and when the meeting was over she began to get well. Within a month she was walking.

I don't know, of course, but it seems like her faith was equally strong when she was a Christian, but when she got on the Indian side of religion and took the Medicine, it seems like that's what made the difference and enabled her to walk again.

—Truman Dailey

Two years ago I fell ill. It was scary because I seemed to have a blood disorder. The doctors were talking about sending me to Minnesota to have me treated for leukemia.

Four days before my birthday my parents put on a prayer service for me. They rolled the sacred tobacco, gave me Medicine, fanned me with the waterbird fan, and prayed for me. Four days later I went to the hospital. My blood count was normal and I gained back the weight I had lost.

I think too of the miracle that happened to my Grandpa

Philip from whom I get my middle name, Afraid-of-Bear. Once when he was home on leave from the service there was a prayer ceremony for a young man who was desperately ill. Grandpa was the Fireman at this meeting and Peyote was placed in the center of the room so everybody could eat as much as they wanted. In ceremonies that are specifically for healing, people ingest as much as they can, and more, to increase the Medicine's power.

As the evening wore on, it seemed clear that the young man was dying. So when midnight arrived and it was time for Grandpa to pray officially as Fireman, he asked the creator to take his goodness, his wholesomeness and his strength, and give them to the younger, dying man. He said he would be willing to lay down his remaining years for the man, who had not experienced as much of life as he had.

The young man survived his crisis and did pull through. But Grandpa! The next day, while chasing horses, he keeled over with a heart attack. That was sad, but it happened the way he wanted it to. Peyote listened to him, and granted what he asked.

—Loretta Afraid-of-Bear Cook, Lakota

New Life and Behavior Change

I started using Peyote when I came back from the army in 1962. I stopped using liquor because it is not right to use it with the Medicine. I told all these people when they came to my meeting, you guys better straighten up. This liquor doesn't go with Medicine. White people say liquor and gas don't go together and it's the same with Medicine. Liquor and Medicine don't go together.

—Irvine Tachonie, Navajo

It's been twenty-three years that I've lived a life of sobriety. I don't smoke cigarettes like I used to. I don't drink alcohol or use any kind of drugs because of the life that I now live in the Native American Church. That's what it's done for me, for my family and my relatives on both sides: my Winnebago in-laws, my Sioux relatives, my Menomini relatives, my Creek and Seminole and Ottawa and Iowa, my people back home in Oklahoma.

—Johnny White Cloud, Otoe-Missouri/Creek/Seminole

I'm into my sixth year of recovery from alcoholism. I'm off the bottle now, but the temptation is still there. When I go to Peyote I pray God to forgive my old alcoholism and to keep me sober.

Over the years Peyote has taught me many things, though actually it is God who taught them to me through Peyote.

—Andy Kozad, Kiowa

When I first used Peyote I became deathly sick. It seemed like I vomited several bottles of whiskey, several plugs of tobacco, and two bulldogs. This accumulation of filth represented all the sins I had ever committed. With its expulsion I became pure and clean in the sight of God, and I knew that by the continued use of Peyote I would remain in that condition. I was transformed – a new man.

—A Winnebago

My heart was filled with murderous thoughts when I first took Peyote. I wanted to kill my brother and sister. All my thoughts were fixed on the warpath. Some evil spirit possessed me. I even desired to kill myself.

Then I ate this Medicine and everything changed. I became deeply attached to the brother and sister whom I had wanted to kill. This the Medicine accomplished for me.

—John Rave

Women and the Feminine

A long time ago one of my aunts, Ethel Blackbird, told me that Peyote can be used for many kinds of healings and that we should take it when we have our babies.

When it came time for me to do the woman's ritual of having my first child, my mother came from Washington State where she had been living. My labor pains started around four in the morning and she immediately got up and gave me some Medicine. This pulled my resources back inside of me and kept me calm. Being calm enabled me to experience the birthing process clearly and calmly, consciously sensing and feeling what was going on.

—Loretta Afraid-of-Bear Cook

Bereavement and Death

Many people have gone on the path of this life and beyond. Our altar, which is in the shape of a mound, is Mother Earth where you come from and where you return. It is the same as the biblical passage, "from dust to dust." As you eat Peyote, the altar becomes a grave into which many a man has gone.

—Lawrence Hunter, Minneconjou Lakota

My grandmother wanted her Last Sacraments. They were going to call a minister, but she said she wanted them in the Peyote way. William Black Bear gave her four Peyote balls, and my father sang four songs. They said the Lord's Prayer, and she said Amen and breathed her last breath.

—Eva Gap, Oglala Lakota

CHAPTER TEN

Something Like a Summing-Up

The tentativeness expressed in the title of this concluding chapter is emphatic, for in rereading the essays in this book I realize that I am no closer today to answering the central problem they circle – given what we now know about brain chemistry, can entheogenic visions be validated as true? – than I was after my first encounter with them almost forty years ago. All I can do is stuff into the duffel bag a few remaining items I think should be included and then pull its drawstrings and knot them.

Three encounters with the entheogens – none of them mine – have crossed my path that seem significant enough to relate. Two of them were positive, the other negative, and I begin with the hell side of these heaven-and-hell substances.

In 1962 I delivered the first of what has become an annual Charles Strong Lecture on World Religions to the universities of Australia. The appointment called for spending three or four days at each university, and (as the Harvard psilocybin research was still in place and had attracted global attention) several of my hosts raised the subject with me. Some months after my return to Cambridge, an Australian who had been party to one of these conversations turned up on my doorstep. He was carrying LSD that was burning a hole in his pocket, but he was reserving it for Niagara Falls, which he had targeted as the perfect place for his initiation. His plan – to ingest capsules of unknown dosage alone in a potentially dangerous physical setting – struck terror in my heart, but there was no stopping him. His mind was made up. He was on his way to what, from hearsay, he was certain was going to be the peak experience of his life.

Several days passed and then my phone rang. It was a collect call – would I accept charges? The operator wasn't altogether clear as to who was calling, but it was someone in Buffalo, and, suspecting immediately his identity, I accepted the call. There was a long silence, and then an agitated voice kicked in. I couldn't make heads or tails of what it was saying. Garbled sentences would be left hanging in midair and replaced by other false starts until the whole dissolved into sobs and wails. There was a desperate, prolonged cry of "Hoooooooston" which trailed off into nothingness and the line went dead.

It wasn't until several days later when two police officers appeared at my door that the story emerged. The police had gotten involved when, on the rim of Niagara Falls, "John" (I forget his actual name) exposed himself publicly. Converging on him and finding him incoherent, the officers handcuffed him, whereupon he snapped the chain – a first in the cops' experience. John was behind bars in Buffalo awaiting a hearing. Somewhere down the line I heard that he had gotten into a serious legal scrape for having imported drugs which by then were illegal. That was the last I heard of the matter.

The other two stories have happier endings.

In Kathmandu in 1976, my wife and I heard that there was an American in a nearby Buddhist nunnery, and, being curious, we looked her up. Her story was this.

Born of Russian aristocracy, she immigrated with her parents to the United States where (after winning a beauty contest in Cannes) she married a Hollywood mogul. At the time when the psychedelics broke over America she found her life completely empty. She felt trapped by her marriage and was depressed to the point of seriously thinking of ending her life. LSD presented itself as a straw she might latch on to, and she ended up having ten sessions in all. I remember her accounts of four of them.

From her initial session she emerged remembering nothing. Zero. Total amnesia. It was as if eleven hours had been excised from her life, with one proviso. She was certain – and remained certain as she told us the story – that it was the most important event of her life.

Somewhere down the line, a male friend persuaded her – pressured is more accurate – to take LSD under his guidance. He wanted her to undress, but she agreed only if he provided her with a sheet to wrap herself in. He was an experienced mountain climber, and the walls of his apartment were papered with giant blow-ups of Himalayan peaks. All she remembered from that session was that she turned into an iceberg and sat upright and motionless the entire night.

In her ninth session, her life passed before her as if on film, and with that rerun behind her, her tenth and final ingestion had no effect at all, despite the fact that the dosage was as strong as in her other sessions. The chemical could have been aspirin for all the difference it made, and she resolved to become a Buddhist nun, a move she had been contemplating for several months. The Dalai Lama asked her to wait for a year, which she spent wrapping up her racy jet-set life, whereupon he acceded to her request, and there she sat before us. Her head was shaven, and her thin robes and sandals made our padded armor against the Himalayan winter seem self-indulgent. Yet she was cheerful and gave every evidence of being at peace with herself.

My third account concerns Bill Wilson, the beloved founder of Alcoholics Anonymous. My acquaintance with him was limited to an afternoon in his hotel room in Kansas City and the lecture he gave to two thousand recovering alcoholics that evening, and I was no more than a bystander throughout, as I had driven Gerald Heard (who was in residence at Washington University at the time) to Kansas City so the two men could update their long-standing friendship. Even so, to have been in the presence of those two men for three consecutive hours was an unforgettable experience. Bill – no one ever called him either William or Wilson – mostly set the agenda. He had recently taken LSD and was under the compulsion (typical in the first weeks following the initiation) to talk about it. The reason I mention that afternoon is to report that he counted his entheogen experience as equal in the conviction it engendered to the conversion experience that led him to his founding Alcoholics Anonymous.

Moving from those three anecdotes into my final wrap-up, I will begin with something that puzzled me for a while but which I think I have pretty much resolved.

Why, when I count several of my entheogen experiences as being among the most important in my life, have I no desire to repeat them? On occasion I have gone so far as to rank them with family and world travel in what they have contributed to my understanding of things, yet – with the exception of peyote, which I took in the line of duty while working with the Native Americans as described in chapter 8 – it has been decades since I have taken an entheogen, and if someone were to offer me today a substance that (with no risk of producing a bummer) was guaranteed to carry me into the Clear Light of the Void and within fifteen minutes return me to normal with no adverse side effects, I would decline. Why?

Half of my answer lies in the healthy respect I have for the awe entheogens engender; in Gordon Wasson's blunt assertion in the frontispiece to this book, "awe is not fun." I understand Meister Eckhart completely when he says that "in joy *and terror* the Son is born" (emphasis mine). I speak only for myself, of course – that durable formula of set and setting again – but if I am honest I have to say (and age may figure in this) that I am afraid of the entheogens. I will take them again if need be, as I did with peyote, but the reasons would have to be compelling.

The second half of my answer is that I have other things to do. This may sound like a limp excuse for foregoing ecstasy, so I will invoke the Buddhist doctrine of the Six Realms of Existence to explain the force it has for me.

Metaphysically, that doctrine posits six kinds of beings and the realms they inhabit. (The doctrine can also be read psychologically as six states of mind that human beings keep recycling, but I will stick to its metaphysical reading.) The two populations that are relevant here are the demi-gods, who are always happy, and human beings, whose lot is harder but who are actually the best off of the six kinds of beings because they alone possess free will with its power to change things. (The four I haven't mentioned are instinct-ridden *animals*, fiercely envious *jealous gods*, insatiably greedy *hungry ghosts*, and *hell beings* who are ravaged by rage.) Blissed out on

Cloud Nine, the demi-gods are still subject to time, which means that sooner or later their holidays will end and they will find themselves back in the form of life from which they were granted temporary leaves. Only the human state opens onto *nirvana*, which is why one of the three things that Buddhists give thanks for each day is that they have been born into a human body.

I will not try to separate what is literal from what is figurative in this account; only its moral teaching interests me here for supporting my second reason for having no desire to revisit the entheogens. The Sufis speak of three ways to know fire: through hearsay, by seeing its flames, and by being burned by those flames. Had I not been burned by the totally Real, I would still be seeking it as knights sought the Grail and moths seek flame. As it is, it seems prudent to "work for the night is coming," as a familiar hymn advises. Alan Watts put the point more directly: "When you get the message, hang up the phone."

The downside of swearing off is, of course, the danger that the Reality that trumps everything while it is in full view will fade into a memory and become like Northern Lights – beautiful, but cold and far away. The problem besets all epiphanies; the psalmist's lament, "restore unto me the joy of my salvation," has already been quoted. During the three years of the Harvard experiments the entheogens were the most exciting thing in my intellectual life, but at this remove I have to work to get my head back into those years and revive the excitement. I suspect that there are thousands of people out there, possibly millions, who would have reached passionately for a book such as this had they come upon it soon after their first ingestion when they thought the world would never be the same again, but who at this remove find its subject interesting but no more than that.

The question comes down to which experiences we should try to keep in place as beacon lights to guide us and which we should let lapse. The intensity of the experience doesn't give us the answer, as this final personal anecdote of the book bears out.

When the first UFO craze swept America in the 1950s I was teaching at Washington University in St. Louis where the president of McDonnell Aircraft gave Chancellor Arthur Compton a grant to convene a conference on Science and Human Responsibility. Having himself headed the team of physicists at the University of Chicago that produced the first chain reaction of splitting atoms, Compton was able to attract world-class scientists of the order of Werner Heisenberg to the conference. I was enlisted to manage arrangements and produce a record of the event.

On the evening of the conference I was in my dean's office reviewing the checklist of things to be done. He was speaking, when suddenly, midway through a sentence, a look of horror swept over his face and he plunged for the window behind me. I whirled to follow him, at which point my account becomes embarrassingly corny because what we saw fit the UFO stereotype so exactly. Five illuminated saucers were sweeping in a semicircle across the leaden clouds of the late November evening sky – astonishingly close to our window, it seemed. They were moving so fast that they were out of sight almost before we saw them. We bolted into his secretary's office hoping to follow them from her window, but they were gone.

Without exchanging a look, we retraced our steps and resumed our chairs in his office where we sat in total silence without looking at each other for about five minutes. The reason I am telling this story is for what we experienced in those minutes. We felt shaken to our foundations. Finally, the dean bestirred himself, looked at me, and said, "Well, Huston: I'm a dean and you're in religion; they'll never believe us." We went our respective ways and never mentioned the matter again.

In its immediate force, that experience rivals that of the entheogens, but, unlike the latter it had no lasting impact, the obvious reason being that I don't believe in invading extraterrestrials. That disbelief leaves me suspecting that a naturalistic explanation for what we saw exists, even though I don't know what it might be – an aircraft's wings that were reflecting rays from the setting sun, perhaps? What the account points up is the way our basic beliefs adjudicate what we make of our experiences, a point the Victorian poet Ella Wheeler Wilcox captures beautifully in the following quatrain:

> One ship drives east and another drives west
> by the self-same gale that blows,

'Tis the set of the sail, and not the gale,
that determines the way she goes.

I can think of no better way to close down the ruminations of this book than to describe as succinctly as I can the set of the sail that has vectored them.

I believe that when "set and setting" are rightly aligned, the basic message of the entheogens – that there is another Reality that puts this one in the shade – is true. There is no way that the prevailing view of the human self (which depicts it as an organism in an environment that has evolved purposelessly through naturalistic causes only) can accept that claim, which means that its Procrustean anthropology must go. That it will go, has been the critical (as distinct from constructive) burden of all my writing, for it rests on assumptions that are too arbitrary to escape scrutiny indefinitely.

Endings, though, are not the place for argument, so I will let Robert Frost deliver my parting shot. I do not see how anyone can deny that the traditional, theomorphic view of the human self which the entheogens endorse is nobler than the one that common sense and modern science (misread) have replaced it with. Whether the theomorphic view is true or not cannot be objectively determined, so all I can ask of the opposition is that it not equate noble views with wishful thinking. They can be as demanding of us as are their opposites, as Frost suggests in his poem "A Cabin in the Clearing."

In it he has two wraiths, Mist and Smoke, talking about an old woodcutter and his wife huddled together in a cottage in a small clearing it the forest.

No one – not I – would give them up for lost
Simply because they don't know where they are,

says Mist. To which Smoke replies,

If the day ever comes when they know who
They are, they may know better where they are,
But who they are is too much to believe . . .

APPENDIX A

Secularization and the Sacred: The Contemporary Scene

This first appendix steps back from the entheogens proper to situate them in their modern social context. The essay was written thirty years ago as a chapter in The Religious Situation, *edited by Donald Cutler, and I have rewritten it considerably to adapt it to the concerns of this book. I have retained references that link it to the sixties for the concreteness they give to the piece.*

Religion has lost ground startlingly in the last decade, at least in the eyes of the general public in the United States. In 1957, 14 percent of a Gallup national sample was of the opinion that "religion is losing its influence." By 1962 the figure had risen to 31 percent, by 1965 to 45 percent, and by 1967 it stood at 57 percent. In ten short years the proportion of Americans who see religion as in retreat has quadrupled, jumping from one-seventh to over half the population. Yet squarely within this decade of seeming collapse, an astute sociologist, Robert Bellah, has observed that "the United States is experiencing at the moment something of a religious revival."

The cognitive dissonance occasioned by Gallup's statistics and Bellah's perception sets the problem for this paper. Is one of the two mistaken, or are sacred and secular related in ways that are sufficiently multivalent to allow both to be correct?

"Secular" characterizes regions of life that man understands and controls, not necessarily completely but (as the saying goes, and here it is exact) for all practical purposes.

Thus defined, secularization has increased steadily with the advance of civilization. Primitive man obviously managed or we wouldn't be here, but there was little in his life that he understood sufficiently to effect clean control over it. As a consequence his society was whole and sacred throughout. For tribal peoples, hunting and husbandry are sacred activities; Toda priests tend sacred buffaloes, and totemic rites bind Arunta hunters to their prey in ties that both expedite and legitimize their killing. Likewise with healing: Medicine men are both physicians and priests.

In civilizations the situation is otherwise. Whether we read "In God We Trust" on our currency as vestigial or as precise index of the extent to which the Almighty Dollar has become an object of worship, Western economies have become completely secular. We do not fully understand how economies work and so can only marginally control them, but we certainly don't charge their vagaries against God, any more than (short of desperation) we implicate him in health. Similarly with Medicine. Priests now enter the picture with the approach of the inexorable – death – and even the cure of souls has become, through psychiatry, a secular pursuit. Politics has gone the same route. In China the process is currently passing through a decisive phase as the pragmatics of nation building challenge Mao Zedong's "theologically" vectored Cultural Revolution, but in the West the battle is over. The priest king has gone the way of the medicine man. The claimed discovery (by an Indian astrologer during the heat of the 1964 presidential campaign) that Lyndon Baines Johnson had practiced austerities on the banks of the Ganges in his previous life "staggers belief," as an Indian newspaper commented.

It might seem natural to suppose that (with man's progressive mastery of domains of his existence and the attendant emancipation of these from the custody of religion) the sacred is shrinking and becoming residual, but we should be careful about jumping to that conclusion. It is true that the sacred is less evident in contemporary life than it was in the past, but to this perception three that are less obvious should be added. (1) More of it persists than meets the eye. (2) What remains of it is durable, sufficiently so that it is not likely to decline much further. (3) On the contrary, the sacred is likely to make a comeback – there are signs that it is already doing so.

These are controversial claims, so they require support. I begin with some theoretical considerations and proceed to ones that are a bit more empirical.

If "the secular" defines regions of life that man *controls,* the view that it is displacing the sacred entails the corollary that life is becoming more manageable and (if one carries the view to its logical limit) will eventually become completely manageable. In actuality, however, it is far from clear that man's control over life is increasing. *Parts* of life are coming under control – infectious diseases, infant mortality, and dental hygiene, for example – but it is a far cry from that fact to the conclusion that life as a whole is more under our control.

Item. Human problems tend to be mercurial; solve one and another moves in to take its place. Frequently the new problem is created by the solution to the original one. We wipe out infectious disease and face the population explosion. We solve our energy shortage by splitting the atom, and with precisely that act we are handed the atom bomb and nuclear waste. We invent pesticides, and "silent spring" awaits. Technology used to be considered our servant, but there are ominous signs that it is carrying us to no one knows where. To the extent that this fear is justified, technology dethrones *deus ex machina* to ensconce itself as *deus in machina*, and man remains as much a creature as he was in the former. We are back to the same general problem we faced when we asked about the relation of secularization to the sacred. There it took the form, Does increasing secularization diminish the sacred? Here it reads, does solving life's problems reduce their number or merely change their specifics?

Item. Beyond being unattainable, the idea of complete control isn't even coherent. Man is a social animal and as such lives as one will among others. No one of these wills could become omnipotent without ceasing to be social, for social existence necessarily involves both give and take. This positions "omnipotent" and "social" as logical alternatives.

Finally, not only is the notion of total control contradictory, it isn't even attractive. To enter a friendship, to say nothing of marriage, with intent to control it is to soil the prospect from the start.

Life calls for balancing the rewards of control with gifts that come to us through openness and surrender. The more we resolve to have things our own way, the more closed we become to the virtues in alternative ways. If we cannot perceive the virtue inherent in the capacity to surrender – to surrender to another person in love, or to obligation in the sense of feeling its claims upon us – cynicism awaits us.

These three points relate to our inquiry as follows: Assuming that the sacred lies somewhere within what exceeds human control, there is no reason to think that technology has shrunk that domain. Life continues to dangle over seventy thousand fathoms; only the regions of the deeps have shifted.

Are there new quarters within what we do not control where the sacred might be bidding to show itself today? To sharpen our search we can note that the sacred exceeds not only our control but our comprehension. We can't control the weather, but now that meteorologists have banished the nature gods we no longer find it mysterious.

One candidate is unconscious regions of our minds. By definition, we don't know what goes on there, and clearly we don't control it. "Where id was, let ego be," Freud counseled, and he would have been the first to insist on how far we are from reaching that goal, if indeed it is reachable in principle. To these two features of the unconscious we should now add a third condition of the sacred which the unconscious satisfies: its importance. Recent experimental studies of sleep are reported to reveal that approximately 80 percent of the time male subjects are dreaming they have erections. If the report is accurate, it provides experimental support for what I suspect is true in any case; namely, that the unconscious lies close to the wellsprings of our vitality.

Incomprehensible, indomitable, and important – these are authentic marks of the sacred, and the unconscious possesses them all. Following a lecture my psychologist wife gave at the Graduate Theological Union in Berkeley, the first question she was asked by a theological student was, "Now that we have Jung, do we need religion?" And Ronald Laing, Director of the Langham

Clinic for Psychotherapy in London, argues that in certain cases psychosis is a state of mind in which religious resources of the unconscious can surface with unusual directness. Our unconscious minds hold over us the power of life and death, psychic health and insanity, which is a kind of psychic death, with severe depression falling somewhere in between. We stand in relation to it as creature, a sufficient reason being that (as remarked) it is unfathomable: its thoughts are not our thoughts nor its ways our ways, "our" referring here to what we are in conscious possession of. Human beings are not noticeably open to the divine in any quarter today, but angle for angle, the unconscious provides one continuing locus for its entry. This partly explains the striking impact that Hinduism and Buddhism are having on the West, especially on Western youth, for Asian religious symbolism – "Atman is Brahman," "each of us possesses the Buddha-mind already" – accommodates epiphanies from our unconscious minds more readily than does Judeo-Christian imagery, which tends to picture God as residing outside ourselves.

In addition to the unconscious, there is a second frontier on which the sacred retains its rights in the face of continuing secularization; namely, interpersonal relations, that highly elusive, treacherous, and potentially sublime terrain where two or more persons meet and exchange words and feelings.

The importance of these relations needs no arguing. They create and sustain us, but we have seen that importance by itself doesn't make something sacred. Personal relations have always been important, but in the past they have provided less of an aperture for the divine because they were less problematical. In traditional societies most men and women live their entire lives in a single primary group whose members know one another intimately and for life. Such abiding communities don't prevent unhappiness, but they do spare their members the deeper disturbances that appear when communication breaks down seriously: loneliness, withdrawal, identity confusion, and existential dread.

Thanks to the fact that our society has become impersonal, these maladies are with us. American families now change residence across state lines on the average of once every five years, and wherever they live, they usually work elsewhere. As a consequence they know one another only (a) in stages of life and (b) in aspects of self – in roles, such as husband, professional, customer, voter, client.

These developments have produced in industrial societies a kind of crisis in communication; personal relations, while continuing to be essential, have become precarious as never before. This combination of importance and fragility provides a clearing for the divine, particularly when to those two attributes we add unpredictability and the mystery that attend deep encounters between persons. A perceptive soul like Martin Buber sensed this and wrote a tract for our times, *I and Thou,* which argues that God lives precisely in the intersection of lives at depth level. Encounter groups (T-Groups, Sensitivity Training Groups) provide another evidence that the interpersonal has become an area where the sacred can surface today. Last summer I directed in India a seminar composed of university women, sixteen American and sixteen Indian. The Americans found encounter-group exercises both more threatening and more significant than did the Indians, who (still nurtured by relatively tight and stable primary associations) chattered merrily along as if this were their normal fare. Westerners may like or feel threatened by encounter groups, but almost invariably they find the experience to be different. It also tends to be *important,* for some, the most important experience of their lives. With a dozen or so independent and emotionally charged souls affecting the course of the group, it is unpredictable and *uncontrolled.* And the trajectory of the group *defies comprehension;* there come points when so much is going on, at so many levels, interpreted so divergently by everyone present, that one looks out at the group with eyes of insanity, so totally lacking is anything that approaches an objective grasp of what's happening. All three prerequisites of the sacred are present. It is, therefore, no surprise to find coming from the pen of one of the most practiced leaders of encounter groups an article titled "Sensitivity Training as Religious Experience." When, writes James Clark,

> for forty or fifty hours one is confronted in a complex and deeply human way by nearly a dozen other people in a circle

> with no imposed task to fly into, no hierarchy to bind, contain and ritualize, more often than not one expands. A person experiences the limits he and his environment have imposed on him, and expands beyond them. Knowing what one cannot give, one is able to offer what one can give. And knowing what one can give leads to a sense of where one is in the universe and a genuine experience of the prizing of all men, the mystical, deep, religious, expansive experience of knowing that "no man is an island."

Of an encounter group at Tavistock Institute in England, Margaret Rioch writes: "Everyone, I think, suffered a sea change into something rich and strange, something that borders on the sacred."

In targeting the unconscious and interpersonal regions of human life as two places that are open to the sacred today I have no intention of overlooking or belittling the place one thinks of first, namely churches, synagogues, and their like which came into existence to make room for the sacred. It is, rather, that, driven as they are by crosscurrents of innumerable sorts, institutionalized religions present such a tangled scene today that I have deliberately chosen to exclude them from consideration in this essay. Some day I may address their case, but at this point my thoughts are too unformed and conflicting to be committed to print.

With that proviso in place I turn in the balance of this essay to the regular mode in which the sacred announces itself – revelations – to ask if they are occurring less today than formerly.

Alternatively known as epiphanies or theophanies, revelations are unheralded manifestations that cast things in a different light. As such, they conform to the three marks of the sacred I have been working with: they are not subject to our control, they wear the stamp of importance, and they outstrip our understanding. The genuineness of what they reveal – are they authentic revelations or only subjective inventions? – cannot be objectively determined, and images used to describe them are various. Some recipients report being transported to another, more momentous world, like the man Saint Paul tells of who was transported to the third heaven where he was told things he was not permitted to repeat. Others find messengers from the momentous world turning up in this world as angels. Still others don't speak of other worlds at all, but rather of the pieces of this world being gestalted anew in ways that now make sense.

Revelations can be terrifying, but as conveyers of new gestalts they bring happiness of a sort that differs from the ordinary kind for carrying noetic claims – they ask us to upgrade our estimates of life and the world. The most emphatic epiphanies are those that arrive in times of darkness and despair. The happiness these bring has earned for them a distinctive name, blessedness, the distinctive feature of which is its paradoxical character. Blessedness is paradoxical for shining in darkness that remains visible while light permeates it. To fully comprehend this fact is to find it not just astonishing but astounding; it is an affront to human logic. The peace that comes when we are hungry and find food; when we are lonely and find a friend; when we are sick and feel health returning – these happinesses are understandable. But blessedness is the peace that arrives when these resolutions do not occur. It comes – or can come; there is nothing inevitable about it – when all options are exhausted and life has us by the throat. It could be the death of a child, or an incurable disease, or the onset of blindness. Such things happen, and they make us wish that life were different, that it would approach us in a different guise. But this is the way it has approached us, and we have no choice but to accept its visitation as our identity from that point on.

The question for this essay with its focus on the contemporary scene is whether our secular age is visited less frequently by epiphanies than were former ages, but that they continue to occur is incontrovertible. On my desk is the report of a New York housewife who, in the midst of a deep depression caused by her sister's suicide which left three motherless children, found her depression suddenly lifted and replaced by an inexplicable serenity that embraced rather than eclipsed clear-eyed cognizance of all that had happened. I find such resolutions also in Sidney Cohen's reports of terminal cancer patients who under LSD continue to feel their pain but in a way that no longer matters, so completely is it set in cosmic perspective.

> The pain is changed. I know that when I pressed here yesterday, I had an unendurable pain. I couldn't even stand the weight of a blanket. Now I press hard – it hurts, it hurts all right – but it doesn't register as terrifying.

And these:

> I could die now, quietly, uncomplaining – like those early Christians in the arena who must have watched the lions eating their entrails.
>
> I see that the hard deaths, too, must be borne. Like difficult births, they are a part of life.
>
> When I die I won't be remembered long – there aren't many friends, and hardly any relatives left. Nothing much accomplished – no children, nothing. But that's all right too.

Several swallows don't make a spring, of course, and I know of no way to move from spot-checks like the ones given to statistical conclusions about our times. In his empirical studies of peak experiences, Abraham Maslow found both that such experiences are virtually universal – almost all his subjects reported having had them – and that to an extent that surprised him as an investigator, they tended to characterize them as religious in their character and feel. But what are we to make of that finding?

Despairing of arriving at a quantitative conclusion, I will fall back on surmises, which begin by noting that revelations have two poles: a sender (whether personified as God, or limited to the state-of-affairs that is disclosed) and a receiver. Secular modes of thought could handicap receivers today, causing them to discount intimations they might otherwise take seriously, but there is another way to view the matter, and I will devote the remainder of this essay to sketching it.

If man is in fact a theomorphic being, as the historical religions all claim, the image of God is the most important fact about us. It stands to reason that this component will not submit without resistance to being caged in a secular outlook. If it actually exists, it will be like a jack-in-the-box, the spring of which constantly pressures the lid to open and let it out.

I am impressed by the evidence I see that supports this view of the self. Equating secularism with skepticism, we think of ours as a skeptical age, but I have come to question the equation. With New Age credulities busting out all over, our scientific, high-tech, and in ways secular age may turn out in fact to be one of the most believing ages in history. The objects of belief have changed, but faith remains securely in place.

Item. Last year while teaching for a quarter at Santa Barbara, I was invited to the home of a family I had chanced to know in Mexico several years earlier. In the course of the evening it developed that a visiting sister was completely wrapped up in the Los Angeles chapter of Maharishi Mahesh Yogi's International Meditation Society, while the high-school-sophomore daughter was a moving spirit of the local chapter of the Sokagakai chanting sect. Driving home I stopped for gasoline and found the filling station attendant completing D. T. Suzuki's *Zen Buddhism.* In a single evening I had met six persons, three of whom were involved – two very deeply – with religion in forms that elude the usual opinion polls.

Item. Lest the preceding episode be discounted on grounds that it is precisely what one might expect in California, I choose my next one from staid New England. Two years ago a group of eight M.I.T. upperclassmen formed a preceptoral group and asked me to be their instructor. It was to be an independent study project which the students were to conduct themselves, my role being limited to that of adviser and consultant. Ostensibly on Asian thought, it began respectably enough with standard Chinese and Indian texts, but as the weeks moved on and the students' true interests surfaced, the original syllabus began to lurch and reel until I found myself holding on to my mortarboard wondering whether to continue in the role of sober professor or turn anthropologist, sit back, and observe the ways of the natives. For natives in thought-patterns they were; far closer to Hottentots than to scientific positivists. In the end the anthropologist in me triumphed over the academician, for I found the window to this strange and (in the technical, anthropological sense) quite primitive mentality fascinating. I cannot recall the exact progression of topics, but it went something like this: Beginning with Asian philosophy, it moved on to meditation, then yoga, then Zen, then Tibet, then successively to the *Bardo Thodol,* tantra, the kundalini, the chakras, the *I Ching,* karate and aikido, the yin-yang macrobiotic (brown rice) diet, Gurdjieff, Meher Baba, astrology, astral bodies, auras, UFOs, Tarot cards, parapsychology, witchcraft, and magic. And, underlying everything, of course, the psychedelic drugs. The students weren't dallying with these matters. They were *on* the drugs; they were *eating* brown rice; they were meditating hours on end; they were making their decisions by the *I Ching,* which one student considered the most important discovery of his life; they were constructing electronic experiments to prove that their thoughts, via psychokinesis, could affect matter directly.

And they weren't plebeians, Haight-Ashbury flower children. Intellectually they were aristocrats with the highest math scores in the land, Ivy League verbal scores, and two to three years of saturation in M.I.T. science. What *they* learned in the course of the semester I have no way of knowing. What I learned was that the human mind stands ready to believe anything – absolutely anything – as long as it offers an alternative to the desecularized mechanomorphic outlook of objective science. Some may see the lesson as teaching no more than the extent of human credulity, but I read the matter differently. If mechanomorphism is the truth, then indeed the students' gropings reveal no more than human unwillingness to accept its strictures. But if reality *is* in fact sacred, the students' frenetic thrusts suggest something different. In matters of spirit, subject and object mesh exceptionally – no faith, no God; no response, no revelation. It follows that the sacred depends heavily on man's nose for it. Given noses as keen for the chase as my students' were, if the sacred lingers anywhere in the interstices of contemporary life, it is going to be flushed out.

Item. Credulity isn't confined to the young. Returning recently by plane from Chicago to Boston I found myself seated next to a physics professor. She discussed physics with interest, but flying saucers with passion – passionate belief. UFOs are not, to be sure, ideal exemplars of the sacred, but they impinge on it by virtue of awakening numinous feelings, for if they should exist, they are probably manned by creatures more advanced than we. So they do more than puzzle. They are strange.

Item. Turning from personal anecdotes to the world at large: that Japan is the most literate nation in the world, and the most industrialized in Asia, has not prevented more than five hundred new religions from springing up there since World War II.

Item. Science fiction is booming. Only in part does space exploration account for this; we can as readily turn the matter around and say that the will to transcend the mundane has fueled space exploration. In the past this will, working on the imagination, produced the "Gothic" novel with its preternatural incursions. It also produced spiritualism and its interest in phenomena that are neither clearly natural nor clearly supernatural. Today science fiction, especially where located in space as in *Star Wars,* accomplishes the same imaginative transcendence by envisioning encounters with natural but transterrestrial beings. Mythological and psychological supernaturalism are replaced by a transterrestrial and uncanny naturalism.

To conclude: the polarity of the sacred and the secular continues to hold its place among the mighty opposites that crisscross life: systole and diastole, action and repose, freedom and form, centrifugal and centripetal, yang and yin. If man is in fact *homo religiosus,* he is by nature vulnerable to transcendental visitations – susceptible to intimations of Otherness at the thin places that separate us from an encompassing More. By this reading, when 57 percent of the American public say that religion is losing its influence, they should be saying that institutionalized religion is losing its influence in certain areas of life where its presence used to be more evident. But as institutionalized religion isn't the whole of religion, the fact that the sacred has withdrawn from certain spheres does not prove that as a free-floating phenomenon it is diminishing. Robert Bellah could be right in detecting in the United States at the moment something of a religious revival.

APPENDIX B

Thinking Allowed with Jeffrey Mishlove: A Televised Interview

I *tack on this second appendix as a kind of coda to provide an idea as to how the thoughts that this book has ranged through might be summarized for a general audience. It consists of the edited transcript of Jeffrey Mishlove's 1998 interview with me in his television series, "Thinking Allowed."*

JEFFREY MISHLOVE

Hello and welcome. Our topic this evening is the psychology of religious experience, and my guest tonight is one of America's scholars of religious traditions, Dr. Huston Smith. Dr. Smith is a former professor of philosophy at M.I.T. He's the author of a classic study, *The World's Religions,* which has sold over two million copies, as well as eight other books on philosophy, religion, and psychology, most recently one called *Beyond the Post-Modern Mind.* Welcome, Dr. Smith.

HUSTON SMITH

Thank you. It's good to be here.

MISHLOVE

It's a pleasure to have you here. Your background in philosophy, psychology, and religious studies is extensive, and the topic that we're going to discuss – the psychology of religious experience – is broad: there are innumerable religions and they are very diverse. And yet ultimately they all seem to reflect the human mind. Would you say that being a scholar of religions has made you a more religious person yourself?

SMITH

I certainly don't feel that I've become less religious, and I also feel that my studies have deepened and broadened my – what? – my beliefs. In that sense I guess one might say that my studies have made me more religious. I might prefer to say that they have matured me religiously, for my religious bent dates back as far as I can remember.

MISHLOVE

I suppose it's always a little delicate for a scholar, who is supposed to be objective, to study something as intense and passionate as religion can be.

SMITH

Well, some see it as a problem, but I've been fortunate in that I have never felt I had to choose between passion and objectivity. Quite the opposite. *Not* to have the two conjoined puts you at a disadvantage, it seems to me, for if you are studying something you are not in love with, what are your chances of getting deeply into it and seeing it from inside? It has been my good fortune to have been able to devote my life's work to dealing with what I most love.

MISHLOVE

My first encounter with the psychology of religion in a deep and personal way came from reading William James's classic, *The Varieties of Religious Experience* . . .

SMITH

. . . a wonderful book . . .

MISHLOVE

. . . in which he described his experiments with nitrous oxide and other then-known drugs.

SMITH

That was an adventuresome, courageous move and fully in character with the man that he was.

MISHLOVE

Then in the mid-sixties, I read a book by Timothy Leary and Ralph Metzner called *The Psychedelic Experience,* in which they attempted to create the analogy between the pantheon of gods in the

Hindu and Buddhist traditions and the dynamic forces working in the subconscious mind.

SMITH

Again, a venturesome hypothesis. But it seems to have stood up. There does seem to be a correlation between chemically induced paranormal experiences and spontaneous ones such as those that produced traditional pantheons.

MISHLOVE

An overlap, at least.

SMITH

A marked overlap. We can trace the religious use of mind-altering substances back at least three thousand years, and we now know a good deal about how the brain processes them.

MISHLOVE

You were involved in some of the early work on the connection.

SMITH

Close to the eye of the cyclone, actually. Those were the 1960s. I was teaching at M.I.T. and acted in effect as Aldous Huxley's social secretary during the semester he was with us as a visiting professor. He had published *The Doors of Perception,* which opened the public's eyes to the visionary possibilities of the entheogens. It turned out that Huxley's M.I.T. semester coincided with the fall that Timothy Leary moved from Berkeley to Harvard by way of a summer vacation in Mexico where, on the side of a swimming pool in Cuernavaca, he ingested a handful of psychoactive mushrooms that opened up his mind in ways that took him completely by surprise.

MISHLOVE

Psilocybin mushrooms, I presume.

SMITH

Yes. He had a three-year research appointment at Harvard's Center for Personality Study where he could name his own project, and he decided to see if the dramatic experiences that entheogens can produce have the capacity to change behavior. He had read Huxley's book and drew Huxley into his project as an adviser, and I got drawn in through Huxley.

MISHLOVE

At that time the drugs being studied were perfectly legal.

SMITH

Not only legal; they were respectable – this was officially authorized research at the top university of the land. The first thing that Leary did was to mount an open-ended study in which people would simply describe their experiences under the drugs, and he found that a large proportion of the descriptions had a mystical cast to them. That was the finding that interested me most.

MISHLOVE

You had been studying mysticism long before this, I presume.

SMITH

A fair statement.

MISHLOVE

Had you thought about the relationship between mysticism and drugs prior to your encounters with Leary and Huxley?

SMITH

Academically I had. I had read Huxley's *Doors of Perception,* and his contention that phenomenologically, which is to say descriptively, drug-occasioned visionary experiences are indistinguishable from ones that occur naturally. I conducted an experiment on that point in which I took accounts of classic mystical experiences and intermingled them with descriptions that Leary's subjects provided and asked knowledgeable judges to separate them into their original piles. They were unable to do so.

MISHLOVE

From the accounts they gave, your judges couldn't separate the classical mystics from Leary's subjects.

SMITH

Exactly. Their guesses didn't pan out.

MISHLOVE

That sounds similar to a study Lawrence LeShan conducted. He took statements of mystics and statements of physicists and compared them, and they too seemed almost indistinguishable.

SMITH

That's interesting and pertinent as well, but I won't go into how. The two cases are strikingly similar. But I think it's important to add another point. The evidence I cited shows only that drug and non-drug mystical *experiences* are alike while mysticism includes much more than mystical experiences. Its real concern is with mystical lives, including the compassion and other virtues such lives embody. It think it was Robert Ornstein who put this point graphically when he said that the object of mysticism is not altered states but altered traits. Experiences come and go, whereas it is life's sustained quality that counts. So we have to ask not only whether mystical experiences feel the same but also whether their impact on the lives of their subjects is the same.

MISHLOVE

Good point. And now that we have a twenty-year perspective on the original experiments you refer to, I think it's quite obvious that psychedelic cults don't have the staying power of authentic religious traditions.

What about Leary's claim in *The Psychedelic Experience* that the gods that people tend to project onto the world actually exist inside us, as parts of our own psyches? He seems to have been saying that the pantheons of the ancient pantheistic religions are forces that actually exist, but exist within us. I think he would hold that the same principle holds for monotheisms.

SMITH

We live in a psychological rather than a metaphysical age, and I see no harm in putting things the way you attribute to Tim. Whether we go the psychological or the metaphysical route is a fielder's choice because the important points can be stated either way. We have that option because geography doesn't apply to things of the spirit which elude spatial matrices. The spatial imagery that we attach to spirit is metaphorical only, and not literally accurate. It follows that the distinction between out there and in here, which in everyday life holds categorically, is relativized when religious objects are at issue.

If I can carry the issue one step further, it's natural to position values (good and evil, better and worse) on a vertical axis – superior and inferior, thumbs up, thumbs down.

MISHLOVE

Heaven?

SMITH

Yes. All peoples position their heavens above them and their hells in the bowels of the earth. Gods invariably dwell on mountaintops, and angels sing on high. But we need to add still another point. When we talk about the external world, good is imaged as up and bad as down. But when we introspect and attend to what's inside us, that reversal of attention causes the vertical (value) axis to invert as well. Deep and profound thoughts are better than superficial ones, and a fundamental truth is more important than a shallow one that only skims the surface. I say these things to support my point that the distinction you brought up – between out there and in here – cannot be taken literally when speaking of spiritual matters. It is relative and hence negotiable. Traditionally it was customary to think of God as out there, but in our psychological age there is a move in the direction you pointed to – to think of him/her/it (the pronouns never work) as within ourselves.

MISHLOVE

Another related notion, I think, is the one originally developed by Durkheim, the French sociologist, in which he suggests that religions are really reifications of society's group mind or collective consciousness. God, he argued, is actually a personification of a society's shared values. That sounds like Jung's collective unconscious to me.

SMITH

I can understand why you say that. I find both Durkheim and Jung's hypotheses useful when they are not pushed to their limits.

For one thing, both oppose the all-too-common notion that the mind is reducible to the brain, and therefore (so this reasoning goes) since the brain can be positioned, the mind can be located too. But I remember in a weekend conference in Tucson some years ago where Gregory Bateson asked the psychologists present – people like Carl Rogers and Rollo May – where their minds were located. His question took us aback, but when we saw what he was getting at, we saw it as a deft way of making his point, which was that it's wrong to think that our minds are located in our heads. That's where our brains are, but our minds are at large.

MISHLOVE

And of course we can always go back to the argument of Bishop Berkeley that the entire physical universe, that everything we experience – pots and pans and TV sets – exist only in your mind.

SMITH

Yes, though in fairness to Berkeley, we have to add that the durability of the physical universe derives from the fact that its contents reside in God's mind as well as ours. Otherwise the world would depend on human beings and vanish if human beings became extinct.

MISHLOVE

Right. A needed amendment.

SMITH

We've gotten into ecology here. Everybody's concerned about the ecology of nature, what about the ecology of mind? We're just beginning to get used to that idea, yet it's a daily experience. You can walk into the room and (in current terminology) feel vibrations. We can sense what feels like a wall of anger and hostility, and also sense an ambience of peace. And of course with the move toward field theory, physics too is becoming profoundly ecological. Dig deep enough and we see that what present themselves to us as free-standing, independent objects actually float on webs of relationships. Networking is the way the world works.

MISHLOVE

Picking up on the God within, or "the beyond within" as you title the section on Atman in your chapter on Hinduism, I notice in contemporary religions, particularly in evangelical Christianity, a resistance to God-within language. I assume that they are afraid of self-deification – usurping God's position. How do you feel about that?

SMITH

They have a point and you nailed it. I mean, if someone comes along and says, "I am God," it's perfectly reasonable to counter, "Well, your behavior doesn't seem to bear that out." God by definition is perfect, and what human being can make that claim? So I think the teachers that you refer to have a valid point, but it doesn't annul the concept of the divine within, which remains in place. Hinduism has been the most emphatic of the religions in saying that in the final analysis Atman is Brahman. That translates into saying that the deepest component of the human self is divine. But they accommodate the point you've raised with an analogy. A lantern can be coated with dust, and dirt, and finally caked with mud to the point where the light within it is totally concealed. So both sides of the matter are true. We are indeed divine in essence, but that essence is so clouded with human frailties that it is only marginally detectable.

MISHLOVE

In a previous program you told us about some of your experiences with primal peoples, such as the aborigines in Australia. As I recall your accounts, such peoples can access the divine quite regularly.

SMITH

I think that's so. The aborigines distinguish between our everyday world and what they call the Dreaming. The Dreaming is a transcendent state, or place, wherein the living participate in the lives of their ancestors, and indeed the creation of the world. I suppose we might call it a trancelike state, but they can be in it while going about their daily routines. There's another way in which they're in touch with a transcendent reality or state, and this has to do with parapsychology as we know the word – telepathy, specifically. I was in Australia for a summer (their winter) lecturing at its universities, and I spent my free time with the aborigines and the anthropologists who study them. Those anthropologists were unanimous in believing that the aborigines have telepathic powers. They told me story after story of cases when on a walkabout someone would suddenly announce that someone back home had fallen ill or died. They would retrace their steps, and invariably (according to the anthropologists' reports) the intuition would be validated.

MISHLOVE

That's a strong statement coming from anthropologists, who tend to be quite skeptical.

SMITH

Quite so. Their theory, or presumption, was that these are normal human powers, but like any power it can atrophy if it is not used. It can also be blocked if our conceptual mind disbelieves it.

MISHLOVE

Do you see some religious traditions encouraging the cultivation of psychic capacities more than others do?

SMITH

Well, it's interesting. I'll put it slightly differently. Most religions believe that paranormal powers exist – *siddis,* the Indians call them . . .

MISHLOVE

. . . meaning powers? . . .

SMITH

. . . yes, and that they tend to increase as one advances spiritually. However, religions tend to treat them gingerly. If you make their acquisition your goal you are settling for too little, and if you master some of them your wonder-workings can breed spiritual pride. Also, the domain is treacherous, so if you don't have a competent teacher you can go off the deep end. So authentic teachers tend to acknowledge the *siddis* while advising that we not get caught up in them.

MISHLOVE

But aren't there traditions – the shamanistic traditions in particular – that place great importance in these powers?

SMITH

Actually, in various ways all traditions do. To pick up on the shamans that you mention, the Mayans call them "spirit-lawyers," that is, men or women who go to the spirits and try to argue them into giving benefits of various sorts to human beings. They perform valuable services for their communities, but one doesn't associate sanctity with shamans.

MISHLOVE

Well, as our program winds up, I wonder if there's a message that you would like to leave with our viewers this evening? Something you might pass on to them from your studies?

SMITH

Let's see. Let me try this:

Religious institutions, though they are indispensable, are a mixed bag, but the basic claim that they put forward is true. And what that claim asserts – I'm paraphrasing William James here – is that the best things are the eternal things, the things in the universe that throw the last stone, so to speak, and say the final word.

NOTES AND REFERENCES

Chapter 2

1. A. R. Vidler, ed., *Soundings: Essays Concerning Christian Understandings* (Cambridge: Cambridge University Press, 1962). The statement cited appears on page 72, in H. A. Williams's essay "Theology and Self-Awareness."

2. Edith Hamilton, *Mythology* (New York: Mentor, 1953), p. 55.

3. Quoted in Alan Watts, *The Spirit of Zen* (New York: Grove Press, 1958), p. 110.

4. George Mylonas, *Eleusis and the Eleusinian Mysteries* (Princeton, NJ: Princeton University Press, 1961), p. 284.

5. Henri Bergson, *Two Sources of Morality and Religion* (New York: Holt, 1935), pp. 206–12.

6. Mary Barnard, "The God in the Flowerpot," *The American Scholar* 32, 4 (Autumn 1963): 584, 586.

7. R. C. Zaehner, *Mysticism, Sacred and Profane* (New York: Oxford University Press, 1961), p. 12.

8. Quoted in William H. McGlothlin, "Long-Lasting Effects of LSD on Certain Attitudes in Normals," printed for private distribution by the RAND Corporation (May 1962), p. 16.

9. Ibid., pp. 45, 46.

10. Timothy Leary, "The Religious Experience: Its Production and Interpretation," *The Psychedelic Review* 1, 3 (1964): 325.

11. Walter N. Pahnke, "Drugs and Mysticism: An Analysis of the Relationship Between Psychedelic Drugs and the Mystical Consciousness," a thesis presented to the Committee on Higher Degrees in History and Philosophy of Religion, Harvard University (June 1963).

12. The first account is quoted anonymously in "The Issue of the Consciousness-Expanding Drugs," *Main Currents in Modern Thought* 20, 1 (September–October 1963): 10–11. The second experience was that of Dr. R. M. Bucke, the author of *Cosmic Consciousness*, as quoted in William James, *The Varieties of Religious Experience* (New York: Modern

Library, 1902), pp. 290–91. The former experience occurred under the influence of drugs; the latter did not.

13. James S. Slotkin, *Peyote Religion* (New York: Free Press of Glencoe, 1956).

14. G. M. Carstairs, "Daru and Bhang," *Quarterly Journal of the Study of Alcohol* 15 (1954): 229.

15. Michael Polanyi, *Personal Knowledge* (Chicago: University of Chicago Press, 1958).

16. Sigmund Freud, *Totem and Taboo* (New York: Modern Library, 1938).

17. William James, *The Varieties of Religious Experience* (1902; reprint ed., New York: Macmillan Publishing Company, 1961), pp. 305–6.

18. "The Hallucinogenic Fungi of Mexico: An Inquiry into the Origins of the Religious Idea Among Primitive Peoples," *Harvard Botanical Museum Leaflets* 19, 7 (1961).

19. Margaret Prescott Montague, *Twenty Minutes of Reality* (St. Paul, MN: Macalester Park, 1947), pp. 15, 17.

20. "The Current Scientific Status of Psychedelic Drug Research," read at the Conference on Methods in Philosophy and the Sciences, New School for Social Research, 3 May 1964.

21. Quoted by Dr. Sanford M. Unger in the paper just mentioned.

22. Albert Camus, *The Myth of Sisyphus* (New York: Vintage, 1955), p. 38.

23. William James, *op. cit.*, p. 379.

24. From an early draft of a manuscript by Philip Kapleau that was published as *The Three Pillars of Zen* (Boston: Beacon Press, 1967).

25. Slotkin, *op. cit.*

Chapter 3

1. Quoted in *A Zen Forest: Sayings of the Masters*, trans. Soiku Sigematsu (New York: Weatherhill, 1981).

Chapter 4

1. R Gordon Wasson, *SOMA: Divine Mushroom* of *Immortality,* Ethnomycological Studies, no. 1 (New York: Harcourt Brace Jovanovich, 1968, 1971).

2. I have not been able to find the statement in his writings, but I am confident of my memory. Frithjof Schuon says of the Vedanta that it "appears among explicit doctrines as one of the most direct formulations possible of that which makes the very essence of our spiritual reality" (*Spiritual Perspectives and Human Facts*, p. 95).

3. Recounted in his book co-authored with the greatest living mycologist, Roger Heim.

4. "Fly" from the fact that the mushroom attracts flies and sends them temporarily into a stupor (Europeans used it as a flycatcher until quite recently); "agaric" from an error in classification. When Linnaeus came to mushrooms he found the whole domain so frustratingly complicated that he grew careless and used "agarikon," which is actually a tree fungus, to designate the gilled, fleshy-capped mushrooms.

Publications by R. Gordon Wasson Cited in the Text

1. "The Hallucinogenic Fungi of Mexico: An Inquiry Into the Origins of the Religious Idea Among Primitive Peoples," *Botanical Museum Leaflets* (Harvard University) XIX, 7 (1961).

2. With Roger Heim, *Les Champignons Hallucinogenes du Mexique* (Paris: Archives du Museum National d'Histoire Naturelle), Serie 7, Tome VI, 1958 (1959).

3. With Valentina Pavlovna Wasson, *Mushrooms, Russia and History,* 2 vols. (New York: Pantheon Books, 1957).

4. *SOMA, Divine Mushroom* of *Immortality* (New York: Harcourt Brace Jovanovich, 1968, 1971).

5. "*SOMA:* Comments Inspired by Professor Kuiper's Review," *Indo-Iranian Journal* XII, 4 (1970).

6. "*SOMA:* Mr. Wasson's Rejoinder to Professor Brough," in press as a monograph to be published by the Botanical Museum of Harvard University. Page numbers are those in the typescript.

7. "The Soma of the Rig-Veda: What Was It?" *Journal of the American Oriental Society* XCI, 2 (April–June 1971).

Reviews of Wasson's SOMA in English, French, and German

8. Andre Bareau, in *Journal Asiatique* (1969): 173–76.

9. G. Becker, in *Revue de Mycologie* XXXIV, 1 (1969): 84–87.

10. John Brough, "Soma and Amanita muscaria," *Bulletin of the School of Oriental and African Studies* (University of London) XXXIV, 2 (1971).

11. P. Demieville, in *T'oung Pao* LVI, 4–5: 298–302.

12. Robert Graves, "The Divine Rite of Mushrooms," *Atlantic Monthly* (February 1970): 109–13.

13. Catherine R. and Kate Hammond, "The Mystery of Soma," *Sunday Herald Traveler* 2 (August 1971), Book Section, pp. 1–2.

14. Daniel Ingalls, "Remarks on Mr. Wasson's Soma," *Journal of the America Oriental Society* XCI, 2 (April–June 1971): 188–91.

15. Daniel Ingalls, in *The New York Times Book Review* (5 September 1971), p. 15.

16. Jacques Kayaloff, in *The Russian Review* (April 1970): 233–39.

17. Stella Kiamrisch, forthcoming in *Artibus Asiae.*

18. F. B. J. Kuiper, in *Indo-Iranian Journal* XII, 4 (1970): 279–85. Rejoinder by R. Gordon Wasson, pp. 286–98.

19. Weston La Barre, in *American Anthropologist* LXXII (March 1970): 368–71.

20. Claude Lévi-Strauss, "Les Champignons dans la Culture: A propos d'un livre de M. R. G. Wasson," *L'Homme* X (1970): 5–16.

21. Portions of the above, translated by Alfred Corn, "Claude Lévi-Strauss: Mushrooms in Culture," *University Review* 12 (1970), unpaged.

22. B. Lowy, in *Review Mycologia* LXI, 4 (July–August 1969): 849–51.

23. M. M. Payak, in *Indian Phytopathology* XXII, 4 (December 1969): 527–30.

24. A. Pilat, in *Schweizerische Zeitschrift für Pilzkunde: Bulletin Suisse de Mycologie* XLVIII (November 1970): 133–43.

25. Winthrop Sargent, "Mainstay of the Sky, Foundation of the Earth," *The New Yorker* (30 May 1970): 90ff.

26. Richard Evans Schultes, in *Economic Botany* XXV, 1 (January–March 1971): 111–12.

27. Richard Evans Schultes, in *Journal of Psychedelic Drugs* 111, 2 (September 1971): 104–5.

28. Michael Sullivan, in *Journal of the American Oriental Society* XCI, 2 (1971): 346.

29. Unsigned, "Ariadne," *New Scientist* (3 September 1970): 494.

30. Unsigned, "Daily Closeup," *New York Post* (19 August 1971).

31. Unsigned, in *Times Literary Supplement* (22 May 1969). Correspondence: 21 August 1970; 11 September 1970; 25 September 1970.

32. S. Henry Wassen, in *Saertryk af Friesia* (Copenhagen) 330–32, and in *Svenska Dagbladet* (8 August 1969).

Chapter 5

1. Irmgard Schloegl, *The Wisdom of the Zen Masters* (New York: New Directions, 1975), p. 21.

2. A. K. Coomaraswamy, *Hinduism and Buddhism* (New York: The Philosophical Library, 1943), p. 30.

Chapter 6

1. The flyleaf of Rank's book which served as almost the bible for Grof's work in one of its stages carries a quotation from Nietzsche: "The very best . . . is, not to be born. . . . The next best . . . is . . . to die soon."

2. Dr. Grof's studies did not stop a quarter century ago when the foregoing report of it was written. His recently published book, *The Cosmic Game: Explorations of the Frontiers of Human Consciousness* (Albany, NY: State University of New York Press, 1998), picks up the story where it leaves off here.

Chapter 7

W. N. Pahnke, "Drugs and Mysticism: An Analysis of the Relationship Between Psychedelic Drugs and the Mystical Consciousness." Unpublished doctoral dissertation, Harvard University, Cambridge, 1963. Summarized in W. N. Pahnke, "Drugs and Mysticism," *The International Journal of Parapsychology* 8, no. 2 (1966): 295–320.

R. Doblin, "Pahnke's 'Good Friday Experiment': A Long-Term Follow-Up and Methodological Critique," *The Journal of Transpersonal Psychology* 23, no. 1 (1990): 1–28.

Chapter 8

Hillary Jenkins, *Newman's Mediterranean Voyage* (Dublin: Irish University Press, 1974).

D. H. Salman and R. H. Prince, eds., *Do Psychedelics Have Religious Implications?* (Montreal: McGill University / R. M. Bucke Society, 1967), pp. 1–12.

INDEX

C

D

E

Q

R

T

U

V

W

BLACK MONDAY

R. SCOTT REISS

SIMON & SCHUSTER
New York • London • Toronto • Sydney

SIMON & SCHUSTER
Rockefeller Center
1230 Avenue of the Americas
New York, NY 10020

This book is a work of fiction. Names, characters, places, and incidents either are products of the author's imagination or are used fictitiously. Any resemblance to actual events or locales or persons, living or dead, is entirely coincidental.

For information about special discounts for bulk purchases, please contact Simon & Schuster Special Sales at 1-800-456-6798 or business@simonandschuster.com

Designed by Dana Sloan

Manufactured in the United States of America

10 9 8 7 6 5 4 3 2 1

Library of Congress Cataloging-in-Publication Data
Black Monday : a novel / R. Scott Reiss.
p.   cm.
1. Terrorism—Fiction. I. Title.
PS3618.E5726B58 2007
813'.6—dc22   2006051228
ISBN-13: 978-0-7432-9764-6
ISBN-10: 0-7432-9764-4

For, in order of appearance,
Ariel, Chayim and Samantha

BLACK
MONDAY

ONE

OCTOBER 27TH. 6 HOURS BEFORE OUTBREAK.

A plague that will cause the death of millions. A plague that will destroy countries. A plague that will plunge the world into a dark age.

A plague that will make nobody sick.

Lewis Stokes—or so the false name on his Nevada driver's license reads—feeds another dollar into the Wheel of Fortune machine in the lobby of hotel New York–New York in Las Vegas and feels his heartbeat pick up, but not because of the game. The onetime beggar boy—whose mother was publicly beheaded—has just spotted the twenty-year-old University of Nevada English major that he's flown six thousand miles to kill.

The boy—slovenly-looking and dark-haired—is weaving toward him, past the single-deck blackjack tables, heading for the reception desk. He's drinking from a foot-long glass beaker filled with bright red liquid, probably a Singapore Sling or mix of rums and fruit juices. The boy looks tipsy, unaware, alone.

The kid must be killed by 12:14 tonight.

"Not one minute later," Lewis's mentor had said when he'd provided the usual range of perfectly made false IDs.

Lewis tenses to stand, to follow. But he realizes that the boy is too tall to be Robert Grady.

He just *looks* like Grady.

Lewis curses under his breath and puts another dollar in the machine.

Normally a handsome blond, Lewis is disguised as a balding dark-haired man today. Normally lean, he looks heavy and clumsy from the belly-extender bladder and black-framed glasses. His posture is slumped. He walks with a limp. The few people who notice him register a nerd in a box-cut sports jacket. A cheapo off-the-rack design.

Playing slots enables him to sit within view of the reception desk, invisible to the bellboys, desk clerks and house detectives. One more gambler among hundreds. But *this* gambler conceals a Glock under his jacket and a serrated K-bar knife in the pit of his back. Lewis killed his first person at age twelve, in self-defense, in a tent.

"Wheel . . . of . . . Fortune," shouts a chorus of tinny mechanical voices in his machine as the wheel spins on top and multicolored lights flash, and potential amounts of winnings, $800, $100, $20, rotate in pie-wedge shapes on the wheel.

He hates Las Vegas, the brashness, the noise, the anarchy that reminds him of the refugee camp where he grew up. The damn ground floor is the worst. It's like Fellini designed the place. It's a madhouse of rock music, kids running, machines clanging, drunks laughing. No windows to the outside world. No glimpse of anything except the asylumlike gaming area, laid out in a maze through which flows a never-ending human jackpot. People spilling like coins from the elevators and heading out toward other local traps; the Riviera and the Paris, the Monte Carlo, the Gold Coast, none of them remotely resembling the romantic spots for which they've been named.

More to the point, where is Robert Grady?

"Make it look like robbery if possible," Lewis's mentor had said. "But if that boy is standing in a crowded lobby at 12:14, walk up and shoot him in the face. Can I depend on you to sacrifice yourself if necessary, my old and special friend?"

"What happens at 12:15 if he's still alive?"

"The world may—unfortunately—stay the same."

"Why will killing a college student make so much difference?"

"I want you to know his exact role. You *deserve* to. But if the Americans catch you, if they figure out who you are, they will do anything to make you tell."

Five hours and thirteen minutes left.

Lewis arrived in Las Vegas two days ago. Plenty of time to work. But he's been unable to locate Robert Grady. The boy has not gone home. He's not attended class. His telephone answering machine is so filled with messages that it refuses to accept new ones. Does he know Lewis is here? Who the hell is he, anyway? His girlfriend, when Lewis phoned her apartment, pretending to be from the school, said she'd not seen Grady in a week.

"He's a degenerate gambler, and I'm through with him," she'd snapped. "He only applied to your stupid school so he could play craps in casinos. When he disappears it means he won money. He'll keep playing until he loses it back."

Finally, an hour ago, Lewis had made a fourth round of calls to casinos that the kid frequented and learned that a Bobby Grady had a reservation to stay here tonight. So Lewis reserved a room too. The file said the kid always stays on the eleventh floor, Century tower, because he considers that tower "lucky." So Lewis checked in to that tower too. It was the only way to obtain key-card access to the elevators leading upstairs.

Lewis checks his watch, takes a break at the machine and calls the hotel operator on the house phone.

"Mr. Grady just called. He said he'll be checking in a little late," she tells Lewis.

"How late?"

"He didn't say."

"Did you talk to him?"

The operator seems offended that he's asked. "I'm telling you all I see on my screen, sir."

Lewis curbs his irritation, slumps his shoulders to remain inconspicuous, ambles back to the Wheel of Fortune machine.

A white-haired old lady in a wheelchair now sits beside him at another machine. She balances a plastic cup filled with quarters on her skinny lap.

The lady smiles at him. "This place is *so* exciting!"

He doesn't answer. She'll remember him less accurately that way. He's remembering his last visit to see his mentor, in August, transported in his mind to a more quiet, beautiful place. They'd sipped orange juice in a cool green garden. Mist-shrouded oaks had rimmed the vast lawn. The crash of the nearby ocean had mixed with the cry of wheeling terns as the mentor and younger man sat on nine-hundred-year-old stone benches. Everything around them, the private forest and green mountains and the sprawling home beyond the sculpture garden, had been solid, lovely, old.

"Actually, Robert Grady is one of several people I'm hoping you'll visit in America," Lewis's mentor had said, conveying orders as if they were requests, as always.

Lewis flashes back to the last murder, three weeks later, after he'd flown to Washington, D.C., bought a car and driven up I-95 to the Taconic Parkway, the Berkshire hills and the village of Becket, Massachusetts. There he'd located the isolated dirt-road home of a fifty-nine-year-old custom kayak maker. He'd entered through an unlocked door. People in the area did not fear home-breakers. He'd disemboweled the man when he returned home on a Friday night from a Savion Glover dance performance at the Jacob's Pillow Dance Festival. Lewis had worn latex gloves on the job. He'd used his left hand during the attack to fool forensics experts, analysts of angles of attack.

Lewis was a righty except when on jobs.

After the killing, he'd rifled the medicine cabinet for pills, stolen the cash in the man's wallet and taken some antique silverware. He'd dumped the loot in the waters of a nearby quarry, a deep green lake.

"ROBBERY MOTIVE IN MURDER," the local *Berkshire Eagle* had proclaimed.

As his mentor had said, "Deception is success. Disinformation is deception. Always make Americans blame others for what you do."

"By 12:14 Robert Grady will journey to the other side," Lewis had promised, recalling words that his great-great-grandfather had written after World War One. Words he carried with him on travels, in a dog-eared edition published in 1927. "*Blood is always on our hands, but we are licensed to it.*"

And now, finally, he spots Robert Grady.

The kid passes within two feet of him on his way to the reception desk. At first glance Grady seems like one more easygoing college boy. Open-necked button-down white shirt, slightly wrinkled. Faded Levi's. Worn Avias and an Eastern Mountain backpack over his right shoulder. The face is young and scruffy with a brown beard. The boy has baby blue eyes.

But Lewis also glimpses something raw beneath the soft surface. The eyes aren't exactly clear and innocent but fixed on something invisible. Lewis grew up surrounded by desperation. He knows its forms: need, terror, obsession, greed. This boy is haunted by premonitions. By the pull of compulsion. The slavery to odds.

Lewis watches Bobby Grady turn away from the reception desk. But instead of going upstairs, Grady hands his knapsack to a bellboy, points at the elevator and gives the kid a tip.

Robert Grady wants to go gambling right now, it seems.

Lewis sighs, feeds one last dollar into the machine and lets the guy pass and draw away again, into the casino. He pushes the Wheel of Fortune machine button one last time and rises silently to follow.

But the machine goes berserk. Bells clang. Wheels spin. Everyone within a hundred-foot radius gapes at Lewis. Bellboys. Guests. Kids. A prostitute. Security cameras in the ceiling will be recording the scene. Hotel guests, people checking in, lined up with their luggage, crane to see. The Wheel of Fortune machine has been programmed, on super-rare occasions when it pays off big, to make a commotion as loud as an air-raid siren on an American military base. The noise almost drowns out the rock and roll music blaring through the lobby.

Blingblingblingbling!!!!

The old lady in the wheelchair gasps. "Good Lord! It's not *stopping*! Five thousand and . . . oh . . . oh *my*!"

Robert Grady, who has not turned to watch, is drawing farther away, heading for the sports betting area.

A flashbulb pops. Someone has taken a photograph of the big winner.

"Hotel newsletter," the woman holding the camera announces to Lewis with a grin as one of the casino attendants, a Hispanic-looking man in a brick-colored jacket, approaches with an immense smile, holding a clipboard that probably contains a form to be filled out for the IRS. You cannot deduct gambling losses on tax returns in the U.S. But you pay taxes if you win. Is life fair?

In the fraction of a second during which Lewis decides what to do, the process comes so fast he experiences it as instinct. If slowed to logic, his thoughts would be:

Don't worry about the witnesses. The photos won't show what I really look like. No one will connect what happened here to what happens to Robert Grady later tonight.

He turns to the woman in the wheelchair as the stupid machine keeps clanging, as numbers spin, zeroes flash, nineteen turns to twenty . . .

"I have to catch my flight!" he gasps with Lewis Stokes's "southern accent." He's been schooled in Austria, Bahrain, Tunisia.

The woman gapes at him. This is not what she expected to hear.

"My wife doesn't know I'm in Vegas," he says. "I'm here with her best friend. If I miss my flight, I'm dead!"

The old lady's eyes go wide. She understands now, all right. She's probably getting more shock and titillation material in the last minute to pass on to the knitting club back in Houston than she normally experiences in a decade.

"I can't *believe* it," he says. "I lost almost all my money at Circus Circus and now . . . *You take it.*"

"Whaaaaa . . ."

Ms. Wheelchair's mouth opens so wide you'd think she could swallow the whole damn slot machine. Lewis pushes through the crowd, ignoring tourists snapping shots. The woman from the newsletter looms close to get a profile portrait. The overhead surveillance cameras undoubtedly follow his every move as he quick scopes the gaming area and limps toward the faux cobblestone restaurant row, toward the sports betting area, hurrying because Bobby is gone!

He hears astounded voices behind him, saying things like, *"He's leaving without the money!"*

"He told the lady to take it!"

"Why her? Why not me?"

Nobody coming after him though. They're glued to the Wheel of Fortune machine, rapt to see who will get his winnings. He moves faster, pushes a heavyset man aside and catches sight of the kid standing in the sports betting area, staring at the screens showing today's trotter races from Aqueduct.

Robert Grady turns away from the screens and heads outside. Apparently he's seeking a different casino.

Stokes jams his hands into his pockets and follows onto Las Vegas Avenue, the famous strip, in the always-moving crowd. At their backs rises the phony Manhattan skyline of hotel New York–New York. The false Statue of Liberty. The black towers made up to look like high-rises jutting incongruously into the sucked-out pastel desert sky. The perpetually racing yellow hotel roller coaster roaring and twisting, carrying screaming passengers.

I need to change appearance but can't lose sight of the kid.

At dusk the casino lights are coming on. In the distance, beyond backed-up traffic on I-15, the mountains look hazy, lavender. The city's glow blocks out early stars. It's so dry here that even at a hundred degrees, Lewis doesn't sweat, or rather, his sweat dries before he notices it. He relishes heat. Out beyond the garish hotels—temporary edifices—is the timeless desert. He's spent years working in deserts, but this one is different, harder at the surface, less white with sand, spotted with razor-needled cactus and flinty rocks. But it is clean like a good desert, blasted by heat and nightly cold from nature's purification process. The desert is a testing ground for human capability, luck and mercy, a place where those who lack survival skills perish as if they had never lived.

Lewis wills away human smells, the dirty odors of tar, hot dogs, perfumes, bus exhaust. He blends in perfectly.

For as his great-great-grandfather had written, in the stilted prose of the World War One era, "If I cannot assume the character of strangers, I can at least conceal my own, and pass amongst them without friction."

The book, a gift from Lewis's mentor, gives him purpose. When he's lonely, it provides comfort.

"We lived for the day and we died for it," great-great-grandfather had written, knowing well the secret life.

Bobby Grady ambles into the first casino he approaches, the Monte Carlo. He changes dollars for chips at a blackjack table. The kid hits blackjack on his first try.

Lewis takes a chance and ducks into a nearby men's room. He hadn't planned on morphing until later but can't risk being recognized after the big payoff at the Wheel of Fortune machine. People who walk away from $20,000 in winnings might land up on the news, he thinks. He envisions some tourist providing his photo to a local TV station, imagines a story broadcast later tonight.

Have you seen this man?

He finds an empty stall, his pulse racing. He's gambling that Robert Grady will stay at the blackjack table for a few minutes more. Lewis forces himself to move slowly to keep from making a mistake. Off comes the headpiece, changing him from a balding brunette to a full natural blond.

Hell, how many wigs are made with gigantic bald spots on top? Not many, he thinks.

Off comes his bland and ill-cut blue jacket, off comes the shirt and tie to reveal a white knit tennis shirt beneath. Off comes the belly extender, a simple deflatable bladder. Presto, off come the black glasses. His vision is perfect. He straightens his posture, adding two inches to his height. The limp is gone. The rubber cheek wedges are gone. The mustache is gone.

From his hip pocket he produces a scrunched-up reinforced knapsack of the thinnest, strongest polymer fabric. The discarded props go into the bag. The trousers stay the same, unfortunately, and so do the rubber-soled shoes. There are limits to making quick changes.

His Glock 9mm pistol now lies nestled against his back, beneath his shirt.

Less than three minutes after the balding forty-something "tourist" ducks into the second stall, a healthy, tanned, handsome blue-eyed beach boy type—the real Lewis—emerges and reenters the casino to see Robert Grady getting up from the blackjack table, scooping up chips, walking off.

He follows Bobby outside again, slows his pace to match the man's and drops the knapsack in a trash can, already spotting a scavenger—a Vegas bum—veering over to take it, and probably sell the props inside.

I wonder what happens after 12:14.

Ten fifty-nine now.

Bobby Grady is still walking.

It's unbelievable. Doesn't the kid rest? He's been in and out of casinos for hours, losing thousands of dollars at blackjack, craps, baccarat, sports betting. He's dropped in at the Palace Station casino and the New Frontier.

How does a student get all this money?

The kid's in a trance. He gambles and drifts. Gambles and walks. He's even paused to watch outdoor shows between bouts of losing. At the Venetian Hotel, and its re-creation of the canals of Italy, he'd sat for a long time eyeing hotel "gondoliers" pushing tourists around a "canal" as they imagined that their absurd imitation approximated reality.

Water again, Lewis thinks in disgust, following Grady into the Bellagio casino, past a line of sprouting jet fountains outside. The fountains sway back and forth like coordinated dancers, to the strains of "The Star-Spangled Banner." It's funny. In a desert city dedicated to expense, the most valuable tourist draw, the most fabulous part of so many shows is plain water.

Eleven ten.

One hour and four minutes left.

Robert Grady strolls into the hotel's mist-cooled tropical garden. He pauses to gaze stupidly at the gigantic mechanized bald eagle, whose huge head swivels back and forth, an allegedly patriotic incarnation whose jerky movements remind Lewis of robotic dinosaurs in 1950s films. He used to watch old American movies in the refugee camp, when he was a boy.

Wait a minute. *Grady's walking into a men's room.*

Lewis follows.

His hand goes behind him, feels the knife.

He hears heavy laughter in there, though, and when he walks in sees half a dozen large men—police types from the build and haircuts—at the sink. They're wearing name tags. They're part of a convention.

"When did you join the FBI?" one man asks another.

Unbelievable!

At eleven twenty-four, Bobby comes out of the men's room, rubbing his belly as if he feels a stomachache.

Eleven thirty.

If I have to do it in public, I will.

Robert Grady turns and pushes back out of the casino, turns south toward his hotel, passing the pizza and burger places serving the lower-rent tourists, and stepping over hundreds of discarded business cards—phone numbers of prostitutes handed out by illegal aliens, Mexican workers—lying on the sidewalk.

New York–New York coming up again.

Lewis follows the man back into the hotel. The lobby echoes with Irish music, rock music. Robert fishes in his wallet for his key card, matched to a specific elevator.

He walks into the elevator behind the boy. They're alone. He can hear the kid's rumbling stomach.

Bobby shrugs at him and asks, "How'd you come out tonight, guy?"

"Ahead," says Lewis in his southern accent. "You?"

"Hey, luck has to change, right?"

Lewis precedes the man off the elevator, onto the eleventh floor, half a step ahead so as not to alarm him. Both men turn left at the hallway junction. Lewis notes that there's no one else in the hall. Dirty room service trays sit outside a few rooms. The AC is arctic level here, probably to keep the guests from sleeping, Lewis thinks.

"Good luck tomorrow," the kid remarks over his shoulder, inserting his card into his door slot.

It's all pretty easy after that, pretty routine.

Lewis's hand comes out in one smooth movement. He hits the boy from behind, right hand over the mouth as he drives Grady into the room, left hand shoving the knife in between the third and fourth ribs. He's carried out the motion a thousand times over the years, on sandbags, on dummies, on prisoners, on his mentor's targets of choice.

Only seconds have passed. He shuts the door. Robert Grady lies facedown on the carpet, facing the bathroom as blood spreads on his shirt. He's voided himself in death. He never even struggled. The AC will cut the smell that will eventually reach the hallway and attract a maid, despite a DO NOT DISTURB card on the door. The TV is on. The window is open by half an inch, maximum space. Hotel windows in Vegas only open that far, probably so depressed losers won't jump out, Lewis thinks.

Phony evidence time. Latex gloves on!

Lewis steals the wallet and stack of hundred dollar bills inside. He slips from his pocket a small Ziploc bag from which he removes one gold earring, a prize he'd spotted beside a slot machine. A bit of woman's dried blood discolors the post. He drops the earring on the carpet.

He also opens a small bottle of cheap perfume and knocks it over on purpose. That way the smell will stay in the carpet, even after he puts the bottle back in the bag.

Then he makes sure he leaves two partial haunch impressions on the stiffly made-up bed, as if two people had been sitting there, side by side. An investigator will conclude that Robert stood up from the bed, and got hit from behind while on the way to the bathroom.

Finally, Lewis lets himself go crazy with the knife, using his left hand, of course, stabbing and slashing the back and neck and base of the head so that the death blow will be considered just one more thrust, not the first.

Lewis Stokes—who will as of later this morning become Clayton Cox—decides that one day he too will record his memoirs, like his great-

great-grandfather. Perhaps he will begin formulating them over the next few weeks, when he's been ordered to go to Washington, to monitor certain people at the Pentagon and wait for instructions from the man who changed his life.

Deception is the key to success.

On the night table, the red digits of the clock turn over. It is 12:14 A.M.

I wonder what will happen a minute from now?

Time to get out of here, to leave Vegas. The mentor has supplied a safe route.

For as Lewis's great-great-grandfather had written of his secrets, for history, for governments, for scholars, "We had on our heads prices which showed that the enemy intended hideous tortures for us if we were caught."

TWO

OCTOBER 28TH. OUTBREAK.

Midnight in Las Vegas is 10 A.M. in Qatar, the small, fabulously wealthy oil-state located on the southwest shore of the Persian Gulf. The recently relocated headquarters of Al Jazeera, most popular television network in the Arab world, occupies the top three floors of a spanking new seventeen-story glass tower overlooking the shark-infested water. Below, rich pleasure boats roar past outgoing oil tankers and lumbering incoming container carriers bringing new computers, large-screen TVs, gas-guzzling SUVs and Italian designer furniture to the nouveau-riche millionaires whose condos line the overbuilt shore.

On the sixteenth floor, filled with disquiet, Hassan el Kader sits in the air-conditioned control room of studio C, sweat forming in the small of his extremely large back as he wonders if he's about to get himself murdered over the upcoming show.

Have I gone too far this time, he wonders?

Hassan is executive producer of the award-winning weekly news show *Faces and Places.* He's also a CIA spy, who passes along tips—things his reporters learn—to the Americans. And the Americans were very interested to find out about today's taped segment two on the show, which will begin in minutes and reach over one hundred million viewers across the Arab world.

Out on the set, beyond the control room window, the makeup girl finishes brushing face powder on show host Leila Shaalan.

It had all started three weeks ago, in the same way many of his top stories had begun over the years, with Hassan's cell phone chiming at 2 A.M., with a familiar whispered voice in the darkness. "I have a terrific story for you," one of his best sources, a man he had never met, had said. "It is about an old man in the mountains of Pakistan."

An old man who will probably be hunted down by the Americans or British by tonight, Hassan tells himself now as the QUIET ON THE SET light comes on and the hosts turn toward the cameras, ready to begin. He thinks back to the late-night call.

"A holy man who makes predictions," the voice had said.

"Everyone makes predictions," he'd responded.

"But his come true. He knows things that will happen far away from his village, even in Washington. He has a huge prediction to make. Or shall I offer the BBC the opportunity to tape it instead of you, my fat friend?"

With thirty seconds remaining before showtime, the pace in the room quickens Hassan's heartbeat. The 36-year-old Kuwaiti—ex–University of Kuwait BS degree, ex-Columbia University School of Journalism, ex-prizewinning reporter—sits before banks of screens lining the walls, providing him with simultaneous scenes from all over the world. A protest march against high fuel prices in Chicago. A shot of rioting rail workers in Paris. A parade of goose-stepping troops in North Korea. The floating debris marking the wreckage of an Alitalia Airbus that crashed into the Mediterranean Sea last night on its way to Rome.

Human sensory systems are not designed to support secret lives, Hassan sometimes thinks, remembering the way the Americans had recruited him back at U Kuwait. They'd saved his country from the Iraqis. His own father—a pilot—had been killed in that war. It had been clear to the idealistic student that Arabs had to participate in the fight against terrorism.

We need someone at Al Jazeera, the American journalism "professor" who had recruited him had said back then. *We need a pipeline into the terrorist world. We can save Arab lives if you pass along names, addresses, sources. Only once in a while, on important stories. Will you do it?*

Since then he's helped stop bombings in Madrid and Baltimore. And thanks to his warning today, terrorism alert levels are raised all over Europe and America. Extra police are out. Extra security helicopters are aloft. Extra airport and harbor security is in place.

"And now," says the voice of an announcer as a globe spins on half the screens in the room, "welcome to *Faces and Places.*"

Hassan thinks, *If the man who called me realizes what I've done, I'm dead.*

Segment one, an interview with the King of Jordan, begins. The king's face fills ten screens in the control room. The segment was taped at his palace in Amman.

Hassan reaches into his shirt pocket and removes a small foil packet of Tums.

"Tell me more about this holy man," Hassan had asked the caller two weeks ago, sitting in bed.

"He gathers a growing following. Tribesmen. A few military people. Students. His reputation grows."

"What predictions has he made that came true?" Hassan had demanded like any good journalist. He helped the CIA out of idealism. He'd refused money over the years. He'd hidden his clandestine activities from his wife and children.

And on the night of the call, it was clear to him that one of the high-profile organizations—Red Brotherhood, Al Qaida, Black Baghdad—wanted the Imam on TV, probably planned some event around the broadcast. Or maybe they sought to build the Imam's reputation. Extremists often used Al Jazeera to get out messages.

Hell, if it's news we run it, not because we sympathize, Hassan thinks, casting back to the late-night phone conversation again, and the caller's response.

"Just last month the holy Imam told us that death would come from the sky within twenty-four hours in an Algerian valley named Bar El Kab. It came to pass. American fighter planes wiped out a training camp there for mujahideen. The Imam sees the future, hears the intent of God."

"How did the Imam know what the Americans would do?"

"Send a reporter to ask. Check his earlier predictions. They really occurred. Isn't this news, knowing the future?"

"What else did he predict?"

"Two months past he warned all true believers in the American city of Tulsa to leave that country and return home. He said American police would make their lives miserable within the week, arresting many."

"Two months ago," Hassan had said, doing the math, "was when the FBI rounded up Tunisian immigrants in Tulsa."

"The very next day. Send a reporter. Share with the world what the holy man says. Do it by November first."

"Why that day?"

"By then, he says his next prediction will come true."

"How does a poor holy man in Pakistan know what happens in Washington?"

"Have your reporter ask."

"Is the Imam connected to a political group?"

"He will answer all questions."

"What is his new prediction?"

"I have to wait, like you. Do you want the story or not?"

"Who are you?" Hassan had asked the caller for the hundredth time over the years.

"A friend whose advice has never been wrong."

Now Hassan pops another Tums as on the set, the lovely Leila Shaalan tells viewers, "For our next segment today, something really incredible. *Faces and Places* reporter Fauzan el Harith journeys into the rough and dangerous northwestern mountains of Pakistan to meet a holy man."

The map on-screen shows the region topographically, with little red dashes showing Fauzan's trip. Ah, Pakistan, Hassan thinks with a journalist's fondness for trouble spots. If the government isn't being overthrown there's a nuclear showdown with India, or a crackdown on militants, or accusations of atomic secrets for sale.

"To reach the holy man, we traveled for two days," Fauzan's voice-over tells viewers around the globe, in Cairo coffeehouses, Damascus police stations, Lebanese villas, Palestinian refugee camps. Western analysts in London and McLean will be watching too, of course. They grind their teeth each time Al Jazeera reporters interview someone the CIA would like to arrest, like the subject of segment two.

Fauzan dragging out the anticipation as he tells the story, waxing eloquent about the sun blasting down and the air cooling as he climbed higher into the Katar Range. About bedding down with guerrilla escorts beneath the stars, and watching a meteor shower surge crosswise below an Arabic crescent of white-hot moon.

On-screen, the trudging journalist reaches the snowline. The men top a ridge and begin descending into a brown, harsh, rocky valley.

"That's when I saw the caves," Fauzan says.

We all use each other. The CIA uses me. I use the callers. The terrorists use the network. The network uses the news, Hassan thinks as the Imam hobbles into view. He is a short, crippled, elderly man who leans on a wooden crutch, and whose useless left foot dangles six inches off the floor of the cave as he moves.

"Imam Suleiman has never traveled more than ten miles from the village where he was born. He's been crippled since age six," Fauzan's voice tells Al Jazeera's viewers.

Ten screens in the control room show a ring of somber men in loose-fitting pants and vests sitting cross-legged in the cave, on rugs, as sunlight shafts in from a natural hole in the roof. Dust motes dance in the column. Hassan, who in his own days as a reporter covered Pakistan, imagines the smell of unbathed men, sweet tea, curried lamb, charcoal bits, oiled weapons.

"Word of Imam Suleiman's predictive powers have spread and many now make their way to this valley," Fauzan says.

The camera focuses on the old man's ravaged face and milky white left eye. Hassan's whole staff had been riveted the first time they saw this tape.

Fauzan begins his interview by asking the Imam about his public announcement last month warning someone named Abu Gabra, far away in the south, that men wearing hawk feathers would attack him. Fauzan asks if Suleiman knew that "Abu Gabra" is actually the name of a Sudanese oil field. And that the French paratroopers who dropped into it a week after the prediction was made—to rescue hostages—wear hawk patches on their sleeves.

"I only repeat the words that God puts into my mouth," Imam Suleiman says in his oddly high voice.

Fauzan asks whether the Imam's inspiration actually comes from information provided by extremists, terrorists.

"If that were true, my son, if they knew what was planned, why would not our fighters at Abu Gabra have been ready to repel the French soldiers when they arrived?"

His disquiet growing, Hassan checks the foreign networks, in case a terrorist attack is timed to take place as *Faces and Places* airs. Everything looks normal in the TV news world, whatever normal means. A segment on a nuclear submarine launch shows on French TV. A big fire rages in Berlin on German TV. Two ships have crashed into each other in Tokyo. One disaster flows into another, as unreal as in a Hollywood film.

Yet even to jaded Hassan, there is something compelling in Imam Suleiman's delivery. Knowing what is coming on the tape, Hassan experiences a primitive dread that pierces through his cynical armor as a journalist.

His prediction is impossible, Hassan thinks. *He's merely using flowery language. What he says cannot become real.*

"I understand that you have a prediction to make," young and handsome reporter Fauzan asks the Imam.

The camera seems to look long into the old man's face. The milky eye. The craggy cheeks. The scar across the forehead, allegedly from a lightning bolt the man survived at age six. God's mark, the local herdsmen say.

Imam Suleiman's voice coarsens into the voice of a younger man. It is as if another soul has taken possession of his body. Straightening up, he seems bigger.

"Rejoice, my brothers, for I do not mourn today for fallen comrades. I predict not the destruction of friends this time, but the annihilation of the great enemy. God has had enough of their murderous arrogance, their corruption, their greed, their heretical ignorance and mocking of the proper ways and commandments, and so I issue a fatwa."

The men in the cave lean forward as if pulled by invisible force. Even Fauzan seems captivated by the power of the Imam, the certainty in his voice.

"Let their great towers fall into darkness in Washington and London and Bonn. Let their people rise up against their rulers. Let their homes burn, their food disappear, their machines cease to function and drop from the sky, *starting by the end of this month.*

"Let their soldiers rebel, their governments topple, their religious houses crumble by their own hands. Let nature refuse them bounty, *starting by November first.*"

No prediction has been this specific, this vast, Hassan thinks. No threat has had this kind of timetable.

"His presence in person is formidable, but on TV he's awesome," Fauzan had said.

I wonder if the Americans have attacked the cave yet.

The Imam is saying, "And their great illicit fortunes will wither to uselessness. Their corrupt friends in Muslim lands will answer to oppressed millions. This is not an idle wish. The Lord's retribution is to begin *not* in the distant future but perhaps even as soon as my words die away."

October will be over in three and a half days, Hassan calculates. *So whoever is behind this threat has 86 hours to attack.*

Imam Suleiman adds, "From the wreckage, the survivors will build a beautiful new world."

He refuses to talk anymore. He subsides back into an old man. He hobbles away, helped by aides.

Millions of people across Arab lands, Hassan knows, in homes, coffee shops, businesses and palaces, are looking at each other and wondering whether by some miraculous power the old man has actually seen a future of celestial retribution.

Fauzan looks deeply into the camera. "We leave you with a question. The clock is ticking. Will the Imam's prediction come true?"

The credits roll.

Out on the set, the host is removing her lapel microphone.

Hassan feels the tension in his back begin to ease. It's always that way after a show he's told the CIA about. He thinks, as always, *Maybe I'll get away with it again.*

"Hey, that's not the Alitalia crash," the associate producer says, looking at screen nine. "That's a different crash."

Hassan glances right. He sees the Golden Gate Bridge on it, and gawkers lining the railing. The caption reads, "Delta Jet Hits San Francisco Bay."

"What the . . . *wait a minute,*" one of the technicians says, minutes later, pointing at monitor six, at BBC TV. "That one's in the English Channel! That makes three crashes, including Rome last night."

It seems that the British Airways 747 went down north of Le Havre, on a blue-skied, cloudless afternoon. It broke up on impact. Just fell into the water too.

How many other planes did they rig to fall, Hassan thinks, horrified, feeling his pulse beat in his head.

Animals. How did they time this to happen so soon?

"Give me sound on London," Hassan orders, and a moment later hears the BBC announcer. "The pilot reported trouble on all four engines. So did the crew of the Air India flight that fell 32,000 feet into the Arabian Sea twenty minutes ago. We repeat, there were *no* explosions. Authorities are shocked by four crashes in one day. The planes are of different makes. Terrorism has not been officially blamed so far. But most major airlines are canceling service until their fleets are checked for explosives or sabotage."

Hassan staggers to his feet. The control room seems confining now, airless.

I shouldn't have put the damn story on the air. I helped the people who did this.

He feels his heart pumping crazily. He pushes away from the TV screens and walks into the hallway, wanting to be closer to sunlight, ocean, something real. He makes his way to the northeast side of the newsroom, toward the wall of windows overlooking the harbor, but it seems that everyone else here has beaten him to it.

What are they gaping at down there?

Hassan praying silently with terrible fear, Don't let it be another plane down there.

One of the camera girls is crying.

He doesn't want to look. But he gazes down. What he sees at first looks normal. It's the new coast highway, the fine ribbon of tar that proudly winds past the modern buildings lining the shore. For a moment his dread lifts.

But then he realizes that he's wrong.

One of the secretaries is shaking. "Out there. See? Far out . . . the smoke. It was in the air. It just fell . . ."

"Let their machines cease functioning," the Imam had said.

Hassan asks himself, How did they coordinate it? There were hundreds of people in those planes. I was in a plane myself last week.

The phones go crazy in the newsroom, but the staffers won't leave the window.

Someone saying, "It's just like the Imam said . . ."

Someone else saying, "No, he said it would be *more.* He said it would be *countries,* not just a few planes."

Hassan finds himself gripped by an almost primal terror. *He said it would be governments. He said it would be armies. He said the collapse would start within days.*

One era in the world becomes another, but people—most of them—don't see that at the time.

"There will be a logical explanation," Hassan assures his staffers, to try to calm them. But at the same time he's thinking, perhaps there will be *no* logical explanation. Perhaps God *is* a logical explanation. Perhaps a Koranic dark time is beginning, and this day is the sort that happened in the Bible.

And when the next phone call comes for him, minutes later, his terror rises and crests and hits heart attack level.

"It's your old friend," whispers the familiar voice in his ear. "Did you pass along the little story? Did you tell your friends overseas about the cave?"

THREE

OCTOBER 30TH. 2 DAYS AFTER OUTBREAK.

By the evening before Halloween, 63 hours after the five air crashes, no new incidents have occurred. Terrorist alert levels remain high but Washingtonians are breathing easier. Planes are flying again. Extra security is in place. The Imam in Pakistan, intelligence experts believe, overstated the threat, as terrorists often do.

A 39-year-old epidemiologist named Dr. Gregory Gillette grills franks and onion burgers for his neighbors lining up in his northwest D.C. driveway.

He's relaxing after two nervous days of playing war games at the Pentagon, planning responses to germ attack should it ever come, waiting for the wave of terror predicted by Al Jazeera. Arranging for medical supplies, hotel and office evacuations, labs to handle anthrax spores, hospital bed space.

Now Gillette wishes his two best friends would stop arguing about the crashed jets. The argument is valid. But his friends are scaring young kids in the line.

"I'm telling you," says Les Higuera, "experts checked every plane in the country. Every engine. There was nothing wrong with them. So they were allowed to fly again."

Les is executive producer of ABC's new show *Newsline.* Heavy, dark and excitable, he often learns important news before other residents of the block.

"The government should have kept *all* planes grounded until divers reach the ones that went down, and someone figures out *why,*" snaps Bob Cantoni, twenty-nine, former Marine, National Rifle Association lobbyist and political liberal about everything except gun control. He's the only resident of the block who owns firearms. A Remington shotgun. And a 9mm Sig Sauer. "That nut in Pakistan predicted bigger attacks."

"You want terrorists to run our lives?"

Greg says softly, "We can talk about this later."

The annual Marion Street Sunday-night fall block party is in full swing and Gillette is surrounded by people he loves, in a place far from the Georgia trailer park where he was born, the foster homes where he grew up and the detention centers where he spent much of his youth.

Only four miles from the White House, Marion Street is short and leafy, lined with eight small homes. A mini tudor. Gillette's brick federalist. A couple of converted cottages and townhouses. There's nothing visibly special about it. He supposes that's why he loves it. He's worked so many trouble spots—drought areas, malaria outbreaks, cholera panics—that Marion Street seems like heaven to him.

At the moment, both ends of the block are sealed from traffic by wooden sawhorses. Orange and blue paper lanterns hang from the oaks and pines, and pumpkins lit by candles grin from doorsteps. The "Mexican food line" occupies the Higuera driveway. The "bar" is in the Cantoni driveway. The "dessert table" sits in the driveway belonging to Eleanor Holmes, a D.C. judge, and her husband, Joe, a builder.

Swing era jazz—"In the Mood"—pumps from speakers, softly enough not to bother neighbors on Ingomar Place, but loud enough so people dance in the street. Gillette spots his wife, Marisa—a teacher—jitterbugging with their twelve-year-old adopted son, Paulo, and their fourteen-year-old adopted daughter, Annie. Annie's pretty good.

I am a lucky man, he thinks.

Les starts in again about the jets. "Why did they all have to go down in deep water? No black boxes to find."

Temperatures are pleasantly warm for the season. The stars are out, the leaves fringed with gold. Gillette's a slim man, ex–Georgia Tech on a ROTC scholarship, ex-Georgetown med school. He's tall and lightly freckled, with a small scar beneath his left eye, suffered when he crashed a stolen motorcycle as a teenager, during a police chase. He has thick black hair and eyes of intense blue—the deadly combination, Marisa likes to say. On the average-looking side, he turns handsome when he smiles, even when wearing an apron. Like now.

Gillette's life turned around at age sixteen when tuberculosis had hit the detention center in which he'd been residing. The people he'd feared or respected until then—inmates or guards—had been terrified. Only the CDC doctor who showed up—Wilbur Larch—had remained calm.

I want to be like him, the boy had decided, and stayed behind when the healthy boys evacuated, to offer aid. Dr. Larch refused it but began helping the boy after that, arranging summer jobs at CDC, easing entry into college and later med school. Larch had become his mentor.

Now the worry over the crashes starts up again. Judge Holmes says, "All the jets that crashed laid over in Riyadh. Terrorists must have monkeyed with the engines."

"I understand that the Riyadh airport is closed, and three workers are in custody," says Chris Van Horne, pastor of St. Paul's Lutheran Church, who lives three doors down. He's a small, bald, loud man with hearing aids in both ears.

People on line, kids too, glance at the sky. The moon is full, the night cloudless. Gillette makes out red or green running lights on jets heading toward Reagan Airport.

Bob Cantoni shivers. "Les, your own network keeps broadcasting those last words from the cockpits. 'Engine one is out. Engine four is out.' Christ, it could have been *us* on those planes. Over a thousand passengers."

"That's enough," Gillette says firmly. "Who are you picking to win tomorrow night? 'Skins or Giants?"

"Any team that doesn't go on strike," Bob says.

Les snaps, "Why shouldn't football players be allowed to strike like everyone else, you fascist?"

Arguing as always, the men move off.

Next in line is little nine-year-old Grace Kline, from two doors down. She's had a crush on Paulo ever since he punched out a bully threatening her at school. Her parents, Neil and Chris, are both lawyers with the Environmental Protection Agency.

"Are you really in the Navy?" she asks Gillette as she holds out her paper plate. "That's what Mom says."

"In a way."

"I thought you're a doctor."

"A special kind, honey. See, when people get sick, sometimes even doctors don't know how to make them better. I try to find the reason. I work for a part of government that's technically part of the Navy. We hunt germs."

"Do you wear a uniform?"

"Sometimes."

"Are you an admiral?"

"A commander." He laughs.

"Do you have a gun?"

"A microscope. Come over to the house if you want to talk more, and I'll explain. But there are hungry people behind you. Hot dog or hamburger, Grace?"

"Mommy says everyone on the block is friends because of you. You made us into our own village."

The kid walks off, munching a mustard-smeared frank, and Gillette pushes his unease over the crashes from his mind. The girl was right about Marion Street being a village. Marionville, as residents call it, lies off Nebraska Avenue near Connecticut Avenue. The nearest fire station is two blocks away, across from the CVS drugstore. The nearest restaurants, shops and gas stations occupy a one-block strip a four-minute stroll from his brick home.

Gillette's job, in a way, is to keep this place safe.

Marionville's smaller kids have always walked to the Montessori or public elementary schools behind the fire station. There are three churches close by, a couple of old-age homes, a movie theater, a bookstore, doctor's offices.

People here help each other, look out for each other, watch each other's kids.

A few problems, sure, but mostly it's a comfortable life. The best of

America, *Washingtonian* magazine called the neighborhood. An easy drive to downtown even during rush hour for residents who work in government, including the most highly placed one of all the neighbors.

Gillette.

"Oh, Dr. Gillette. Someone stole my new leaf blower from my garage," frets Alice Lee, next in line, a retired violinist from the National Symphony and widow and part-time clerk at the Politics & Prose bookstore. "It just stopped running. I went inside to call Sears. When I came out it was gone. Who would steal a leaf blower?"

"Someone from 5110 Connecticut," Gillette guesses with a sigh, naming the trouble spot of the neighborhood.

Riinggggg, goes the phone inside his kitchen, through his open window. He ignores it. Emergency calls come on the special Pentagon cellular phone, not the landline.

"Do you mind if I ask you a personal question? I don't usually pry until after three gin and tonics, not just two," giggles a tipsy blonde who doesn't even bother to bring a plate. Gail Hansen owns an art gallery on P Street, and comes on to Gillette regularly. She wanders over to the house when he's in the yard, phones for help if something breaks in her house, offers to cook him and the kids dinner when Marisa visits her parents in Vermont.

"I know you hate to intrude, Gail."

"I've always wondered, why did you adopt children instead of having your own? I mean, with that beautiful wife of yours?"

"When I met Annie in Sudan and Paulo in Brazil, I fell in love," he says, but the other half of the answer is, *We can't have kids naturally. We're blessed to have those two.*

Burgers and questions. Franks and jokes.

"Is it true what *Smithsonian* magazine wrote about you in the 'Microbe Hunter' story? That you stopped an epidemic at an Air Force base in the Philippines? That you got a presidential citation?"

He remembers a hospital ward filled with dozens of sick men and women; shaking, moaning, disoriented. He thinks, *Major Novak and I traced the brucellosis to an infected cafeteria worker, and ended the outbreak with doxycycline.*

Any thought of Major Theresa Novak usually makes him uncomfortable, and this one does now.

"I'm just a boring bureaucrat," he says. *Seconded from the Centers for Disease Control to Biological Defense.*

The phone in the house doesn't stop ringing. Perhaps he ought to answer, hand the spatula to someone else . . .

But then he spots the cop walking toward him. The local police are de facto citizens of Marionville. Officer Danyla is a single mom whom Gillette helped out last summer, getting her son into an experimental treatment program funded by the National Science Foundation for kids with multiple sclerosis.

"We found Paulo's bike, Greg. Want to meet the suspect? We'll have to walk, though. Our squad car just broke down. Half the cars at the precinct are out. The Captain thinks someone put sugar in the tanks, but I can't figure out how vandals got into the lot."

In the last hours of the ending era, these are the worst problems confronting Gillette.

A stolen bike. A stolen leaf blower. A cop on foot because a car broke down.

"I didn't steal the bike. I found it lying on the road and I went for a ride. I was going to look for the owner."

Einstein never formulated the notion as a law of physics, but Gillette knows there are certain pockets in the universe where assholes congregate. Five-one-one-oh Connecticut Avenue—"The Oasis" is its name—looks nice from the outside. It's prewar, dark weathered stone, with gargoyles on the roof and bricked arches over windows. It seems little different at first glance from the better-kept condos, co-ops and rentals lining the avenue. But Gillette knows that the building is deteriorating. The new landlord has been trying to drive tenants out by doing as little repair work as possible, so he can convert 5110 into condos. There are problems with electricity. Some apartments are empty. The elevators need work. Two fires have broken out in vacant apartments in the last year.

The fate of the building is in court.

Obviously, residents who put up with these conditions are stuck there, like the elderly, or can't find decent housing elsewhere or stay because the court has frozen rents. Sometimes they're too depressed to care. Or their tenant records make it difficult to find other housing.

Recently the building's problems have started to affect the whole neighborhood. First police arrested the Senate aide from Tampa, a 5110 resident who had been sending e-mail bomb threats to the White House. Then there were the two American University coeds who'd been paying for their parties by selling drugs, and attracting clients at all hours. When loud music plays at 3 A.M. in the neighborhood, odds are it's coming from 5110. When there was a rat problem in July on the streets, health inspectors traced it to poor sanitary conditions in the building's basement. As 5110's reputation deteriorates, its residents seem to spiral down.

And now Gillette, Paulo, Annie and Marisa stand in the small courtyard outside 5110, with a couple of uniformed cops and Theodore "Teddie" Dubbs, who is big for fourteen, shrewd, mean and backed up by his dad, Gordon. Gordon runs security at "Three Faiths Charities Warehouse," near the airport and has "problems with his temper," various police and employee reports have read over the years, Gillette knows from the cops. Dubbs in fact used to be a cop.

"I saw you carrying a chain cutter, Dubbs," Paulo tells the bigger boy. Last month the two kids gave each other bloody noses when Teddie tried to rob one of Paulo's friends after a church basketball game. Paulo is fearless, despite his size.

"You got the chain cutter out of a van," Paulo says.

"That's ridiculous," snaps Gordon Dubbs. "Teddie can't drive. He's too young." Divorced, Gordon's a broad-shouldered, tawny-haired, handsome ex-detective, wearing a clean white T-shirt beneath red suspenders. In his forties, he's still bitter over his firing from the police.

"I bet you stole Alice's leaf blower too," Paulo says to Teddie, not backing down. "I bet you're selling things."

"If Paulo says something, it's true," Marisa tells the cops, fists clenched.

Teddie Dubbs turns disdainfully to Marisa, who looks extra gorgeous tonight to Gillette. Her face is flushed, her lean body in tight jeans and a T-shirt, her long legs ending in cute tennis sneakers, her blond hair hanging down to brush her lust-inducing ass.

Teddie smirks. "If he's your son, how come he's a different color, Mrs. Gillette? Been screwing around on your husband?"

Gillette takes a step toward the kid and Dubbs senior is suddenly standing there, eager for a fight. The cops separate the men. From the direction of downtown comes the sound of sirens. There must be an accident there, or fire, Gillette thinks.

"Fuck you and your whole family," Teddie says.

"Don't curse," Gordon says, holding back a smile.

"Touch my brother and I'll bash your head in," says Annie, who is Teddie's age and gorgeous. "Come to think of it, he'll do it himself."

"He's not your brother. You're black and he's like from Mexico or somewhere yard workers come from."

"Brazil," Paulo says, his round face angry. His curly copper hair tangled. His chest starting to come in, and his muscles to elongate. "We *chose* each other as a family," he adds, as Gillette proudly hears the words that he and Marisa have taught their children.

Officer Danyla asks Paulo, "Did you see him take your bike, son?"

"Nobody ever sees him. He's too smart for that."

There will be no charges filed tonight. The families draw apart. From a sixth-floor window of 5110 comes blaring rock and roll music.

Teddie calls after Paulo, as they walk away, "Just wait, midget. I'll get my chance."

The way Dubbs senior eyes Gillette, he seems to be thinking the same thing.

"Remember when we worried about my parents catching us making love," Marisa says, "and not our children?"

The moonlight shafts through the window to fall on the woman in bikini panties straddling Gillette. Hands on his chest. Haunches on thighs. Hair brushing his forehead. Eyes huge and breast tips hard.

Eleven-thirty now, and the house is quiet, the party over, the street clear below Gillette's second-story master bedroom. The phone messages on the answering machine turned out to be an automated real estate pitch. Gillette dimly hears the drone of a television playing, from elsewhere on Marion Street. His neighbors do not usually stay up this late, so there must be something interesting on.

The pleasure is so great that he's only vaguely aware of horns sounding from out on Connecticut Avenue. But on Sunday night, the road should be clear.

The last day of October is only thirty minutes away.

"Stop that," Gillette tells Marisa.

"You mean this?" Marisa says innocently, reaching and cupping his engorged penis in both hands. She taps his shaft gently with oiled fingers. She moves her fingers up and down. The bottle of lotion stands on their night table.

It's an old game. Whoever comes first, loses. Whoever loses has to cook dinner tomorrow night.

"Yes, that," he says hoarsely.

"Maybe you'd rather I do *this,*" she says, reaching up to push aside the bottom of her silk panties. She's so wet inside that when she lowers herself onto him, he slides right in. He loves the sweet feel of her, the glide of fabric against his skin. He finds her crazy spot on her spine, between her shoulder blades. He runs his nails softly along the spot.

"That's not fair," she says.

They start rocking slowly. He takes her right nipple gently between the tips of his teeth.

"I . . . will . . . hold out . . . longer than . . . you," she says.

"You're tough." He moves faster.

"Keep that up and I'll start screaming and wake the kids," she whispers.

He grips her thighs, feels sweat sting his eyes.

"Damn you," she says. "I won't ask Neil Kline for Redskins tickets if you don't slow down."

The window is open, the night still warm. Last time Gillette checked, a half hour ago, Paulo was sprawled on his bed two doors down, beneath a Lance Armstrong poster and beside the desk where he'd printed out his science class report on the Irish Potato Famine of the 1840s.

"Are you really twelve years old, or a doctoral student?" Gillette had asked the boy yesterday, scanning draft two with pride.

"Look at these old drawings, Dad. A million people starved because a microbe wiped out their only crop. I want to be like you when I grow up, and hunt bad germs."

Ah. Paulo. The kid's real father had died of malaria.

Annie is asleep too, in the corner bedroom, beneath a blown-up photo of herself—an actual ad currently running in the Metro—showing her feeding a baby cheetah by bottle at the National Zoo. A Friends of the Zoo volunteer, she's recently been allowed to help care for the new cubs.

Ah. Annie.

Annie is shooting up, growing tall, like a Dinka. She's as thin as a model. Gillette had first seen her in a southern Sudan feeding camp, as a baby, when he was combating a cholera outbreak. Her mother was dying. The baby's weight/height ratio had come in at numbers that meant the eighteen-month-old would not be fed.

My kids.

Even now, making love, he knows that Marisa, like him, is never far away in her head from those sleeping kids down the hall. Probably every parent in the world has been worrying more about their kids lately. A plane crashes halfway across the world and your first thought is, *I wonder where the kids are.* A half-crazed religious fanatic ten time zones away predicts disaster, and you can't stop thinking about your kids.

Thunk.

"What was that?" Marisa says, slowing, looking up at the ceiling.

"Raccoon on the roof again," he says.

Thunk . . .

He pulls her close but the bed suddenly slides four feet sideways. The mirror shatters over the chest of drawers. The windows blow in and the house rumbles. Annie the war orphan starts screaming down the hall.

"What is ittttt?"

The air seems to be sucked from the room and the walls actually bend and contract with the force of the second explosion. The night goes orange outside. He hears a rumble dying away, a branch breaking off beside the house.

The bed has been thrown against the night table.

Gillette's up, running for the kids . . .

Paulo stands in the hallway, hands stiffly at his side, eyes huge, looking like a four-year-old.

"A man fell out of the sky, Dad."

Annie grips him and screams. She'd been only an infant when the militia rode down her father in Sudan, but Gillette has always thought the memory is there. Even the sound of gunfire on TV has always unnerved her.

"A *hand* fell on my windowsill."

The family holds on to each other, making sure they're all right. And then he's pulling on his shorts and sneakers and, shirtless, heading downstairs at a run. He leaves the door open behind him. Flames are visible beyond Bob Cantoni's roof, coming from Ingomar Place. In the middle of Marion Street, Gillette sees a smoking airline chair toppled sideways, with a charred body in it, still wearing a seatbelt.

He can't tell if the body is a man or a woman.

Car alarms blare. Windows have been shattered in homes and autos. Neighbors stumble from houses in bathrobes, a few people bleeding lightly from flying glass, all of them terrified. But they're moving. They're all right.

"Christ, what *was* that?"

"A bomb?"

Gillette cuts through Bob's backyard to reach Ingomar and the fire, seeing where the jet tail had crashed into what had been a small brick home. Gillette shields his eyes, fighting off heat, trying to get through the front door past the collapsed portico. He does not know the owner of this house. In the living room he spots a limp hand protruding from the fallen-in ceiling, and the child's toy dump truck at the foot of the stairs. He hears sirens closing as he's driven from the building by smoke.

Two other houses are on fire.

"Bob! Les! Help me!"

Marionville's men and women are responding now. Judge Holmes runs for blankets. Annie calls 911 on her cell phone. Les Higuera props an injured man against a tree, wiping blood off his face.

"My wife! Help!"

By the time the cops arrive Gillette is going body to body among the injured lying on lawns. He's directing firemen and neighbors. Doing triage. Just like in an epidemic. Ignore the dead. Tend to the injured. Figure out who is fatally wounded and who might be saved.

"Mommy!" a little boy screams, thrashing on a lawn as Alice Lee tries to hold and soothe him. "I can't seeeee!"

Gillette checks pulses and asks the same questions he asks in epidemics.

"Do you know your name? Does this hurt? Can you tell me what day it is?"

Hours seem to pass but he knows that probably it's only minutes. He leans back on his heels after doing mouth-to-mouth. Marisa is holding out his open cell phone.

"For you, Greg."

It's his encrypted Pentagon unit, and when he answers, sure enough, the voice on the other end identifies itself as belonging to tonight's duty officer at Emergency Response. He should stay where he is, the voice tells him. Officers are on the way to take him to the Pentagon.

"Is there an outbreak too? Not just crashes?"

"The officers will answer your questions."

"Are aircrews getting sick? Is that causing the crashes?"

"Sick, sir?" The duty officer seems confused. He lets out a deep breath. The man's voice is scared.

"How many planes went down this time?" Gillette says.

The duty officer sounds southern, not deep south, not a long drawl. Maybe Virginia. Or Maryland.

"Haven't you been watching TV, sir? It's spreading. Europe five hours ago. Then New York."

"I don't understand."

"It's not just planes, sir," the duty officer tells him. *"The machines are stopping. All over the world."*

FOUR

OCTOBER 31ST. 1:30 A.M. 3 DAYS AFTER OUTBREAK.

The national Biological Warfare Defense Program was formed in 1996, during the presidency of Bill Clinton. It created a quick-response command center at the Pentagon to identify and neutralize germs or chemical weapons that officials feared would be used against America in attacks.

"The intelligence community believes it's not a matter of *if* it's going to happen, but *when*," Clinton's national security adviser told the president at the time. "We're assuming that over the next few years terrorists will get access to some of the world's worst diseases. The list includes anthrax, cholera, smallpox, plague and viral or hemorrhagic fevers, like Ebola.

"Such a catastrophe will kill millions of innocent men, women and children in the U.S. and around the world, if not checked."

Two major assumptions lay behind the Rapid Response Program command structure: that different government agencies would cooperate if teamed together, and that if an attack came, it would be directed against humans.

Sorry.

"Why come with two cars, just to get me?" Gillette asks the Air Force captain beside him in the rear seat of the black Defense Department Taurus, as they head toward the Pentagon over the Memorial Bridge.

"In case one breaks down, sir. Look at this mess on the road," Captain Heidi Ross replies. She's small, fit-looking and soft-spoken, and intense and professional despite the anarchy around them.

At this hour traffic should be moving quickly over the Potomac River.

But the drivers of Gillette's miniconvoy need their red dome lights to help them maneuver around stalled, crashed or abandoned vehicles. At least a third of the cars and trucks on the road have simply stopped running. Other traffic has backed up. Bits of still-burning small plane float, down in the river.

With sirens blaring, they weave onto the breakdown lane, back onto the road, and even onto grass as they hit the Virginia side.

Drivers stand around, not knowing what to do. The Taurus passes a broken-down ambulance, a Safeway food delivery semi, a "Gospel & Blues Band" bus. Gillette sees a man pedaling a one-speed bicycle on the highway, beneath vapor lights, a sight he associates with third world cities. An escaped pet—a Russian wolfhound—runs between cars.

"When did this start?" he asks, aghast.

"The first reports came in about six this evening. Cars stopping in Marseilles and Barcelona. Mediterranean cities. Planes down. But then Russia also. Russia has it the worst so far. Their trains are out too."

"What about the U.S.?"

"It's mainly the northeast, so far. New York. Philly. The highways are a mess. Nothing at Fort Myer is moving. Half the Maryland State Police vehicles are disabled. Planes burning on runways at Kennedy and Boston. The FAA's grounded all other flights. And the president went on TV, asked everyone but essential personnel to stay home. They're broadcasting the appeal every few minutes. Hopefully people will stop moving around until we figure out why some cars run, others don't."

The Ford swerves around a two-car accident, just a fender bender this time. A Chevy van ran into a pizza delivery truck. Ahead, Gillette sees the glow of flames in the northbound lane. He's trying to get his mind in gear. The landscape—downtown Washington and the roads—have turned into some horrible Hieronymus Bosch depiction, filled with wreckage and lines of soldiers walking toward the bridge, away from their stalled troop carrier.

Part of him is still back on Ingomar Place, bending over burn victims, thanking God that the catastrophe spared his family.

So many fires, he thinks. *As gasoline burns.*

Of course!

"Fuel. That's the common denominator with the crashes four days ago," he says, turning from bystander to command center analyst, spectator to doctor. It's the way Dr. Larch trained him to function during outbreaks. You seek the source of infection. You track it back to where it began.

What did Eleanor say back at the party? That all the downed jets overnighted in Saudi Arabia. She blamed sabotage. Well, those jets would have all gassed up there.

He asks Captain Ross, "The gasoline supplied to the places you mentioned. Did it come from Saudi Arabia too?"

Ross looks impressed with his ability to make connections. "The Saudis were sending information on that as I left to get you, sir. That's all I know."

Gillette can't help but remember the face of the old Imam on the news for the last few days. The craggy features and milky eye. The English translation, broadcast repeatedly across Gillette's screen.

"Let their machines cease to function. From the ashes, the survivors will build a beautiful new world."

Sure. Gasoline. It has to be. But Gillette asks himself again, *How could terrorists get to the fuel supply in so many places? Or was the tainted fuel distributed from one pipeline? One oil field?*

"We're almost at the Pentagon, sir."

He frowns, knowing from briefings that oil supply has always been a potential target of saboteurs. That oil companies regularly test oil for contaminants before sale.

So why didn't they catch the problem, unless the people doing the testing are the terrorists, or the contaminant is something new that they didn't detect.

Is it possible?

Of course it's possible. It's why new infections break out in hospitals, antiseptic environments, all the time.

He hopes he is wrong. The Imam's warning would become pretty damn accurate if there was a problem with oil. The immensity of the possibility takes his breath away, but he tells himself he doesn't have enough information yet.

Captain Ross shows ID to get past a cordon of armed troops ringing the Pentagon lot. More cars than usual are here at this hour, despite the breakdowns on the roads.

"If the problem started hours ago, how come we weren't called in until now?" Gillette asks Ross.

The merest flick of hesitation precedes the answer. "Actually, sir, the Center has been busy since seven. My original orders were not to bother you. Just to bring everyone else."

"But why . . ."

Then he understands.

I deal with attacks on people, not machinery. They left me out.

Gillette feels a flush of anger.

Damn you, Hauser, he thinks, envisioning the man he believes to be responsible. *The whole purpose of using people from different agencies is so we don't miss things.*

"Welcome to consequence management, doctor."

In his head, he flashes back to his first day at the Pentagon, his meeting with Major General A. L. Hauser, chief of staff at the Rapid Response Command Center. Hauser runs the center day to day and answers to a political appointee, an under secretary of defense.

"We're a fusion organization, Gillette. Our other advisees come from CIA, Homeland Security, FBI, Energy and Health." The general, cold toward Gillette from the first, was a handsome man in his late forties; ambitious and talented at currying favor with higher-ups, Gillette would learn. Hauser's silver star had been awarded after he took fire for ninety seconds in Iraq, while on a fact-finding tour. He'd spent years at a desk. Hauser, it turned out, had been the Pentagon official embarrassed when Gillette proved the Philippines outbreak was unrelated to terrorism. He dreamt of new stars on his shoulders and feared embarrassment more than enemy gunners. He'd never wanted to work with Gillette. But rejecting Gillette for the team would have looked bad.

"We're not a democracy," Hauser had said on that first day. "You advise me, and Under Secretary Ames, and he reports to the secretary of Defense. In the event of chemical or biological attack, Rapid Response has the lead in any investigation. Authority over local law enforcement.

Priority access to labs around the country. By the way, Commander, your maverick way of doing things was lucky for you in the Philippines. Here you'll follow the chain of command. I'm responsible for what you do."

Now, Captain Ross escorts Gillette past the turnstile and metal detector in the lobby. The paintings lining corridor walls—burning battleships at Pearl Harbor, fallen towers of the World Trade Center—remind him of the scene back on Connecticut Avenue.

Reaching Rapid Response Command, Gillette sees that the big bullpen area is alive with activity. In partitioned-off cubicles, permanent military staff—intelligence, operations and logistics crews—man computers, answer phones, study maps or agency reports.

Ross leads Gillette toward the walled-off corner conference room. Its door is closed. Her voice grows softer.

"Sir, I know you're the last one we picked up, but your neighbors said you saved half a dozen lives tonight."

"It's *what? An oil-eating bacteria?*" Under Secretary of Defense Dennis Ames gasps as Gillette walks in.

"Oh my God," someone else says, in the dark.

The teleconference is in progress between the command group and—via encrypted video—a lab-coat-wearing scientist at Fort Detrick, Maryland, which houses some of the nation's top germ containment and chemical-warfare labs. *It's Theresa Novak,* Gillette thinks, surprised, and flashes to a beach in the Philippines. A full moon. An empty wine bottle. A celebration that the outbreak was beaten. A long kiss broken off and only mentioned once after that.

Theresa—promoted to lieutenant colonel now, and another probable irritant for Hauser—is saying, "The Saudis found this stuff in one Riyadh jet fuel tank on the 28th. But they sat on the information. They didn't want to cause panic. They thought they had the problem contained. You're looking at nanobacteria, smallest known life on earth. A colony, magnified 35,000 times."

Gillette takes the only unoccupied chair, the one furthest away from

Under Secretary Ames, at a long T-shaped conference table. Beside Ames is Hauser, whose eyes flick to Gillette. The lights are off. Other Rapid Response advisees—also dressed in Sunday night casuals—stare at the flat-screen TV.

"Hell, can something that tiny really be alive?" asks the FDA rep, a skinny, balding former Marine named Bob Haskell.

Colonel Novak shrugs. "Remember, before Louis Pasteur discovered common bacteria under his microscope, life that small was considered impossible too."

The scene on-screen reminds Gillette of comic-book science fiction landscapes. Long, undulating greenish translucent rods—cylinder shapes like twisting denuded trees—rise out of a bulbous, gelatinous surface that looks like hundreds of melded fish eggs, or sacs.

"I thought oil-eating bacteria were good guys. They clean up spills," says the FBI man at the table. Mark Wallach is an antiterrorism expert who worked on the Oklahoma City and World Trade Center bombings; he's a youthful, fit-looking ex-lawyer whom Gillette instinctively trusts.

"Up until now you're right," Novak says.

Gillette thinks, *They're not anything I've ever seen.*

The colony quivers. The whole alien landscape froths pink for a moment, and goes blurry. A bubbly saliva-like substance has replaced the green. Then the picture crystallizes and the green returns. The rods have grown longer. The surface area looks bigger. Gillette spots small black diamond shapes embedded in the surface. Sometimes the diamond shapes are alone. Sometimes they're in clumps.

"It multiplies faster than Ebola," says Theresa in her low, inadvertently sexy voice. Gillette flashes to her in a surgical mask and gloves, bending over a soldier running a high fever in the isolation ward of the military base hospital. He sees her at 3 A.M., peering into a microscope in a field lab. He sees her for a millisecond in a skimpy black two-piece bathing suit, looking smashing at the base pool, after the outbreak was beaten. He remembers her taking his hand in a Manila airport bar, saying, "You're right. I'm glad we didn't go any further, Greg. Temptation makes for nice memories. But self-control doesn't last."

Now, on-screen, her eyes go left for a moment, seem to see Gillette. Then she says, "Our little friends here are chomping away in a petri dish of refined oil. The Saudis theorize that the darker crystal shapes are waste, and waste is screwing up fuel, fuel lines or engines. That's actually happened with some bacteria in heart pacemakers and industrial pipelines. Riyadh has started sequencing."

She sighs. "But we're dealing with guesswork at the moment, running tests. How bad is it? How widespread?"

The tension in the room is evident in the stillness with which people sit, the regularity of wiped brows and sipped water. The mood is not improved by the artist sketches on the walls, Ames's idea to motivate the team. It's a timeline of germ warfare.

Athenians at Delphi shows ancient Greek soldiers diverting the Pleistrus river into the city of Cirrha, and tossing in excrement and hellebore roots to poison enemies.

Mongols, the next drawing, shows their attack on the walled city of Kaffa in 1345. The Mongols are loading plague-riddled corpses onto catapults and flinging dead bodies over the walls.

Other drawings depict Japanese scientists preparing plague-laced chocolates for air drop into China in 1938 and Saddam Hussein's germ-warfare lab in Baghdad being bombed during the second Gulf War.

Up until now this room has been where the team met to strategize over coffee and poppy seed bagels. How should we handle an attack of anthrax spores at the U.N. or in the New York City subway, through the ventilation system? Suppose terrorists use a private plane to distribute plague in the Midwest? How would victims reach hospitals? How would we quarantine towns?

I've never worked with any of these people in a real emergency, Gillette thinks, *except Theresa.*

Under Secretary Ames speaks up as Novak finishes. "The Saudis think it's a waste product, eh?" He seems to regard Saudis as incapable of rational analysis as Neanderthals crouching over microscopes, unsure whether to smash or worship them.

All eyes go to Ames, an energetic, broad-shouldered fifty-year-old who came to government by way of the Shelby Energy Trading Company of Montana. He's a workaholic, a devoted family man and administration appointee who, Gillette decided early on, trusts kiss-ass Hauser to handle day-to-day logistics. He spends half his time tending to overseas responsibilities: security on military bases.

Colonel Novak speaks again. "Well, sir, the Saudis have been training their microbiologists at Harvard for the last forty years. They've got state-of-the-art labs. They neighbor Iraq, where Saddam spent years developing chemical and biological weapons for use against them. And they know oil. So I tend to take their theories seriously."

Gillette stifles a smile. He can feel Ames turning pink in the dark.

"If the Saudis are so smart why didn't they find these bugs before they downed five jets?" Hauser asks coldly.

"Same reason our Mobil oil quality-control people didn't see them. They're tiny," the colonel says. "Remember, we're seeing a whole colony here. But individual organisms are small enough to get through almost any filter. They don't show up easily on the usual electron microscopes. The Saudis needed a Philips field emission scanner to spot single organisms, and at 100,000 magnification. The biggest bug in here is probably no larger than 0.05 micrometers, one-thousandth the volume of an ordinary bacteria. You don't see them with standard equipment until the colony forms. You don't know they're there until it's too late. Take a look at this."

This time the shot seems to be from the left, instead of from overhead.

"What are we supposed to be looking for? It's the same colony," says Hauser impatiently.

Colonel Novak is petite, with black eyes and olive skin. She has the posture of a dancer. She says, "That's the problem, General. It looks the same, but this picture came in from Moscow an hour ago."

"Oh dear," says a tall, heavy, pretty-faced woman sitting diagonally across from Gillette. Violet Pell represents the Energy Department. "Do the Russians buy fuel from Saudi Arabia? I thought they use their own oil."

"We're waiting for an answer on that," says CIA rep Ed Mallory, a quiet, professorial type who sucks on pencils in meetings, and smokes outside the building when they're over, and has developed a wheeze in his voice.

"If the Russians *don't* get oil from the Saudis, that means the infection is coming from two places."

"At least two," Gillette says unhappily.

The picture changes again. Colonel Novak seems to hover in the air at the foot of the table. "This last specimen comes from my 1999 Honda Accord, which stopped running about 4 P.M. today."

A collective groan goes up.

Theresa says, "I ran tests while we waited to hear from overseas. I've sent for samples from gas stations."

"Thank you for an excellent presentation, Colonel," Ames says, and then Novak is gone, replaced by a Mercator world map, showing blinking red lights to signify areas that have been affected. In the U.S., dots extend inland in diagonal lines from the coast of New Jersey and Delaware. *From oil refineries,* Gillette thinks. In Canada, a lone red dot throbs at Montreal, where Gillette and Marisa had honeymooned fifteen years ago. The map of France resembles a face, with two red dot eyes.

Ames looks around the table calmly. "For the benefit of Dr. Gillette, let's go back over our early recommendations, steps the White House announced. Al?"

Hauser recites, "Airports closed for the time being, and all gas stations. Troops dispatched to protect oil facilities, tank farms and pipelines. Our allies are doing the same thing overseas. Local law enforcement support in protecting infrastructure. Schools and offices to be closed tomorrow. Curfews to go into effect until this thing is over. Merchant marine traffic frozen. Wall Street holiday. Ongoing screening tests on all aspects of the national oil supply, until we know what's clean."

Ames says, "Identify and neutralize. I have to call the White House at five. Suggestions?"

Gillette speaks up. "I think we treat this like an outbreak. Try to find the original source and stop this before it spreads."

"The source is clearly a large organization," says Ed Mallory. "With the resources to hit the fuel system in several places at the same time. Hell, there's a million Muslim drillers, pipeline workers and quality control people working across the Mideast. When someone takes credit, we'll know who did it."

Other people are nodding.

Mallory says, "Meanwhile, ears to the ground. We hit all sources. We continue to ID faces on the Al Jazeera video, men in that cave. I know we blasted them, but *who were they exactly*? And who funded them? Or what country?"

Gillette persists, "I suggest that we not assume yet who did it. That we get someone from the oil industry in here to fill us in on how distribution works."

Ames looks approving. "Actually, an old friend of mine will be here any minute, to help."

Questions come fast and furious.

"How much of the oil supply is infected?"

"Can you kill this thing?"

"How long will facilities be shut down?"

There's a knock at the door. Captain Ross enters.

"Dr. Osborne Preston is here, sir, from the Center for Strategic and International Studies."

She adds, "We're out of gas in one car. Do we risk filling up from a pump, sir? Or siphon gas from a car that runs?"

Ames introduces the visitor, a robust-looking, sunburned man in his sixties. Flattop chestnut hair on top. Shiny, clean-shaven face. Piercing blue eyes. Crisp button-down shirt and pressed tweed jacket. Emergency or not, the man had taken time to groom himself before coming here. He's either vain or doesn't panic, Gillette thinks.

"Os and I met back in '04 at the Geopolitics of Energy conference," Ames says. "All the intelligence agencies participate, every autumn."

"It was an honor to work with you, Dennis."

"The honor was mine. How are your two fine sons?"

Cut the shit. Get to the questions, Gillette thinks.

"Dr. Preston has had a brilliant career in private industry," Ames tells the table. "He was an oil geologist with Texaco, then a VP for energy security. Now he devises strategy games relating to oil supply. He anticipates potential problems. Overall supply decline coupled with terrorist attacks on Mideast pipelines, for one."

"That pushed prices over $4.90 a gallon," says the new man at the table.

Ames says, "You were pretty accurate on your New Orleans hurricane scenario, Os."

"I take no pleasure in that, Dennis. I would rather have been wrong."

Gillette envisions office lights burning all over the capital at 2:28 A.M., that is, in places where people could reach their offices. Men and women will be on phones and around tables in situation rooms, at the Senate Energy Committee, Homeland Security. Trying to get a grip on things.

Ames tells Os Preston, "It appears that some kind of microbe has infected part of the fuel supply."

"A microbe?" Preston looks shocked, but recovers quickly. "We figured sabotage, but a *microbe*? What kind?"

"We don't know, and we don't know how widespread it is, if the thing's been injected into the oil fields themselves or just some tanks, or pipelines."

Preston raises a hand authoritatively.

"It's impossible that the bug is in the ground. Assume it was introduced after refining, during distribution."

"How can you be sure?" Gillette asks.

Hauser glances at Ames as if to see if the challenge bothered him. Preston doesn't mind. "Several reasons," he tells Gillette. "First off, the oil fields themselves. Most people think an oil field is a big underground lake. That's not true. Oil is distributed through porous rock, or in layers over a wide area. The drills crack the rock. The oil migrates up from pressure beneath it, and collects in cracks around the drill. The pipe sucks up oil. So any microbe that managed to live down there in the first place would have to be small enough to move through porous rock."

"Is .05 microns small enough?" Gillette asks.

Preston frowns. "That's pretty damn small."

"*Could* a microbe that size migrate through a field?"

Preston steeples his meaty fingers. "Look, microbes do live down there sometimes, but the refining process kills them. No microbe can survive that."

"The refining process," prompts Ames.

"Well, you probably know how that works . . ."

"Explain it anyway, for those who don't."

I *don't,* Gillette thinks, fascinated.

"Well, Dennis, oil is useless in crude form, the way it comes out of the ground. You have to separate out the different hydrocarbons to get finished products. Gasoline. Jet fuel. Heating and lubrication oil. Plastics. Hell, all the stuff we can't live without."

"How does refining do that?" Violet Pell asks.

"Same way a still does, ma'am. It's like making moonshine. Imagine a refinery if you will, a row of two-hundred-foot-high distillation columns, connected by pipes at the bottom. Round steel stacks rising up."

Preston moves his hands as he speaks, as if outlining the towers, pipes.

"The crude oil gets pumped into the column. Then you heat it to evaporate the oil. I'm talking about temperatures as high as a thousand degrees Fahrenheit, depending on what end product you want. The crude vaporizes and rises through the cylinder. Based on temperature, the different hydrocarbons reach different heights and condense back to liquid. Four hundred degrees separates out gasoline. Six hundred and seventeen gets you jet fuel. Seven hundred for lubrication oil and a thousand for fuel oil."

"That's the whole thing?" Hauser asks, taking notes.

"For forty percent of the oil, yep. For the rest, chemical processing; to break large hydrocarbons into small ones, to combine small ones into large ones, and to rearrange atoms. I can go on, but the bottom line is, no microbe can survive the chemicals and heat."

"So the infection's in the oil delivery system," concludes Ames, thinking out loud. "The tanks. The pipes. The microbes were added *after* refining."

"Give thanks for that," says Preston. "If a bug that destroys oil could migrate through reservoirs *and* survive refining, well, depending on how widespread it was, you'd be looking at what that nut on TV predicted."

"What is that, Os?"

"More or less, in fifty days, by mid-December, the end of our world."

FIVE

OCTOBER 31ST. EARLY MORNING. 3 DAYS AFTER OUTBREAK.

The man now called Clayton Cox—in Washington at least—steers his BMW motorcycle along New Mexico Road, in the northwest quadrant of the city. At 2:51 he is on his way to slaughter a family; both parents, if the father is home, and three children, ages seventeen, fifteen and four, if his mentor's information is—as usual—correct.

The motorcycle runs quietly—its noise will not attract attention—and it uses fuel sparingly. Stalled cars have turned the roads into an obstacle course. In his saddlebags are street maps of D.C., phony "evidence" to leave later, a Marine-style K-bar knife with a six-inch stainless steel blade, a kit of burglar tools and a pack of Chuckles, Clayton's favorite snack. He's holstered his ten-round Glock 17 in the pit of his back, beneath a dark blue windbreaker and Gap jeans.

Chance favors the prepared mind, his mentor likes to say. It is why the man had ordered Clayton to stock fifty-five-gallon drums of gasoline in his rented garage.

He is thrilled by the anarchy around him, filled with pulse-hammering delight at the passing wreckage; the broken-down tow trucks, confused police, the dazed citizens wandering through the streets in defiance of poorly distributed curfew orders. The steady hum of the BMW, the smoothly functioning machinery, is as much an assault on the enemy's sensibilities as are Clayton's orders tonight.

Passing American University, he sees many students wandering off campus, in singles, clusters and couples, in the leafy neighborhood, witnessing history, but unaware that the helmeted man on the motorcycle is its agent.

Clayton has been driving around the city for hours, gathering eyewitness information to report back by encrypted, scrambled cell phone later. He'd seen lines of police and soldiers surrounding the Capitol and White House. Lights burning at the Massachusetts Avenue embassies. Muslim faithful trickling in by foot to pray at the big white mosque off Rock Creek Parkway. They know Islam will be blamed for tonight's assault.

He had also stopped to view the wreckage of a jet that had fallen on Ingomar Place. The fires were out. Police—the ones able to work tonight—stood behind yellow tape that blocked off damage. But it was like trying to stop a crime after it was over. The street had smelled of gasoline and charcoal and a familiar hint of burned flesh. He'd watched ambulance attendants carry a blanketed corpse from a lawn.

Now, closing on the Chevy Chase, Maryland, border, Clayton veers into a leafy neighborhood where the oak-lined streets are named for American presidents. Roosevelt Avenue. Truman Place. Coolidge Way. Reagan Lane. The homes are dark. Cars sit in driveways, where many will stay. Somewhere country music is playing. The house he seeks, which he scouted out two nights ago, sits on a thickly wooded lot. It's a split-level ranch with cedar siding, a screened-in deck and a cute wooden figure of Uncle Sam—a wind-propelled toy—atop the mailbox. Shrubs shield the front of the house.

The name on the mailbox reads, THE NILES FAMILY. BILL, MARY, MIKE, STEVIE AND ALICE.

Bill Niles, Clayton knows, heads a special FBI task force tracking Islamic extremists in the United States. He's on TV occasionally, giving interviews. *Good Morning America. Larry King Live.* Fox News. CNN.

Easy as pie. Isn't that what Americans say?

Clayton leaves the BMW two blocks away, behind Thurgood Marshall Elementary School. He cuts across backyards until he stands at the fringe of the Niles property, his finely trained senses attuned as he listens beneath the trees. He hears a breeze. He hears a squirrel running. He hears no people. No traffic.

There are no normal police patrols out tonight, only emergency response people. And any cars that still run will soon have to fill up with more gas.

Which means another batch will soon break down.

Approaching the house, Clayton is careful to leave one clear partial heel impression, from his size-eleven Reebok, in a spot of soft-seeded sod. His normal shoe size is nine.

He also leaves a front-sole partial in dirt behind the rosebushes.

He's wearing latex gloves.

There's no dog inside, he knows, and no alarm.

He's inside the house in seconds, via the little girl's bedroom window, which is open.

Stupid parents. You'd think an FBI man would be smarter. They lock their doors. They open their windows.

Smelling baby powder, Clayton stands over the sleeping four-year-old. He takes in the limbs sprawled carelessly. The lace nightgown. The poster of a singing group, the Wiggles, above the headboard. An *Animal Encyclopedia* sits on a child-sized table, in front of a red plastic chair featuring Elmo, from *Sesame Street*.

One silenced shot to the little blond head.

In the next room, he sees that the sleeping seventeen-year-old is clearly a big Nats fan, as there are posters everywhere, and lots of baseball paraphernalia. Clayton spots a Wilson fielder's glove. A high school uniform shirt—THE REBELS—lies crumpled over another TV.

The body jerks when he fires. The boy spasms silently for half a minute. The pillow is spattered, but Clayton stands far enough back and stays clean.

The hallway is carpeted, so moving around is easy. Three steps take him to the next split-level up. But the bed in this room is empty. The covers are thrown back.

The sheet is body-warm.

Clayton goes still, listening. He feels his heartbeat pick up. He hears no movement in the house; no breathing, no footsteps, no flushing toilet or running water.

Did he hear me come in? Did he see me in the hall somehow? Is it possible the boy is quieter than me?

The window is open nine or ten inches. He doesn't think that someone could squeeze through so small a slit.

Clayton glances out at the rear lawn anyway, to see garden furniture in the moonlight. A stone fountain. The rear of the property is lined by oaks. No sign of a boy.

The closet?

Clayton glides forward, keeping his Glock 17 in his left hand. There's a poster of the Kennedy Center, all lit up at night, taped to the door.

He pulls open the door.

No kid.

Clayton goes down on his knees beside the bed but he can't see beneath it. The covers block his view.

As he reaches out and pushes aside the fringe of blanket, he hears a toilet flush, down the hall.

Okay, then.

By the time he stands he can hear the muffled padding of bare feet out in the hallway. Clayton presses himself into a dark corner beside the boy's upright cello case and a metal stand stacked with sheet music.

A tall, tousle-haired kid stumbles into the room like a sleepwalker, wearing underpants. He falls into bed. He never sees Clayton.

One shot to the head.

Clayton is uncomfortable with killing children, but recalls as fortification words that his great-great-grandfather had written, that he's read and reread in the tattered volume that he always carries on trips.

"The sorrow of punishment has to be pitiless."

His ancestor had clearly also dealt with unpleasant tasks like this, many wars ago.

The master bedroom lies at the end of the long hallway. When he lets himself in he only sees one form in the king-sized bed. The other half is still made. The man is out. He's probably been called in to the FBI, during tonight's emergency.

No problem, Clayton Cox tells himself, as the woman murmurs something in her sleep. Her eyes start to open.

One shot to the head.

Now comes the next, uglier part, but the one that his mentor said would be especially important.

Out comes the K-bar knife.

He starts with the wife. He uses his left hand only. He carves her up.

Then the kids, which is a little harder. *Do it fast,* he tells himself.

And after that, from the plastic baggie he removes a small stoppered vial and shakes tonight's prime clue on the four-year-old's back. It's a few drops of semen—some pervert's emission—that Clayton collected from a booth in the jerkoff place on Georgia Avenue this afternoon.

Last but not least, he leaves the note he'd written—with his left hand—on the kitchen table. His mentor had told him exactly what to write.

The quote will drive the Americans crazy, especially when they figure out where it comes from.

"Whoever is kind to his creatures, God is kind to him."

"As Napoleon said, it is easier to deceive than to undeceive," his mentor likes to say. "Nothing gets Americans angrier than innocent victims. And when they're angry, they make mistakes."

By 4:13 he's on the BMW, heading back along Reno Road toward Macomb Street and his small rented house. The streets are clear. The houses are dark. The moon is waning. At 5:30 he is to report in.

It's all gone perfectly so far, he thinks, relaxing, but at precisely that moment he sees the bright light in his mirror, and what has to be the only working cop car left in northwest Washington pulls in behind him.

He slows down. So does the cop car. He turns a corner.

The police car does too, and starts to close.

At the Pentagon, meanwhile, no new red lights have appeared on the incident occurrence map. No new jet crashes have occurred in the world.

No one on the Rapid Response Team has gone home, and Gillette—picking at donuts and sipping coffee—is still in the conference room.

Dr. Preston tells the group, "Clearly we're looking at enormous price shocks here. Rationing, probably, short-term. But our strategy scenarios *have* envisioned sudden severe shortages. From fundamentalist overthrow in Saudi Arabia for instance. Or embargo or more hurricanes. Other countries make up for the lost oil. A flow cut is bad, but fixable."

The under secretary has left to go to the White House for a meeting. General Hauser is in control.

"Meantime," Preston says, "once you figure out how to kill your microbe, your big problem will be disinfecting the tanks, pipelines and ships that carried it."

Yeah, all we have to do is kill it, Gillette thinks, amazed at the assumption that things will be that easy.

Preston remains diligent and upbeat despite the hour, a technology-always-triumphs kind of guy. But until Gillette hears proof, he will not blindly accept assurances that the microbe is not—worst-case scenario—in the ground, and powerful enough to survive refining.

Germs find a way around precautions.

"How many oil fields *are* there in the world?" he asks, aware that Colonel Novak is back on the screen. She still looks fresh and alert.

"Over 100,000, including small ones. Checking their infrastructure will be a lot of work," Preston says. "The disinfecting will break a few banks. Heat. Chemicals. Find whatever kills this thing and go tank by tank."

Gillette says, "Assuming it can be killed."

"Everything can be killed."

"Yes, but," Theresa pipes up on-screen, turning a few heads, "we also want to leave the victim—oil—usable."

Gillette thinks, *We've been trying to kill AIDS for twenty-five years so far, without success.*

Hauser wraps up the meeting, gives assignments.

"Energy Department," he tells Violet Pell briskly, as if disaster can be avoided by confidence, "we'll need an inventory of oil supplies across the

U.S. We want a fast, reliable testing system for reservoirs, pipelines and refineries. Infected areas to be quarantined. Uninfected to be guarded. You'll be partnered up with the oil companies on this, and FDA."

Violet nods. "My people are on that."

Gillette thinking, *As soon as we break I've got to call my family. We better stock up on food.*

"Colonel Novak," Hauser says, "what the hell is this bug and where does it come from? Is it toxic to humans? Is it natural? How do we kill it? Does anyone have experience with it? Universities? Defense Department? Could the damn thing have come back from space? Detrick will be our national coordination center. I want thorough testing of all military equipment. Planes, vehicles, even furnaces. We need to know quickly which machines we can still use."

"We may be underestimating the danger," says Gillette. "We're not sure yet that it's not in the fields."

"Oh, I think Dr. Preston eliminated that possibility."

If cars won't run people can't get to work. If heating oil is infected, we need to ration what's left from last year.

The room seems suddenly warmer to Gillette.

He flashes to the scenes of anarchy he recalls, after weather disasters like Hurricane Katrina. The looting. The illness. Police and firemen unable to work.

But in that case, help arrived from outside. And there were unaffected places that victims could go to. Because fuel worked. Transport worked. The emergency was only in one area, not every place, he thinks.

Hauser turns to Mallory. "Intel Services, *who did this*? I want daily reports. You will check out all extremist groups; religious, environmental, anti–oil company. I want information on the microbial warfare programs in Iran, Cuba, North Korea. You will withhold nothing from each other. This is a career breaker or maker. You will impress your people with this fact."

Gillette thinking, He's worried about careers?

People start closing folders, pushing back chairs.

Gillette speaks up. "What about CDC, General?"

"Ah! CDC and Emergency Management. Ve-ry important! Get me lists of energy needs at hospitals and labs nationally. The White House will need to know requirements for civil defense. Power. Transport. Getting doctors to hospitals. You'll liaison with state authorities."

Gillette protests, "I believe my people can also be helpful in identifying the microbe and killing it."

Hauser responds firmly. "It attacks oil, not people. If it turns toxic to people your role will change."

"Sir, microbial behavior is similar. We need to check *all* the oil fields. If I can tell a story, there was a cholera outbreak in London, long ago, and . . ."

Hauser turns away to tell the group, "All of you, get me your travel needs, à la access to flights or fuel. Travel will be on a priority basis once we identify which transport is safe. And anyone misusing privileges that come with your ID will be dealt with in the harshest possible manner, understand? No exceptions."

Gillette breaks in. "My people have experience tracking contagion all over the world."

Hauser flares, "Rapid Response, Gillette. A streamlined team. We can't have interagency squabbling, or individuals trying to hog credit, can we?"

With the greatest control, Gillette holds in his fury. He'll try again in an hour. Hell, he'll order people at CDC into the field anyway. The cholera tale—the first story Dr. Larch ever told him—makes sense as an appropriate point here. For the moment he'll draw up Hauser's lists. Do his part. But he'll also go home and start helping his family cope with tough days ahead.

He hears Colonel Novak say, "General?"

"Colonel?" Hauser's tone toward her is equally cool.

"Sir, I've had some delays in dealing with our people at Biological Warfare. I've asked for anything they have on microbes and oil. They're not getting back to me."

"You'll have written authorization in twenty minutes. Anything you need, let me know."

"In that case," she says delicately, "with all respect, I tend to agree that

Dr. Gillette would add another perspective to the investigation. I've worked with him. He comes up with analyses that elude other people. I also wouldn't mind hearing his cholera story, sir."

No matter how vast the political context, human conflict is always a stopper. People in the room quit gathering up papers and glance from Novak to Hauser, who's gone absolutely still, knowing he's been sandbagged. Gillette feels his pulse quicken.

My people really can help.

"Of course, Colonel," Hauser tells Theresa. "If you think he's important, by all means he is assigned to you at Fort Detrick. Someone else can do his job here."

Gillette says, "When I suggested giving CDC a role, I didn't mean me."

"I meant you," Hauser replies.

From the screen, Theresa's eyes seem to make direct contact with Gillette's. Which is the precise moment when he hears a gasp beside him. Violet Pell says, "The map!"

He knows with horror what he will see before he turns, and there's a new red light on the screen when he does.

"Long Beach, California, is the biggest oil port on the West Coast," announces Os Preston, frowning.

"Louisiana pipelines," says CIA rep Mallory a moment later, when a second light comes on. Over the next minutes more pinprick blinks appear, on the coast of Texas, in Tehran, in Iraq.

Gillette feels his stomach tighten. He understands why the dots are appearing almost simultaneously.

"The White House is releasing all the test results at the same time," he says.

Osborne Preston has gone a bit gray now. For the first time since coming here, he seems shaken.

"Those pipelines deliver oil throughout the South and Midwest. If we can't use pipelines, we can't deliver oil."

Blink. A light goes on in Chicago.

London lights up.

Blinkblinkblinkblinkblinkblink.

"General Hauser?"

It's Colonel Novak again.

"We're getting results on coring tests overseas," she says, and pauses, as if not wanting to say what she's learned. As if saying it will make it more real.

"The microbe is in oil fields. *In the ground.*"

Dr. Preston seems to be pulled halfway from his chair by the announcement. "That's impossible."

Theresa looks grim. "The Ghawar oil field in Saudi Arabia is infected. So is the field at Shayba."

Preston says hoarsely, "But Ghawar produces one out of every twelve barrels of oil consumed on earth."

Colonel Novak reads off names of oil fields that Gillette has never heard of before. "We've got positives from Machete field in Venezuela. Barely there, but it's there."

Preston repeats, barely audible, "Impossible . . ."

Gillette has heard the protest so many times. Every time there's an outbreak, in every real emergency he's studied, someone in authority who lacked imagination uttered the word.

"Samotlor field in Russia is infected . . ."

"Are you sure the tests are correct?" Preston asks.

"It's not a difficult test. You take a core sample and put it under the microscope."

Gillette hears Mallory say, "Fifty days from now is the dead of winter. My God! Cold turkey on oil!"

Dr. Preston puts his head in his hands. He's fumbling for positive thoughts. "It's not *really* cold turkey. *Some* countries must be clean. And the strategic oil reserve has enough fuel to last three months, that is, if other pipelines can be used to deliver it."

Theresa recites, "Forcados Yorki in Nigeria . . ."

The big Mercator map lighting up like a Las Vegas slot machine paying off.

Blinkblinkblinkblink . . .

Gillette has seen this kind of map before. Sometimes in a tent, sometimes hand-drawn, spread over the hood of a hot, ticking jeep in a jungle. Sometimes it's been a computer depiction at the Centers for Disease Control.

Whatever form the map takes, it always signifies suffering. The dots are casualties; the connecting lines routes of contagion. The infected areas keep spreading across countries, airline routes, shipping lanes, camel and trucking routes, widening in confluence with the panic gripping whoever is watching the dots.

Os Preston whispers, "No life on earth can live through refining."

"Whatever it is, it's on earth now," Gillette says.

Each time an outbreak starts, we think, Will this be the time we won't be able to stop it?

The FEMA rep says, "How can we even investigate this if we can't move around?"

They all feel the clock ticking. It's clear that within fifty days, if they don't figure this out, everything they know and value, their families, their loves, their properties and millions of lives will deteriorate day by day and be destroyed.

Gillette says the words.

"It's a new kind of plague."

And General Hauser says, wearily, "Perhaps you should tell the cholera story after all."

"You on the motorcycle. Pull over," the voice on the police hailer finally commands.

Clayton Cox responds immediately, like a good citizen. The police car has been following him for nine minutes now, so he's not gone home. On goes his right blinker. He swings the BMW to the curb. The area of oak-lined Reno Road in which he's been stopped lies less than a mile from Macomb Street, and his house.

The squad car pulls over behind him.

There are no other moving cars on the road here, but several stalled ones have been rolled onto the curb, so they don't block driveways or Reno. Reno is clear. Apparently, some tow trucks have been functioning tonight.

Did some nosy neighbor see me leave the Niles house? Did they phone the police?

In the warm humid air the creak of the opening squad car doors is audible, and looking back, he sees one, two, *three* uniformed cops get out. A white guy and two blacks; a man and a woman.

The police lost lots of cars tonight, so they must have increased staff in the ones that work.

Clayton can feel the warm hardness of the Glock 17 holstered in the pit of his back, the worst possible place to reach it. The cops walking toward him, separating to stay safe, will react instantly if his hand slips under his shirt.

Even if it's just a routine traffic stop, if they look in my saddlebags they'll find the break-in tools. But it won't be routine. The police would not waste time tonight.

"Good morning, sir," the woman says. "May I see your license?"

Clayton looks out from beneath his blue helmet. He's Irish red tonight. Thick red beard. Red mustache. Light blue contact lenses. A real son of the sod.

"I was sure I gave it a full stop back there," he says.

They've asked for the license. That means I can reach around back for it. Do I go for the Glock? If I do and they phoned in my license, other police will come to the house.

He goes subservient, cowed by authority. It's an attitude he'd affected regularly as a boy in the camp, after he'd beaten someone up, stolen something, knifed someone in self-defense.

"Sorry if I did something wrong, ma'am."

"Don't you know there's a curfew tonight?" the woman says, but not unkindly. More like curious. A positive sign.

"Yes, ma'am. I saw it on TV. I was home. I *knew* I shouldn't go out. But Mom ran out of medicine and she phoned me. Her asthma, you know. She needs an inhaler."

The youngest cop, the white guy, walks back toward the squad car, Clayton's ID in hand.

If I'm going to shoot, I better do it before he calls in.

Clayton babbling on about poor Mom's terrible wheezing. Clayton saying, "And when she saw the crash on TV, those people falling out of the sky she got scared and stopped breathing properly . . ."

Clayton thinking, *The damn silencer is not on the gun. If I fire, people may look out their windows.*

But at that moment the lady cop nods sympathetically.

"Where'd you buy your gas?" she asks.

"Ma'am?"

The black man says, "Your gas, Clay. We're asking so we can buy the good stuff when the stations open tomorrow. The captain told us to ask everyone."

"My name is Clayton, not Clay," he says. He hates it when people mispronounce his false names. The names mean something. They're special. They were important to great-great-grandfather.

"Sorry, Clayton," the black man says as Clayton tries to recall a corner where he's seen *any* gas station. As instructed by his mentor, he's been filling his tanks with petrol he'd bought weeks ago and stored in the garage.

"Tunlaw and Wisconsin," he says, remembering a Georgetown station he passed tonight.

The white cop comes back from their squad car and hands Clayton his license. It's real. It's clean. His mentor provided it. *Getting caught from bad ID would be a crime,* his mentor had said.

"Clayton, no more driving after you get home," the woman cop says, wagging a finger. "Nobody is supposed to be on the road tomorrow until we figure out the problem. Tonight we give warnings. Tomorrow, tickets."

"That is, if we can even get to work tomorrow," the black man says.

Clayton hears his own heart beating as the three cops turn back to the squad car. They know his name. They've called in his ID. Is there any danger in that?

A little. But not much. *Leave it alone.*

Clayton kick-starts the bike. He's home five minutes later. He goes into the living room and turns on the television, to one of those usually syrupy dawn programs.

"The president will be addressing the nation at nine," the host says, looking gratifyingly terrified herself.

He reports in and then goes upstairs to sleep, as there will be lots of work to do over the next few weeks. In his dream he is back in the camp, age eleven, surrounded by boys taunting him, throwing rocks and screaming bad names at him, laughing about the death of his mother.

He fights back, lashes out with his feet and feels a boy fall back, but the others surge in. And then the boys seem to fly off him. The boys are pulled away. And in the dream his mentor is standing there, smiling down at him, soothing him and telling him who he really is, whose genes he carries, the genes of a great warrior. It's funny the way dreams work, because in real life he didn't meet his mentor until after he ran away from the camp.

The dream ends well. It ends with Clayton and his mentor and great-great-grandfather, in his white robes, walking by the cliff by the mentor's home.

Clayton Cox, amid the wreckage, sleeps deeply and happily, ready to visit whomever his mentor will send him to next.

SIX

OCTOBER 31ST. MORNING RUSH HOUR. 3 DAYS AFTER OUTBREAK.

Normally Gillette would be on a plane by now on his way to the outbreak, not shunted away from the field. Normally the trouble would be far away. Normally the microbe would be attacking strangers.

Not my family, he thinks.

At eight, he rides the Metro red line home from the Pentagon, having been allowed five hours to "prepare his family" before the Fort Detrick van arrives to take him away. Transferred from Command, he no longer rates a car and driver. Both are needed for "more important people," Hauser had said.

Fortunately the Metro functions electrically, not on petroleum fuel. And engine lubrication oil is synthetic, not natural. Riding, he marvels—as in overseas emergencies—at how normal life exists side by side with disaster. He sees two men in suits trying to get to work. A father sits beside his small son, who wears an oversized skeleton mask. Today is Halloween, Gillette recalls. Two women chat happily, as if going shopping. Why aren't they home, stocking up? Gillette wonders.

OIL SUPPLY INTERRUPTED, reads a top headline on today's razor-thin *Washington Post.* And DEATH TOLL IN THOUSANDS IN PLANE CRASHES AROUND THE WORLD.

And MARINE CONVOY WIPED OUT IN MIDEAST, AFTER VEHICLES BREAK DOWN.

Ads in the Metro show a handsome couple riding in a sports car, a smiling family enjoying their home heating oil, and waving vacationers boarding a jet. Three-quarters of the posters relate to petroleum. Even the public service announcement brags that the Energy Department seeks "alternative fuels" to reduce oil dependence.

You better hurry up, Gillette thinks.

The clattering of rails casts him back for an instant, to the trailer park where he'd lived as a teen, with his uncle, across a weedy lot from an Amtrak line. His father was dead by then, killed in Vietnam. His mom had disappeared. His uncle was usually too drunk to control the boy. One night after Gillette was released from detention, the door chime had sounded, and he'd been astonished to see Dr. Wilbur Larch on the steps.

"It was hard to find you," Larch had said. "Are you serious about wanting to be a doctor? Because there's an internship program for high school students in epidemiology, at my federal agency, the Centers for Disease Control."

"Epi what?" he'd said. And then, as they sat on the steps and the Amtrak Sunshine Express whooshed by, Larch had told him the cholera story, the one Gillette had repeated to Hauser tonight. It was the story of how epidemiology started. The story that Gillette believed might help now.

"Cholera—*Vibrio Cholerae*—was one of the terrors of the ancient world. It still strikes down thousands in impoverished countries. It lives in dirty drinking water, lodges in human intestines, kills with high fever and diarrhea so rampant that a strong man who falls ill at nine in the morning can be dead from dehydration by nine that night."

By the mid 1800s, Larch had told Gillette, a British doctor named John Snow had watched helplessly each time cholera hit London. Poor people locked themselves in their homes to escape contagion. The rich fled the city. Thousands of victims ended up piled on wooden carts, carried off to mass graves.

Then, in August 1854, the disease broke out again, only blocks from Snow's office.

Panic gripped the city. Doctors at the time blamed "bad air" for cholera, or "God's retribution." Microbes had not been discovered yet. Germs did not exist even in surgeons' imaginations.

"Yet Snow *still* figured out how to stop the disease," Gillette had explained at the Pentagon tonight.

"He bought a simple street map of London and marked the locations where victims died. He realized that over five hundred people had perished within two blocks of a single slum intersection: Cambridge and Broad Streets. Visiting the corner, he saw men, women and children lined up at a public water pump, waiting their turn to pump water to drink.

"Could the pump be the source of the disease?" the forty-one-year-old physician asked himself. "If it is, why did some victims die in other neighborhoods? Far from the pump?"

"Snow visited the homes of other victims," Gillette said. "It turned out they all used the water—it tasted better—from the Broad Street pump."

So on September 7, 1854, Snow stood up at a public meeting in the vestry of St. James Parish and convinced desperate officials to shut down the pump. He did not know why the pump was responsible, but within days the epidemic ceased, and the science of epidemiology—Gillette's passion—had begun.

"That's very enlightening," Hauser had said. "But is there some relevant point?"

"That we might be able to slow down the oil bug before we know what it is, if we figure out how it *spreads,*" Gillette had said, and started grilling Dr. Preston as more lights on the oil-field map blinked on. Many fields were officially clean at that point. But the number was dropping as test results came in.

Gillette asking, "Does any single oil company conduct drilling in all or many of the fields?"

Violet Pell of the Energy Department had flushed at that, and snapped, "You're accusing an *oil* company?"

"I'm trying to figure out how the microbe reached so many places unconnected to each other at the same time."

Osborne had shaken his head. "Several companies conduct operations in some of these fields. No single company operates in all of them. Or are you suggesting a plot between *two* companies to destroy their own product?"

Violet Pell had grinned at Gillette like a second-grader watching a stupid classmate ask a dumb question.

Gillette had persisted, "Is there any research project under way that you know of—corporate, academic, U.N.—that involves sending personnel to all these fields?"

"There are always ongoing projects," Osborne had politely replied. "They involve hundreds of people scattered around the globe. Quite a conspiracy, I'd say."

Osborne seemed to think it bothered Gillette if he got a negative answer. But it just eliminated a possibility. Gillette had asked, "Is there any product necessary to oil extraction that's shipped all over the world? A product made by a single company?"

Osborne Preston had frowned, then looked thoughtful.

"Actually, not a single company . . . but you might have an idea there."

The others had stirred at that and Gillette had felt a familiar rise in his heartbeat, a sensation he always experienced at moments of progress in tracking disease.

"Fluids," Preston had announced as a new light popped on in the North Sea. "Fluids are pumped into fields to improve recovery. You see, oil in the ground is under pressure. It's built up over millions of years. In the old days, that's why you had gushers when you hit oil. The pressure drove the oil out. Now we stop gushers with better technology, you know, to keep from wasting oil."

"Get to the fluids," Hauser had said.

"Well, once you start extracting oil the pressure goes down. It's harder to retrieve the rest. So drillers inject fluids beneath a field to increase pressure. Water, mostly, but there can be chemicals or bactericides in the mix. Fluids can be introduced downhole or above ground. Fluids maximize field productivity and increase pipeline flow."

Gillette had envisioned an old lithograph he'd seen of a water pump in old London. The single source of a terrible disease. He'd asked, "What companies make the fluids?"

"Tangier, a French company, is one of the biggest."

"France," Hauser had said with relish. He seemed to like the possibility that the problem came from France.

"Halliburton, in Texas, and Schlumberger. Cougar Energy Services, in Nevada," Preston said, naming the huge multinational energy companies. "But hundreds of smaller companies also make fluids around the world and work on wells."

Violet Pell had bristled at Gillette. "Are you accusing an American company of releasing an infection *in our own fields?*"

"I'm suggesting that accidents happen. Or that there may be research projects of which we're unaware. I'm suggesting that we check these companies—all the companies—for a connection."

"What possible motivation could an *American* company have in crippling our own supply?" Violet snapped. She'd worked for Exxon Mobil before coming to the government.

"Stop talking about motivation," Gillette had replied. "A microbe's only motivation is to stay alive, reproduce."

Violet had asked the table, "Are we actually going to waste time on cockamamie theories?"

"Actually," Osborne Preston had said quietly, "Dr. Gillette might have a point."

After that, Hauser—cooperative for once—had ordered the FBI and Homeland Security to coordinate investigations of the U.S.-based fluids companies, research programs, scientists, facilities. And to get lists of companies making additives or bactericides.

"Send inquiries abroad as well."

"Shouldn't someone from CDC go along?" Gillette had asked.

"Negative," Hauser said. "Homeland Security can do the job. They're trained in this kind of investigation."

"Oh? Have microbes like these shown up before?"

"Turf wars," Hauser flared. "This meeting is through."

Afterward he'd taken Gillette aside, thanked him, and then officially released him from the Pentagon.

"Go home. Check on your family. You work for Colonel Novak now.

I'm sure you'll do a terrific job as a foot soldier, Gillette. That seems to be where your heart lies, eh? Tracking down bugs in the trenches."

Hauser had offered a firm handshake. "Skilled people will handle Nevada and Texas. They'll get answers."

"Send me, sir. I'm a field man, not a lab man."

"Make the best of your famous intuition. Maybe Colonel Novak will find a good use for it. I haven't, Gillette."

His heart beats faster as the Metro escalator approaches the surface. He hears no traffic on normally busy Connecticut Avenue. Walking out into the sunlight, he's surprised to see Les Higuera waiting by the newspaper machines, chubby-looking in a baby blue running suit, propping up a bright red, tandem Schwinn bike. The Higueras bought it last year for rides in Rock Creek Park. A small radio dangles from the handlebars.

Les grins like a limo driver meeting a prearranged fare. He holds up a cardboard sign that says, GILLETTE.

But the smile dies with the radio announcer's words. "The Prince of Wales was killed last night, in a helicopter crash in Scotland. The secretary-general of the U.N. has been confirmed dead in the crash of the Saudi king's private jet."

The sun shines brightly on white brick apartment buildings. The day is warm. Gillette doesn't usually hear birds on Connecticut Avenue during rush hour, but he does today. A lone private car cruises slowly past, with four men in dark suits inside.

"Guess what? I'm nonessential to the network," Les says as Gillette climbs onto the backseat. "They told me to stay home." Moments later they're pedaling up six-lane Connecticut, past broken-down cars and Metro buses. "Marisa got the block together after you called," Les says over his back. "Everyone's at your house to figure out what to do."

They move slowly, it seems to Gillette. Other cyclists pedal along in both directions, many hooked to radios too. It reminds Gillette of rush hour in parts of China. Stores are closed. So is the Van Ness campus of the University of the District of Columbia. Gillette notices an SUV parked at a Mobil station, and recognizes the gold logo as coming from Georgetown University, and a lab seconded to the biological attack program. The two figures in asbestos suits—moving out from behind a pump—will be technicians gathering fuel for testing. The suits protect against infection, in case the microbe jumps to humans.

Could humans be carriers? Gillette thinks.

Teams must be going station to station, identifying which gas is infected, which pumps can be used.

"The president's speech has been delayed until eleven-thirty," Les says. "The White House announced that the oil problem may last as long as a month. Is that true?"

"It could be a lot longer."

More people are walking around than usual in the silence. Others lean from apartment windows to gaze with fear or wonder at the avenue. It will be the nervous people and overly intelligent ones to see the extent of the danger first, no matter what assurances the president gives in his speech.

At Albemarle and Connecticut, Gillette spots a small crowd lined up outside the 7-Eleven where he often buys coffee. People emerge from the store weighed down with shopping bags: food, toiletries, flashlights, beer.

It's not panic yet. Just preparation. The overall sense is of quiet but order, oddness and expectation.

Connecticut tilts uphill. He'd not really noticed how steep it was before. At Fessenden, they pedal toward a crowd shoving in or out of the open rear doors of a broken-down semi. Gillette sees it is a Safeway supermarket truck that must have halted on its way to supply the nearest store, a mile off, near Chevy Chase. People jostle each other in back of the truck, not fighting exactly. Pushing. People hopping off carry away cuts of raw meat, and run toward nearby apartment buildings.

Looting already?

Gillette spots Gordon and Teddie Dubbs in the pack, carrying a plastic cooler between them. He also sees Officer Danyla at the fringe of the crowd, dressed in civvies, watching helplessly. Gordon catches sight of Gillette and gives a long hard look before turning away.

"The driver said they could take the food," Danyla tells Gillette. "It was spoiling. The refrigeration unit won't run without the engine on." But the policewoman is worried. She suspects that next time a truck breaks down a driver's permission may not be sought.

Beyond the truck the road crests and goes downhill past leafy commercial and residential blocks and the turnoff to Nebraska. This is Gillette's neighborhood. At the Sunrise Assisted Living Home the elderly residents sit around on the porch, staring out at Connecticut Avenue. The door to the firehouse of Engine Company 35 is open, except Gillette sees only one truck inside, not two.

He's struck also by the fact that of the dozen stores lining this block, three are gas stations.

Only St. Paul's Lutheran Church is open, set back from Connecticut beside the fire station. A few people trickle in.

"This block hasn't been this dead since before Fort Reno was here," Les says, referring to the fact that long-gone Fort Reno, the capital's largest Civil War signal station, had flown red flags during emergencies at what is now the corner of Connecticut and Nebraska. The flags would have been up today.

Just before turning off Connecticut, Gillette spots an open door to his favorite restaurant, Angler's Feast, a small, intimate seafood place. He and Marisa often eat here on anniversaries, loving the "Big Jack's lobster with dill sauce."

Now he sees the heavyset owner—Jack Angler—loading up the back of a Toyota Tacoma pickup parked outside the restaurant, in a bus zone. Jack's truck is clearly working. The bed is piled with coolers and bags.

Jack waves them over.

"We're out of food at home," Jack tells Gillette. "I've been watching BBC news from Europe. They're saying most oil's unusable. They're saying it's some kind of petroleum-eating microbe gone wild. They're prohibiting private vehicles on the road in Germany and France, to save usable fuel. I've only got a few gallons left, so if I don't take this stuff now I'm afraid I won't get another chance. We'll fill up our basement freezer at home."

Jack brightens. "Want to buy the extra? You can have everything I don't need. Menu prices. I won't mark 'em up."

"*Menu* prices?" flares Les. "*Menu* prices include cooking and serving. You're charging restaurant prices for food we could get at Safeway? Gouger!"

Jack looks stunned. He tries a joke. "I'll throw in salad." Then he loses his temper. "You're a network employee. *You* get paid when you don't go to work. I pay rent on this place. I pay insurance. I have two kids in college. You know what? Fuck you. I changed my mind. Starve or go to Safeway and pray they open the fucking store. By next week people will be clawing out each other's faces for a can of tuna. Happy to pay ten bucks for it."

And with that the rage erupts in Gillette too, at Jack, Hauser, Dubbs, the world. All the old combat instincts jumping in. When he was fifteen he would have hit the man, or thrown a rock through his window. Dr. Larch had worked with him, helped him learn to control his temper. And now he knows that Marion Street needs the food. "Jack, will you take a credit card?" is all he says.

Jack looks from man to man. He wipes his brow. But then he grins. "Gold would be better," he says. "Ah, I'd lose the food anyway, and if MasterCard is going down, so's the whole ship. No hard feelings, amigo," he tells Les. "We can drive the stuff over in my truck. No extra charge for gas. And remember to eat at my place when the fuel starts flowing again. Okay?"

"It's getting worse by the hour," Theresa Novak tells Gillette over his cell phone, twenty minutes later. Jack Angler's food—seven hundred dollars worth—has been delivered. The neighbors have assembled in Gillette's finished basement. The call has interrupted the meeting.

Gillette is to be picked up, Theresa said, and brought with other scientists to Fort Detrick at five. He's to bring a suitcase. He may be sleeping there awhile.

"Saudi Arabian fields are almost all infected," says Theresa. "Iran.

OPEC countries. Even China. Pipelines. Tankers. *It's in there, growing, even if it's not toxic-level yet.* Almost ninety percent of the reserves on earth just went out of reach."

"How did it happen so fast?"

"Maybe Delta-3 got into the fields a while ago. That's what the newspeople are naming it, after the first flight that went down. It gets in and starts multiplying. Or maybe refining triggers growth, Greg. Accident or biological weapon, either way there's a lag time between introduction and impact, so you can't trace it back. I wish Hauser would send you into the field. You see things other people miss."

Gillette hangs up and looks around the basement, at faces that were cheery last night and now look drawn, pale, afraid. Usually neighbors watch Redskins games down here. Today the set is tuned to Les's station. Riots have broken out in Cairo, as the Muslim Brotherhood calls the oil stoppage God's punishment, and cries out for revolution.

"The massive riots seem coordinated," says the announcer, not the regular newscaster, but a substitute. The regular announcer couldn't get to work.

The scene switches to a satellite weather shot showing an immense circular hurricane bearing down on Florida from the South Atlantic. Pinprick black dots indicate half a dozen ships foundering in the storm's path, their engines broken down. Then the scene switches and the caption says, IDAHO SURVIVALISTS READY. Gillette sees a group of small log homes behind a barbed-wire fence. A man holding a shotgun waves it at the camera.

The basement is small, paneled and furnished with stuffed chairs and couches. On the bar are pitchers of water and orange juice. On the walls are photos of the family: on vacation in Rehoboth Beach, baby shots of the kids, honeymoon shots of Marisa and Gillette, an exotic shot of Gillette in a jungle, on an outbreak. It's a time line of family normalcy: Paulo's soccer trophies, Annie's zoo volunteer plaques, Marisa the old college skiing champion, Gillette as a young Navy ensign, standing beside a proud Wilbur Larch, whose recommendation had helped get him into ROTC.

Gillette has spent years sitting in on village meetings during outbreaks, in Zaire, the Philippines, Gabon, Brazil. Advising frightened people how to act.

Now he's chairing a meeting in his own home.

He tells his neighbors exactly what happened at the Pentagon, tells them that they need a group plan for the near future, a block plan to enhance safety.

He advises that they share and ration food, starting now. That they take turns at night patrolling the block, "just in case civil order breaks down."

"Isn't that a little paranoid?" asks Julie Dent, one of the "Sirens," a trio of American Airlines air hostesses who rent the smallest house on the street.

Gillette says, "It never hurts to play safe."

He asks for a show of hands. Do any families have a microscope at home, part of a child's science kit? Paulo does. By tomorrow, Gillette says, if fuel in a car is infected, it will probably be evident under even a weak microscope, since the growth rate of the microbe is so fast.

"Then we'll know which cars we can use, until they run out of gas."

Gillette asks if anyone owns a car that hasn't been filled with gas recently. *That* car, he reasons, should function now. They should take it to Safeway and load up with supplies. Paulo had biked there at eight this morning, and was told by the manager that he was phoning employees, ordering the ones who could reach the Metro to come to work. The civic-minded manager plans to open up at eleven.

Bob Cantoni speaks up haltingly. "We haven't used our Suburban in a month. But Joanna and I figured we'd load up the kids and head out for our country place near Winchester. Worst comes to worst, live off the land."

Les Higuera breaks out laughing.

"Off the *land*? You couldn't grow a tomato plant if you started off with a pack of seeds," he says.

Bob looks down at his belly, which spreads over his belt. "Yeah. I'd be dead in a week. Let's go to the store."

The pastor says, "Surely heating oil will be provided to the church if this trouble goes into winter."

Marisa says, "I can hold classes here for the children, until schools reopen. It will calm them."

Annie looks like she's about to cry. "Who's going to care for the baby cheetahs? They need special care! They *know* me! *I want to go to the zoo, now!*"

Suddenly everyone is talking at once.

"I *told* you we needed alternative fuels!"

"I'm sure the president will have a plan."

"I can teach anyone who wants to learn to fire a gun," Bob Cantoni says. "Just in case."

At eleven-thirty Gillette and Les stand outside the Safeway, in a sea of vehicles and people that spills over the sidewalk, onto Connecticut Avenue, and even the grass on Chevy Chase Circle. Two cops try halfheartedly to keep order. The store's opening has been delayed. The clerks had trouble getting in. The crowd is 800-strong, at least.

Bob Cantoni is two blocks back, guarding the Suburban. Some cars have stalled. Drivers have shut engines to save gas. In the well-heeled crowd, Gillette is being jostled by some of the top lawyers, bureaucrats, lobbyists and journalists in Washington. Lots of radios are on, and the president's speech rises from a hundred open auto windows and even boomboxes.

"We have entered a period of unprecedented crisis."

The manager stands bravely in front of the plate-glass window waving his hands for quiet. Gillette knows the guy. Ali Mohammed is a Pakistani immigrant, a young family man who runs a fine store and clearly fears that he made an error by announcing good-heartedly that he would open up.

Gillette sees employees in there, taking up positions at cash registers. They look nervous too.

The president says, "I am today ordering the release of all oil in the strategic oil reserves, our salt dome storage facilities near the Gulf of Mexico. The oil there is clean. There is enough supply to meet the nation's needs for at least two months if we use it wisely."

Gillette thinks, *But if the pipelines are infected we don't even have two months. We can't move the oil.*

The crowd is restless, pent up.

The president says, "I've instructed top state and local officials to submit lists of essential services and personnel who will receive shipments of clean oil. We will temporarily requisition uninfected tanker trucks and new trucks coming off the line in Detroit to transport this oil. Again, oil will be portioned out by need."

The manager calls out, "We will open in three minutes! Please form an orderly line."

The president says, "Rest assured, our top scientists are working day and night to kill this microbe. Meanwhile, I'm calling on all Americans to make sacrifices and pray we can all do with less for a while."

The manager says, "We are not fully stocked. Our delivery trucks broke down on the way here last night."

A voice from the crowd yells, "Open up, raghead."

That voice belongs to Gordon Dubbs, thinks Gillette.

The president says, "Now is a time for us to show strength, for each family and neighborhood to band together, help each other. Carpool. Conserve heat. Take in elderly relatives. I call upon management and labor to reach amicable agreements limiting work hours and production, at factories that will remain open . . ."

The manager has broken out sweating. He announces, "When I open the doors, twenty people may go in at a time. Each person may buy ten items. As one person leaves, another may come in. This way, everyone can be served."

People in this upscale neighborhood are not accustomed to being told no, or being ordered around without being able to argue. A few men step forward, loudly trying to convince the manager to let everyone in at once. Or let fifty in at once. Or allow customers to buy six items. Or to treat people on a first-come, first-served basis.

Les tells Gillette, "Uh oh. This is how it started in Kinshasa, in '03."

Someone in the crowd shouts, "I've been waiting for this store to open since ten! Other people just *got* here!"

"By the time I get in nothing will be left!"

The president says, voice rising, "Ideally our country could simply purchase more oil. But almost every country—even the big oil-exporting ones—are in the same situation. At the moment it appears that some smaller fields overseas may be clean. In Gabon. Sri Lanka. Indonesia. We are trying to buy oil from these countries. But so is everyone else."

A voice in the crowd shouts back at the president, "So send in the Army! Take it!"

Someone else yells at the first man, "How is the Army going to *move*?"

A handful of people in the crowd start chanting, "Open the door! *Open the door!*"

The president says, "I remind you that America has faced hard times before . . ."

The manager cries out, "If you will not step back, I will not open up!"

Les nudges Gillette. "Let's get out of here."

The president says, "I know we will act responsibly."

A rock sails out of the sea of people, smashes into the plate-glass window and breaks it.

A gasp sweeps through the crowd.

A mob of six hundred people surges toward the door.

SEVEN

OCTOBER 31ST. AFTERNOON. 3 DAYS AFTER OUTBREAK.

The *Pride of Denmark* would have eluded the hurricane easily. But its engines have failed.

The flagship of Florida's top cruise line wallows dead in seventy-foot waves, 100 miles west of the Bahamas, the TV announcer says. There's no way for cameras to reach the scene, so Gillette, in his study, eyes with horror a photo of the *Pride,* as top-heavy and maneuverable as a 79,000-ton saltine box. Her propellers are useless. The five-star floating palace carries 811 crew and 2,031 passengers, most of them trapped in darkness belowdecks.

After all, the ship's electrical generators run on diesel fuel too.

"Mayday. Mayday," says the staticky voice of the captain from the bridge, through a hand-held battery-powered VHF radio whose signal ABC-TV has picked up. Normal ship-to-shore communication is out, as is all steering, radar, pumps and ballast control.

Ten hours ago the *Pride* had been cruising at full speed toward Miami—outrunning the storm—when the engines had started whining, then coughing. The *Pride* had drifted to a halt as the hurricane eye bore down on it.

"We need immediate assistance," says the captain.

"Are you taking on water?" asks the Coast Guard operator.

"Rolling badly."

Seas are black, the announcer says. Weather cups atop the pilothouse ripped away when winds topped 140 miles an hour. Moments ago a massive wave tipped the ship over 34 degrees, almost causing the deadly "roll moment," when a craft goes upside down.

The Coast Guard operator says, "Our rescue helicopters broke down, sir."

Christ, those poor people, Gillette thinks, imagining the ship's officers on the bridge, trying to maintain balance. The panic in staterooms. Luggage crashing from compartments. Crockery breaking in restaurants. Families groping toward each other in the dark, disoriented, screaming, thinking walls are decks.

The second mate, having reached the bridge twenty minutes ago, had reported blackjack tables smashing into slot machines in the casino. And half-drunk passengers breaking into the booze supply in the Starlight nightclub, guzzling scotch in the glow of cigarette lighters.

The captain has ordered lifeboats made ready. But it is unclear how anyone can lower boats by hand in the storm.

A voice in the background shouts, "Oh God! We're rolling sixty degrees! Look at that wave!"

Gillette hears a man cry out, "Mother!"

Then comes only static. And the Coast Guard operator asking, over and over, is anybody there?

Gillette shuts the set off to try to concentrate. For the last two hours he's been packing and helping Marisa make lists of supplies at home. Batteries. Frozen veggies. Bread. Bandages. Now, in his study, he studies his "weird microbe" file while waiting for the transport van from Fort Detrick to arrive. He tries to ignore the throbbing in his jaw where a woman had smacked him during the riot with a package of frozen meat. The bandage is sticky but the wound is clean. Marisa applied alcohol to it. It's uncomfortable, not deep.

That whole scene reminded me of gang fights I fought in, as a kid. People grabbed anything as weapons.

He remembers grown men throwing punches while Gillette and Bob Cantoni pulled the Safeway manager to safety, and Gordon and Teddie Dubbs ran off, loaded down with plastic bags filled with food.

He sees some of Washington's top lawyers and policy wonks, fleeing with boxes of spaghetti and Wheaties. *Just like my gang buddies used to run when the cops showed up.*

Gillette goes back to the material on his desk, books and articles from CDC journals, science magazines, even *National Geographics,* piled in a pool of tensor light.

Maybe somewhere in these files is a clue.

He eyes photos of newly discovered microbes, which do things that scientists thought impossible ten years ago.

"Deinococcus radiodurans showed up in canned meat," he reads. "It survived radiation doses that had previously killed all microbes. This hardy fighter managed to stitch its genome together—actually repair itself—after radiation shredded the membrane of the cell."

He flips a page. "Ammonox bacteria appeared in a Dutch yeast factory. Its cell walls contain sac-like compartments filled with poisonous toxic hydrazine, a form of rocket fuel. Why didn't the poison kill the microbe? No one knows. Not only does ammonox live under conditions that would kill most bacteria, it eats ammonia. Ten years ago no one heard of this microbe. Now researchers are finding it throughout the seas, and rethinking the way the Earth's whole nitrogen cycle works because of this 'impossible' creature."

Gillette has been to plenty of dinner parties in Washington where researchers—academic types who have never watched a person die of Ebola up close—regard microbes as "wonders of nature." But Gillette studies microbes the way a general learns about missiles. He's a gang fighter who replaced his human enemies with germs.

Microbes are living weapons, he thinks. *They keep evolving. Every once in a while they come up with a way to wipe out millions of people. Well, it won't be my family this time.*

The microbes Gillette deals with are not the beneficial kind, that help brewers make beer, or cheese makers produce sharper cheddar. They're the kind that turn healthy people blind and reduce a human brain to bloody mush in less than two days.

Gillette turns pages and stops at a headline: NEW MICROBES RESIST HEAT.

Excited, he reads, "Laboratories use autoclaves to sterilize equipment at 212 degrees Fahrenheit. No life form was thought to survive temperatures above that. Then oceanographers discovered Thermus aquaticus in undersea vents—off volcanoes—that can live in autoclaves."

Gillette realizes that someone has entered the room.

"Even more shocking is the story of scientist Paul Brown, who studied mad cow disease, caused by prions, a kind of protein. Brown freeze-dried a cow brain infected with the disease, heated it for an hour at 680 degrees Fahrenheit and fed it to a hamster. The hamster caught the disease."

"Dad? I figured out how to kill the bug."

Paulo, standing in the doorway in his soccer shirt, clutches a stack of printouts and *Science Kid* magazines. Gillette waves the boy into a chair. Paulo says, "Use bleach!"

"You mean plain old Clorox? From a store?"

The boy nods eagerly, waving papers. "Mom said bleach kills germs. So why not add bleach to gasoline before it goes into a car? The bleach will kill the germs, right?"

Gillette puts his arm around his son's shoulder and wishes the solution was so easy. He tells the boy what he learned from Os Preston when the disinfection question came up. "You're smart, Paulo, but bleach has chlorine in it. Chlorine would damage oil. But there may be some other chemical that kills the bug. The oil companies are trying to find it. So is the government. Everyone is working hard."

"Then the problem will be fixed soon, right?"

"I hope so," he says, careful to be honest. "But even if we find a chemical that kills Delta-3, we'd have to test and retest it to make sure it works. We'd need months to install equipment. Over eight billion barrels of oil stock would need purification. And thousands of miles of pipes."

Paulo looks unhappy. "But the man on TV said we only have like fifty days, uh, forty-seven now, left."

Gillette gazes back at his books.

"That's why you and I have to help figure out how to kill the bug. I'll call you every night, from Detrick."

"Because scientists should share knowledge. Right?"

"I'll tell you as much as I can."

"Oh yeah. I forgot. The lady from Fort Detrick is here to pick you up. Colonel Novak is really pretty."

Gillette stands. He feels warm. He feels guilty. He doesn't want her here.

Why did she come to my house, instead of sending someone else?

"I heard her tell Mom that all the main oil pipelines are infected," Paulo says. "That's really bad, right?"

Microbes.

They're how all important decisions in Gillette's life started. With something that you can't see without a microscope, small enough to escape detection but big enough to end a life.

Microbes are the reason he adopted Paulo and Annie.

"I'm sorry. You can't father children," his doctor had told him after his annual physical—when he was twenty-five.

It had been a March afternoon, cold and snowy. Gillette the Georgetown med student and CDC fellow had just returned from fighting a malaria outbreak ravaging oil workers in the Amazon. He'd avoided malaria but had caught mumps orchitis, a painful jungle strain that brought high fever, swollen glands, painfully enlarged testicles.

"Less than one percent of men who contract orchitis become sterile," a kindly doctor had explained.

"Which doesn't help if you're the one."

Gillette had sat, legs dangling over the exam table at GU, hiding shock, glancing across the room at the edge of a small leather ring box protruding from a pocket of his trousers, which were draped over a chair.

Gillette had stared down at himself. Everything looked the same down there. But it wasn't.

"You can still enjoy a vigorous sex life," the doctor had said.

"Think of the money I'll save on condoms."

"Other than this, you're healthier than a horse. Perhaps you'd like to talk to a counselor, Greg?"

"Nah." Gillette had hopped off the table, reached for his pants. He never showed strangers hurt. He'd learned that when he was a kid, when he got in fights, when his uncle hit him, when he was scared and in a cell. You always act tough. "I always knew I'd come down with something on one of these trips. Funny. I never even got mumps when I was a boy," he'd said.

"Too bad. It would have made you immune."

He'd walked off campus and across M Street, in a raging snowstorm. He'd headed down to the old tow path by the Potomac, feeling the pull of the ring box in his pants. Earlier that day he'd made the restaurant reservation for eight, at Palermo, Marisa's favorite eatery. Her top choices were on the menu tonight. Spinach tonnarelli sauced with yellow peppers and tomato dice. Sardinian sheet-music bread and fava bean spread. Apple timbale flavored with cinnamon. All to be washed down with pinot grigio, topped off with the emerald-encrusted white-gold ring.

Sterile.

And *kids,* he'd thought, an hour later, still on the path, passing a father pulling a sled loaded with two small boys, bundled up and howling with delight.

She loves kids. She made that plain in the beginning.

Well, he'd told himself. She comes from money and I stole it to buy food. Her dad was an Olympic skier and I didn't know mine. She's a northerner. I'm a Georgian. She probably has bad habits I don't know about. Who needs marriage? Half of them end in divorce.

Yep. You move on in life, past problems. You don't waste time moaning about things you can't change.

I'll break up instead, he'd thought, and on the night he'd planned to propose marriage, he'd never even reached the restaurant. *I don't want her feeling sorry for me.*

So he told her that he'd met someone else, told himself he'd get over her. But later—even after months—he was still smelling her when she wasn't there; in a lab, or on his pillow when he woke. A whiff of Marisa would surprise him in a restaurant. On the street. On a damn Air India flight halfway across the globe.

It was more than just the chemistry he missed. It was the confidence and solidness, and knowing that, like him, she'd made the conscious decision to make giving a key value in her life. It was the knowledge that once committed, she was absolutely unchangeable. For the first time, Gillette felt incomplete.

The only person he confided in was Dr. Larch, when the older man flew to Washington for a meeting in June. They shared a customary brew 'n burger dinner at the Guards.

"You should have let her choose," Larch had said.

"It wasn't a fair choice."

"I didn't let someone choose once," Larch had said, and looked sad for the first time Gillette remembered. "I was afraid. Like you. I lost my chance."

"I know Marisa better than you do," Gillette had snapped.

But a week later, when he got home one night, she'd been sitting in the hallway outside his apartment. Legs out. Mittens on. Lovely face purple with fury, brimming with tears.

"Larch called me. How dare you decide for me? There are plenty of orphan children in the world who need parents. You ought to know, asshole. One was you."

Fourteen years of marriage, he thought now. Some great, some bad, but through all of them, he'd never, until Theresa, seriously desired another woman. He had not acted on it. They'd honored his marriage and the institution. Theresa's ex-husband had cheated on her, and she wasn't about to do that to someone else. But chemistry is just that, and during their nights in the Philippines, the urge had taken hold like a stubborn microbe.

"Colonel Novak," he says formally, shaking hands with her in his front hallway, by the door, feeling her warm skin, her long fingers, looking into her black eyes as Marisa watches. "I figured they'd keep you at the fort." His warning means, Do you see what is valuable to me? Nothing has changed.

She smells of vanilla. "You're the one who told me in the Philippines: Get out of the lab. Get into the field. Clues in the field may save weeks in the lab."

Theresa tells Marisa, "Sorry to take away your husband." She asks him, "What happened to your face?"

Marisa tells Gillette what she always does, when he heads off on an outbreak.

"Kill the bastard, Greg. And come home."

He sits in back of the van. She sits in front. He calls her Colonel. She calls him Commander. Beneath their formality he feels tension. The way the attraction had started in the Philippines was, he'd start a thought, she'd finish it. He'd reach for a tool, she'd be holding it. It wasn't flirting or cheating. It wasn't conscious. They were simply in sync.

I wonder what Safeway looks like after the riot? he thinks.

She says, "We passed Safeway. It's destroyed."

I wonder how many functioning vehicles she has at the fort?

"Most cars at Detrick still work," she says, as the van weaves around wrecks on Connecticut Avenue, heading into Chevy Chase. Lights blaze in every dwelling. People are staying home. There's even one cop car on the road.

An all-news station plays softly over the radio. In outbreaks, Gillette knows, missing a key piece of news could have dire consequences. At the moment, Os Preston is being interviewed on the radio, from his home.

"The breakdown will probably happen in increments," he's saying. "First fuel deliveries. Then food. Factories going down from no lube oil. Sporadic power failures early on. You might have one area without electricity and the next block lit up. Phones working in one place but not a block away. Eventually affected areas will overlap. Nothing will work."

The temperature is dropping. Marion street will need heating oil.

"Looks like it might snow. I hope you have heating oil," she says.

I hope she knows what I'm thinking about now.

Because the first snow of the season always reminds him of Marisa, and the story of how they'd met.

He'd spotted her while he was in med school, at a party. One of those Saturday-night Adams Morgan blowouts with rap music blasting and furniture pushed into corners and pizza boxes and wine bottles on tables, counters, hallways, floors. Pine fire crackling in a hearth. The first snowstorm of the year raging outside, shutting in revelers. Best of all, no classes scheduled tomorrow, or duties at CDC.

And there, across the crowded room, as they'd joked about eighteen million times after that, he'd made eye contact with a slim long-haired blonde in jeans and a black tank top. He did not recognize her from the school.

Next thing he knew, it was 2 A.M. They'd been talking for hours. Her date had left. Gillette couldn't stop staring at her. Her slanting green eyes made his forehead throb and his knees go weak. The way she danced made his groin itch. He was experienced with women. He'd been having sex since age fourteen. But he'd never met a girl who seemed—at first—exciting yet familiar, different yet similar, open to possibility but firmly grounded, knowing who she was.

"Want to go somewhere else?" he'd asked at length.

She'd eyed his shoes instead of answering directly. "Size ten and a half, right?"

"What are you, a foot fetishist?"

"Follow me," she'd giggled, and thirty minutes later, after slogging through foot-high drifts, he'd trudged up three flights of stairs to the apartment she shared with two other Teach-for-America girls, off Columbia. What he expected to happen didn't. She opened a hallway closet and pulled out Salomon cross-country ski boots, size ten and a half. And a man's zip-up fur-lined waterproof jacket. A man's ski gloves, gaiters, and long fiberglass skis.

"These should fit you, Greg."

"I preferred you as a fetishist."

"Let's go share a drink with honest Abe. The poor guy's alone in the storm, stuck to a cold marble chair."

"Whose boots are these anyway?"

"Jealous already?" But she liked it and it was true.

Washington shuts down usually at the mention of the word "snow." Its inhabitants plan wars but won't drive in storms. Gillette found himself gliding through fresh powder after the blonde, past the lit-up White House and over the parade ground, toward the white mall by the Potomac. She'd been a cross-country ski champion at Middlebury, he learned. She'd trained for the Olympics but failed to make the team. She'd grown up in Manchester, Vermont, where her dad had taken her skiing in storms all through her childhood.

"Someday I want to do the same thing with my daughter and son. I want one of each," she told Gillette.

Always before, the word "children" had signaled the end of Gillette's relationships. Or rather, confirmed that a particular woman was wrong for him. But flanking the long-legged girl, gliding and pushing, the word sounded different. He was a self-made man who knew what he wanted when he saw it. He'd just spotted home port.

Not to mention, in the magical storm, she was so goddamn beautiful, with her wet ponytail bobbing and snow on her lashes. And the way flakes matted the slim shoulders of her crimson Windbreaker. And those slanting eyes glimmering out from under a pulled-down stocking hat.

"I want to be a teacher," she'd told him on the steps of the Lincoln Memorial, as they sat, got wetter and shared a goatskin bag filled with Dewar's and water under the statue's big sad eyes. "I want to help kids."

"I want to stop outbreaks."

He'd said it before, plenty of times. There was nothing spectacular in the words. But her presence turned the ordinary extraordinary, enriched ambition, accentuated basic drives. Gillette had never wanted to be rich or famous. He'd never yearned for *things.* He sensed that he shared some fundamental outlook with this woman. And he had the guts to acknowledge early on—so as not to make a mistake—that he wanted more permanent things than that.

By the time they returned to her apartment, at dawn, Gillette knew that he'd just experienced one of the great nights of his life. He was famished, cold and happy. They'd cooked omelets in her kitchenette, chopped up mushrooms, red peppers and jack cheese. The domestic rhythm of knives falling together was the music of building memory. They'd washed down sourdough toast slathered with Virginia apple butter with mugs of hot chocolate laced with thick cream.

They'd fallen asleep on a couch, without making love.

But they'd spent whole weekends in bed after that.

You're beautiful, Theresa, he thinks now, *but nothing is going to happen between us. I hope you know that. I hope it won't be a problem. Because you're my new boss.*

The radio carries part of a press conference from Chicago, where the mayor insists that she will not distribute more fuel to all-white neighborhoods than to black ones. Fuel allocated to Chicago will go to essential personnel first so they can carpool to jobs in hospitals, power plants, government or law enforcement. Limited bus service will—until clean fuel runs out—go to areas not served by the El.

"The oil scare is a hoax. Want to bet the Jews did it?" says a man on the street over the radio. "They'll announce they found an antidote and charge an arm and a leg for it."

Theresa tells Gillette, "I know you're frustrated about being stuck at the fort. But Hauser was firm. You can't travel. You'll have access to our labs. And a staff to help you do research."

A news bulletin interrupts the press conference. "FBI official Bill Niles returned home this morning from the White House to find his wife and three children murdered. A Bureau spokesman says a note referencing the Koran was found at the scene, and evidence linking the killings to Delta-3."

Christ. A whole family. My God.

"A staff?" Gillette says. "How big?"

"Originally I'd arranged for six." But suddenly she sounds uncomfortable. "Shipment tracker from Defense. Oil expert from Energy. EPA software maven. A whole team."

"What do you mean, 'originally'? What changed?"

Theresa turns to look him full in the face as the van turns right on East West Highway—a wooded, suburban neighborhood—and makes a quick left into the semicircular driveway of a modern-looking apartment building. The driver goes inside.

Theresa says, "Hauser's diverting people to work on the FBI murders. He wants a massive effort there."

"So how many people are on my 'team,' exactly?"

"The shipments tracker. He lives here."

"One person?"

"He's very good, I've heard."

"Are the agents checking out fluids companies being diverted by Hauser too? That would be a mistake."

"All staffs will need to make adjustments."

The driver reappears, trailed by a pale chestnut-haired man the size of a football linebacker. His canvas overnight bag looks like a toy in his massive fist. The shoulders bulge beneath the Irish fisherman's sweater. But when the man gets into the van, Gillette sees red rims around the pale blue eyes. The formidable-looking guy is sick, or he's been crying. The voice is low and strained.

"Hello, sir. I'm Jim Raines."

The van glides back onto East West Highway, heading for the next pickup, an environmental toxicologist seconded to Detrick from the EPA, Theresa said.

Gillette is accustomed to trusting his instincts about people. Jim Raines is screaming problems from every pore.

"What's wrong with you?" Gillette asks.

"Sir?" The tortured look confirms Gillette's guess.

"If you're working with me and you have a problem, I want to know. Now."

The story comes out. Raines's wife is pregnant, due any day now to deliver a son. But how will she reach the hospital? They've just moved to Washington. They have no friends yet in town. Their Mazda broke down. Taxi companies are closed and suburban buses aren't running, and Mrs. Raines's doctor is stuck in Hawaii, where he was on vacation.

"We tried the police," Raines says. "They said call 911 when the baby comes. But most ambulances and police cars are out of commission. No one knows if they'll be replaced."

"Have you found another doctor for your wife?"

"I was going to ask you to bring her to the hospital, sir, but they said they're not taking patients, unless the baby's on the way. They said they're not a hotel."

Just before Gillette showed up, Raines adds, he'd been going door to door in the building, asking if anyone had a car that worked, for when his wife went into labor.

Gillette orders the driver to turn around. "Pick up Mrs. Raines."

Raines says, "But, sir, the hospital people said—"

"—what they always say during outbreaks. Let's find your wife a place to stay. And by the way, Raines, you better be good at your job. You're my whole staff."

Tears appear in the eyes of the giant, and reappear thirty minutes later, when a nurse at the hospital says that Mrs. Raines can sleep in her apartment, across the street, until she goes into labor. "It's the least I can do to fight Delta-3," she says.

"I know I'm a softie when it comes to family," Raines tells Gillette, back in the van, as a former general on the radio suggests zoning cities into "fuel zones," creating triage areas to keep government functioning, even if it means other areas will get no fuel. "But the rest of the time I'm the best damn shipments tracker in DOD. If there's something in the bad guy's records, I'll find it. If they're lying, I'm the next-best thing to an eyewitness. I'm your man, sir. After what you just did for me and Elizabeth, I'm your slave for life."

EIGHT

NOVEMBER 1ST. MORNING. 4 DAYS AFTER OUTBREAK.

Clayton Cox pays for his hot coffee, two-day-old blueberry Danish and morning's skinny, limited-edition *Washington Post* as the lead FBI assault car careens around the corner a half block away, and races toward him.

Excellent, he thinks. Right on time!

He's on Capitol Hill at 7:59, in a candy / newspaper shop across from the old Eastern Market and within easy walking distance to the Senate and House of Representatives. The hole-in-the-wall store lies close enough to the Metro, so its owner has managed to receive supplies. Or perhaps he's selling stock he had already, at double the price, to those among his regular customers going to work today.

PRESIDENT ORDERS GOVERNMENT TO RE-OPEN, says the banner headline on the *Post.* ON SEVERELY REDUCED SCHEDULE.

The oncoming black Ford careens sideways away from Clayton, at the last minute, and skids to a stop before a modest-looking converted storefront across the street. A hand-painted sign over the plate-glass window reads, in black script, "New Ibrahim Mosque."

Hit 'em hard, Clayton thinks.

The FBI, Clayton knows from last night's video info-pickup, arranged by his mentor, will be simultaneously raiding other mosques across the U.S. at this hour, plus university Arabic studies departments, Arabic language newspapers and apartments inhabited by Middle Eastern immigrants. They'll be gathering up anyone on their watch lists.

They're out for blood after the slaughter at the Niles house.

Seconds into the attack, none of the other customers in Mickie's Market even realizes that the raid has started. They're preoccupied reading their papers, which they'd feared might not even reach newsstands today.

More headlines proclaim, MIDWEST STARVATION PREDICTED WITH PIPELINES OUT and WIDESPREAD MARYLAND POWER FAILURES LAST NIGHT TRACED TO INTERRUPTED COAL DELIVERY.

An FBI assault crew pours from the Ford, blue-jacketed, Kevlar-vested, automatic rifles out. Boots slamming.

Behind Clayton, another customer gasps, "Holy shit! They're attacking the mosque!"

Cox sips his stale, oversugared coffee and misses the stronger kind that he used to brew in the desert as a soldier and drink from battered cups as the sun came up. But you make do with what you have. You never enslave yourself to habits, as great-great-grandfather wrote. You don't eat for days when you're hungry. You gorge yourself on food you hate when you're full. You burn yourself and accustom yourself to pain, change, lack of sleep.

And now Clayton spots snipers with binoculars on rooftops across the street, as a whole second attack wave appears on the block, a veritable rolling showroom of American law enforcement. Black FBI Chevys with five-foot antennas. White Capital Police cars with gold logo shields.

They all want to be in on this particular raid.

Which confirms what he had learned on last night's sex and info tape, that the FBI has identified evidence Clayton left at the Niles killings: dirt from the construction project outside the New Ibrahim Mosque; and vegetable-dyed rug fibers that computers matched to this location, using mosque data banks, which the bureau denies having all the time.

Those worshippers have no idea what's hitting them.

Fine with Clayton. He hates all religions equally. He flashes back for a second, feels thick fingers digging into his shoulder, smells tobacco and testosterone and sees two men in balaclavas holding down his mother in a refugee camp tent, lifting her head up by her long hair. The third man—holding the knife—announces in Arabic, "You defiled yourself with the oilman. You offended God."

Little Clayton can't stop them. He's only eight.

But I killed the last of them on my eighteenth birthday, in Amman. He remembered me, all right.

And now, back to the New Ibrahim Mosque where the Wednesday-morning Koran study group had welcomed Clayton when he'd sat in, three weeks ago, as one more blond refugee from Chechenia or Bosnia. A pilgrim. A friend. Newcomers rotate in and out all the time.

The mosque's Imam preaches against terrorists.

He won't stay so moderate after this.

Other people in the coffee shop make excited cell phone calls about the raid, or snap photos to send instantaneously. A few retreat away from the street fearfully. Others push forward for a better view.

"Get the ragheads!" a white guy in a raincoat shouts.

A little shooting would stir the pot, Clayton thinks like a football fan hoping for a score.

And voila! He's gotten his wish. He hears a shot from inside the mosque across the street. Sounds like an M-16.

It's like fishing. You throw in bait.

"Hey mister!" a woman cries. "Get down!"

Cox turns and wonders where the other customers went. Then he sees that they've hit the floor. One little shot and they think they're in a war zone.

But I just made a mistake. They're staring at me, noticing me because I didn't get down.

So Clayton goes innocent and tells them, "That wasn't a real shot. It was a car backfiring. That's all."

"Mister, I was in the army. That was gunfire!"

"Oh my God!"

Clayton drops to the floor too. Now he's one of them.

Cox the "commuter." Cox the "bureaucrat" today. He's ridden the slowed-down Metro to Capitol Hill like any one of several thousand allegedly essential clerks, lawyers, lobbyists, congresspersons and cafeteria workers who spend their days servicing Uncle Sam, and who have been ordered back to work. At least for a while.

Just like my mentor predicted.

Yesterday's bank and fuel holiday is over. Private vehicles—as of 6 A.M.—are back on the roads, until their fuel runs out. And after that any legal problem will pit owners against insurance companies—but at least nobody will blame politicians for stopping driving.

Cox marvels at the way that his mentor—as usual—predicted the thinking of authorities:

"They'll panic at first and stop all gasoline delivery. They'll ease restrictions when they figure out which supplies are still clean. Because you can't order Americans to stop driving. It's like telling them to stop eating. They think God gave them cars."

Cox hears another shot from inside the mosque.

"That was a pistol," cries the clerky "arms expert" on the floor. "Look! They're heading this way!"

Cox sees two men—civilians—running out of the mosque, toward the line of police and FBI.

One man halts suddenly and puts his hands up. The other keeps going, but starts jerking as gunfire erupts. He flies backward, slams into a fire hydrant and bounces off.

No big deal as far as Clayton is concerned. He's seen men shot hundreds of times. But he gasps. "Sweet Jesus." He feels shocked silence around him. The action is over. White-faced customers start getting up.

Outside, a crowd is gathering as neighbors converge for a better view.

The people in the store hurry out, toward the safety of their offices in the Dirksen or Rayburn buildings. Rushing to get through the protective concrete bomb barriers ringing their pathetic bolt-holes in the department of panic, office of excuses, bureau of incompetency, subcommittee on confusion.

Leaving, Cox asks the shop owner if any other newspapers will come in later. The *Times*. The *London Telegraph*. The global dailies that normally fill his shelves.

"Nothing's being delivered," the man says. "We only got the *Post* because we're near the Metro. Hell, even WTOP is off the air. The staff can't get to work."

Time to go pick up another sex and info tape.

"Things will get better," Clayton assures him, "now that Homeland Security is rounding up bad guys. I have confidence in our scientists. America always wins."

Fort Detrick was built thirty miles outside of Washington, in Frederick and the foothills of the Catoctin Mountains, to protect its labs in event of nuclear war. If a bomb hit D.C., planners hoped, the deadly microbes stored at Detrick—cholera, Ebola, anthrax—would not escape.

"I've burned what we found so far into a CD for Hauser," Theresa says. "Sit with me, Commander. Look."

On the way here they'd passed—on I-270—soldiers protecting food or fuel convoys heading toward D.C. The base looks more like a research campus than a military fortress. Civilians mix with soldiers on the grounds. Labs are housed in red brick buildings, their windows looking out on lawns, a few oak trees whose leaves are turning orange, a soccer field, military housing and lots of new construction. Orderly groups of Humvees leave Detrick regularly to pick up new staffers. Equipment convoys arrive from the EPA or TIGR genetics lab near Baltimore. If fuel runs out, all cooperating agencies must be housed in the same protected place, Theresa had explained before power went out last night. The general in charge of the base had refrained from using its diesel-powered electrical generators to keep the labs operating, saving fuel for the trucks.

Gillette and Raines had spent the night by candlelight, sharing a hastily abandoned bachelor officers' quarters. They'd dined on cold pizza from the refrigerator.

Power had returned at eight.

Now he sees that the main lab is large, with bright fluorescent lights above workstations. A dozen scientists and support staff grind DNA, scan electron or field microscopes, tap on computers, consult other labs by phone.

UNIFORMS WILL NOT BE WORN TO LOCAL BARS, a sign says.

Plunking down his egg McMuffin and cold coffee beside Theresa, Gillette recognizes some of the equipment here from the portable biohazard lab she'd used in the Philippines. There's a Geno/Grinder 2000, a small white box that chops up, realigns and reads bits of DNA. A multitube vortexer, shaking racks of test tubes back and forth, mixing the stuff inside. A freeze-drying chamber and microtiter plate incubator, to save DNA. A fluorescent dye machine to mark it.

He also sees a black shoebox-shaped polymerase chain reaction unit, which allows scientists to extract bits of microbial DNA from samples, and reproduce a single sequence, giving the lab millions of copies to analyze, test, decipher, kill.

"This sample," she says as a dark field containing half a dozen familiar rod shapes swims up on-screen, "comes from an infected refinery in New Jersey, isolated and introduced into clean crude. It shows six hours of growth."

Click.

Now a solid mass of frothing rods blankets the screen.

"Look at the difference when one bug is introduced into *refined* gasoline. Enormous growth," she breathes. "One bug multiplies into twenty million every twenty-four hours."

"You're saying that something in crude oil slows the growth. Refined fuel is easier for it to digest."

"Next," Theresa says, black eyes horrified and excited, "Os Preston asked how the microbe survives heat. Watch what happens when we heat infected crude in an autoclave."

On-screen, time-lapse shots show the rods changing, going dark, oblong, impenetrable. Like little capsules.

"Sporulation," he breathes, eyeing the protective wall.

Their shoulders brush. Raines has come up behind them.

"Transformation starts when temperature tops 150 Fahrenheit," she says. "As you know, normally sporulating bacteria need hours to change. But Delta . . ." She taps a key. Gillette sees no more rod shapes, only dark spheres where rods had been.

"Under thirty seconds," she says in awe.

Several microbes with which Gillette has worked sporulate when threatened. Botulism, tetanus and anthrax also go inert until hostile conditions abate.

Then they turn lethal again.

Spores are the hardiest life forms on the planet. They can remain inert for a thousand years, he knows, and be reactivated by a whiff of moisture, saliva on a pig's tongue, a tiny change in temperature. But Gillette frowns. "I don't know of any spore that can survive a thousand degrees. Botulism can survive boiling for an hour and anthrax can live through explosions, but they only last seconds. What the hell is this thing?"

Her hand brushes his by accident, moving the mouse.

Click. Now Gillette finds himself looking at eight small photos on-screen, showing dark curved surfaces. The upper four are clean of bacteria. The lower are awash.

Theresa's red index finger nail taps photos. "The top shows sections of oil pipeline, gasoline tank and jet engines before introduction. The bottom, seven hours later."

Gillette frowns. "So it gets into an oil field somehow, survives refining as a spore and then reforms, attaches itself to metal in tanks and starts chomping away, fast."

"Our EPA people say it's typical for bacteria that live far underground to save energy by attaching themselves to rocks. There's not a lot of food down there. They wait for it to come to them. We're guessing our ODB—oil-destroying bacteria—is part deep dweller, extremophile. But it was definitely made in a lab. The first phase of sequencing is over. A very small part of the genome matches Clostridium tetani—tetanus. Easy to find in developing countries in soil. Safe to work with if you've been vaccinated. But the dominant organism is unknown. How the hell did they make this thing?"

Gillette, startled by the disease/ODB combo, eyes the swarming microorganisms on-screen. The green tinge is back.

"So refining is actually good for this monster."

A sigh. "Whatever inhibits growth in crude oil is gone after refining. Look, there are theories that microbes may have helped *create* oil. That the oldest oil eaters predate the atmosphere. Ten bucks says the dominant organism in our hybrid is very old, and comes from somewhere very deep."

"Ten, huh? You want to risk that much, Colonel?"

She smiles. It's a lovely smile. It annoys him, as does her tight uniform and nice smell and the fact that they seem to have started joking.

He pushes away the feeling. Now Gillette sees a dark swirling mass filled with crystalline shapes on-screen. The picture is similar to the one he first saw at the Pentagon.

"The waste product?"

"That initial idea was wrong." She shakes her head. "The bug doesn't produce much waste. It's efficient. It *changes the fuel.* Normally refining oil lengthens its chain of hydrocarbons, makes it easier to burn. The ODB shortens the chain again. In layman's terms, it takes two steps back, reverses refining and turns the clean stuff back into tar."

"How much do you need to stop a car?"

"Hell, a handful of sugar will do that," she says. "Three hours in the tank and Delta-3 will turn your new BMW into a lawn ornament. So! Ideas? Questions? Intuition?"

They stand by the window. The building is too low to provide a view of Frederick. But Gillette recalls streets empty of cars. And people lined up in parking lots to buy firewood at four times the normal price, wheeling it away in baby carriages, flyers, carts. He's even seen someone riding a horse on I-270.

"Maybe we'll get lucky when we finish sequencing," he says. "Maybe we'll identify the other bug in the mix."

"Or bugs, plural," Theresa says.

"What exactly *is* sequencing?" asks Raines.

The TV in the corner shows a food protest in Moscow. Troops fire over the heads of rioters. Tear gas erupts.

"A sequence," Gillette begins, "is a blueprint of a cell's DNA. All cells are composed of DNA and all DNA comes from the same building blocks, nucleotides. But different organisms code nucleotides in different combinations. Sequencing tells us the exact combo in an organism."

Raines asks, "How can that help us?"

Theresa picks up the thread. "Well, we take what we learn to GenBank, a database of most known sequences. Tetanus has already been sequenced so the computer ID'd that part right away. The computer gives us 99 percent in a day."

"Will sequencing tell us how the bug destroys fuel?"

"We'll know where the genes are, but not necessarily which one helps it attack oil," Theresa concedes.

"Will sequencing tell us how to *kill* the bug?"

"If we learn what else it's related to, hopefully," Theresa says. "Otherwise, no."

Raines keeps at them. "What percentage of microbes on earth are in GenBank, to compare it to?"

Theresa sighs. "Less than one hundredth of one percent. And they tend to be ones that attack humans. But we've sequenced members of most *families,* so probably we'll get clues. Plus, oil companies have been sequencing bugs that corrode their pipes, trying to figure out how to kill them. We have to trust they're sharing what they know."

Gillette doesn't like all the qualifiers. Maybe. Probably. Hopefully. Sharing.

"Two weeks, maximum, should give us a complete picture of the genome," says Theresa.

"And after that?" Gillette asks, "how long could it take to ID the gene causing fuel destruction?"

Theresa frowns. "Well, even in genomes we know, we *don't* know what half the genes actually do. With a basically new genome we'll have to isolate individual genes or combos to figure out which one wrecks oil."

"How long?"

Theresa sighs heavily. "We'll have to do it by hand. It could take years, unless, as you said, we get lucky with an instant match."

On TV, a press conference is in progress with the head of the FBI, who is answering questions about the Niles family murders.

"The perpetrators may be linked to Al Qaida," he says. "The case is top priority. We're closing in on terrorists responsible for this tragedy and the attack on oil."

"I just hope you're still looking at fluids companies," says Gillette softly, to the screen.

NOVEMBER 4TH. 7 DAYS AFTER OUTBREAK.

Halloween decorations are still up in the fort. THE MONSTERS ARE COMING, says a billboard for a horror film, showing clawed hands clutching a throat. Frederick—when Gillette takes walks outside—reminds him of other military towns he's visited. He sees strip malls, fast-food joints and lots of motels; the Dutch Girl, which is open, and the Quality and Holiday Inns, which are not. A Shell gas station is closed. An Amoco is open, and guarded by MPs servicing military vehicles, ambulances, cars with MD plates.

Why do some companies have clean gas, and others infected gas? Do the shipments come from different countries, refineries, or both?

Now Gillette stands looking over Raines's shoulder at his computer screen, in their subbasement office beneath a steam pipe. The walls peel greenish paint. The better offices, Theresa had said apologetically, are going to scientists.

To lab people, who belong here. Not field people like me.

On-screen is a list of top oil-producing countries, showing their reserves and the number of estimated barrels of oil pumped daily in each nation, until recently. Some fields remain clean. But pipelines or refineries servicing them are infected, blocking delivery.

Country	Reserves	Daily Output
Saudi Arabia	261 billion barrels	7.5 million barrels
Iraq	112 billion	5.0 million
Kuwait	94 billion	1.7 million
Abu Dhabi	92 billion	1.6 million
Iran	89 billion	3.5 million

Country	Reserves	Daily Output
Venezuela	72 billion barrels	2.8 million barrels
Russia	48 billion	6.0 million
Libya	29 billion	1.3 million
Mexico	28 billion	2.9 million
Nigeria	22 billion	1.9 million
USA	21 billion	5.9 million

"Try Halliburton again," says Gillette.

A drop of water hits the desk. The pipe above is slick with condensation and the room smells of mold. For the last three days they've been checking company after company, using Energy Department data to compare supply or fluids shipments to oil field outbreaks, and track the microbe back. Gillette demanded original reports from fluids and drilling companies. Raw and processed data. Filings from intel services, showing the date and hour each infection was detected. And the estimated time—based on infection levels—that the Energy Department believes it was introduced.

"We've got a twenty-seven percent correlation between wells that Halliburton serves and initial infections," Raines says, eyeing his screen. "Usually I track arms shipments, money transfers, bank accounts or offshore payments. But the system's the same."

Gillette says, "Try Schlumberger again. And Tangier. *Some* factor will stand out. *Some* common point had to start the outbreaks."

Raines seems tireless and Gillette feels confidence in the man. "Twenty-six percent correlation with Schlumberger. Same as Cougar Energy Services and BP," Raines announces an hour later. "They've all been hit just as hard."

"I'm sick of this fucking basement," Gillette snaps.

On television, Congress is shown passing the new emergency powers act, enabling the president to seize private property and declare martial law if necessary, in an effort to maintain civil order. Homeland Security is drawing up contingency plans to take over distribution of the national food supply, an announcer says.

* * *

Marisa answers on the first ring. She just got home with the kids from a block lunch—bread and margarine—at the Cantoni house.

"Everything is fine," she says too brightly.

"What's wrong?"

She sighs. "Gail's roaring drunk. Bob's still freaked out by the riot. He said it was much worse than you said."

"The cops calmed everything down in the end."

"He said the mob might have killed that manager if you two hadn't pulled him away."

"They just would have shoved him around a little."

"Now who's sugarcoating the truth?"

"You're lovely when you're mad."

She breaks off for a moment and he hears her arguing with Paulo, "I don't care if your friend lives only two blocks away," she says. "You're not leaving this street at night again until police patrols are back to normal."

"Aw, Mom, it's safe. And school's closed."

"No, it's open and in our dining room and afternoon session starts at two. So get your textbooks, buster."

Paulo shouts so Gillette can hear him. "Dad, she's not a mother. She's a slave driver."

"I'm glad you figured that out," Marisa tells Paulo.

Then she's back. "How's Colonel Novak," Marisa asks. "She looks more like a swimsuit model than a soldier."

"I only have eyes for you. What are you wearing? We haven't had phone sex in two months."

She laughs. "Quick thinking." Then she gives more news. Alice Lee is too frightened to sleep alone in her house, so Marisa invited her over.

"Good. She can stay as long as she wants," he says.

Annie insists on going to the zoo tomorrow, to look in on the baby cheetahs. She wants to take the Metro.

"It will relax her," he says. "She loves those cubs."

People on the street are worried about family members elsewhere, she says. College kids can't get home. Elderly parents are stuck in other cities. Marisa's trying to convince her parents in Vermont to move off the mountain, into Manchester, because if food runs out, they're stuck.

"I'll phone them and put the pressure on," he promises.

"By the way, do you know someone named Clayton Cox? He called and said he knew you at college. He's in town, stuck at a hotel."

Gillette thinks back. He went to college with lots of people. Clayton must have been one of them.

"I told him you'll be gone awhile," she says. "He said he'll call back, asked what you're up to. Maybe I'll invite him over if he's alone in town. He told me a funny story about college. Something about you playing sock baseball in dorms."

"Get his number if he calls again. But don't invite him over," he says. "I don't remember Clayton Cox."

At 5 P.M., he hears reveille outside and knows vehicles and staffers are halting out there, standing at attention, facing the lowering U.S. flag as reverently as Muslims turning toward Mecca. The trumpet music playing over loudspeakers heightens the sense of history at stake, of being under attack.

From an enemy we can't even see without a microscope.

He tells Marisa when she calls, "I'm useless here. I can't help with the science, and Raines and I are duplicating research being done at a hundred other places. I need to think of a different way to go at this thing."

"You will. You always do."

He phones Theresa to ask if he can go home for an evening, help out, talk to his kids, calm them.

"There's no fuel for personal rides. Let's go to dinner. At least Uncle Sam provides three meals a day at the fort."

NOVEMBER 7TH. 10 DAYS AFTER OUTBREAK.

"We've identified a second bug in the mix," Theresa says triumphantly, over the phone. "Leave your dungeon, good knight. Come up and take a look."

On his way to the lab, he passes offices in which radios or TVs broadcast news of widening power failures in New England. A freak subzero snowstorm has hit the Rockies. Thousands are trapped without heating fuel, or ways to leave their homes.

In the lab, Theresa stands with scientists studying the sequence charts, two printouts, each showing a single large circle divided into pie wedges. The first circle represents Delta-3. The second, the newly identified bug. All pie wedges on both charts are marked with the letters ACBG—representing ribosomes—arranged in different combinations. The computer has compared the combos and in several shaded areas, pie wedges match up.

"It's a hydrocarbon eater, from Uzbekistan," she says. "They found it in an oil well seven years ago, but didn't tell us about it until today. They were afraid that they'd be blamed for Delta-3. Finally reason prevailed. Bug two is still not the dominant piece. But it's another part."

"Who owns the oil well?" Gillette asks.

"Crescent Oil."

"Did they own it seven years ago?"

"I don't know."

Gillette calls Raines, names the well and orders him to find out who operated and supplied fluids or bactericides to it seven years ago and up until now.

"Uzbekistan, boss? Well, let's hope we get the promised instant cooperation."

One of the scientists behind Gillette asks, "Uzbekistan is an Islamic country, is it not?"

"Don't jump to conclusions," Gillette says.

Ten minutes later Raines calls back. "Exxon owned it. Cougar Energy operated it and supplied bactericides. Now the fluids are locally supplied. Mostly just water."

"Cougar," Gillette says flatly. But Raines says what they both already know unfortunately to be true.

"That bug could have been transported to a thousand places the last seven years. You can link it to Cougar. To Exxon. To Crescent, or the

Uzbeki oil ministry. To a thousand scientists and workers. We know where it started. But not where it went."

Gillette feels the battle raging between two end products of evolution. The smartest humans, and the toughest microbes on earth. Or maybe the smartest humans are working *with* the tough microbes, mixing one from a deep well in Asia and one causing tetanus.

"The big mystery—the major part of the genome—is still missing," Theresa says, sighing.

"Send me into the field," he says.

"Hauser ordered me to keep you here."

"He did that to keep me out of the way."

"I know. Make the most of it. Hungry? Let's go eat."

They're standing alone out in the hallway. The forecast is for freezing temperatures, the radio said. Gillette wonders how much heating oil is left on Marion Street.

"You didn't have to ask for me to be assigned here," he says.

"What's that supposed to mean?"

"Why did you come to my house?" he demands.

At first he thinks the flush suffusing her face is embarrassment. Then he realizes it is fury.

"Why is it that men," she says, starting off slowly, voice dropping into the arctic zone, "always assume they're why women do anything? You know, Greg, my mother is visiting her sister in Montana. There's no heat in the house. It's snowing. Last time I phoned, state troopers were going by on foot with loudspeakers, urging everyone to keep calm."

"I didn't ask about your mother."

Her fury is rising. "My brother is a diplomat, stuck in Algeria. I don't know how he will get home. And me? I have a horse at home and need to feed her. I'm depending on a neighbor to check my house. But I really *didn't* come to D.C. to eyeball the problem, to get into the field like you always say. I only had thoughts for you."

"Sorry," Gillette says, ashamed.

"Believe it or not," Theresa says, "I don't want you here either. You're wasted here. You're right."

"Then do us both a favor and transfer me."

"Hauser just sent an Air Force major to Leavenworth prison, for thirty minutes of personal use of a staff car. Hauser is not a man you want to antagonize."

She stomps off, leaving Gillette feeling like a fool.

NINE

NOVEMBER 8TH. 11 DAYS AFTER OUTBREAK.

The woman is beautiful and sexy but not to Clayton's taste. Naked, she's too soft. Too white and round. All hip and breasts. But other men find her irresistible.

That much is evident from the tape.

"I was afraid you wouldn't come," she whispers to the man in the king-sized luxury bed beside her. "That they would keep you at work."

Her penthouse apartment is filled with Chinese vases, lacquered tables, Oriental silk screens, Japanese prints. The skimpy black dress crumpled on the carpet is by Armani. The black diamond necklace is custom-made by Katherine Wallach of New York. Her tongue runs up the chest of the fit-looking dark-haired man propped up by an elbow.

If you like girls, she's really hot.

"You look tired," the woman says, her voice dripping with compassion, caring, hunger for sex. She gazes into the hidden camera / recorder. It's top-of-the-line too.

Clayton's mentor never spares expense or precaution. Someday, Clayton knows, he will kill this woman like he killed the man who recruited her to make these tapes.

"My wife's spending the night in Reston, with her mom," the man says, as if his presence is a gift. He traces a line with his finger down the woman's flat belly, into the combed out black V of pubic hair.

"So I can stay longer than usual," he says.

I'd rather you talked more, thinks Clayton, disgusted at the unclean bodily contact, yet aroused, and feeling himself—eyeing the man—becoming a bit bestial himself, sitting in the living room of his rented house.

It's been a while, he thinks, unzipping his fly.

"You will stay with me all night!" the woman says. She claps her hands in feigned delight.

"You know I can't."

A pout. "You never stay."

Great actress, Clayton thinks, stroking himself, knowing from previous tapes that the second the man leaves, she'll scrub herself under a steaming shower. She'll clean the sheets in the wash. She'll brush her teeth and throw away the toothbrush. She'll rid herself of his sweat, dead skin, jism, touch.

"I explained it, honey," the man says. "If I'm not home when Ann calls, she gets suspicious. She calls on the house line for a reason. She thinks I keep a lover in D.C."

"Suspicious? Of you?" The woman laughs and cups his engorged penis. "But you are such a *loyal* husband . . ."

She pushes him down, onto his back.

"A *caring* husband . . ."

She licks him, takes him in her mouth. She stretches one leg over his hip and, groaning with phony pleasure, lowers her jiggly rump onto his muscled thighs.

"A husband denied his rights," she says, rocking back and forth as the man's panting grows audible. The man's face twists in delight.

It's funny how joy and pain look the same sometimes, Clayton thinks.

Well, Great-great-grandfather had written of sex, *"Our men were young and sturdy, and hot flesh unconsciously tormented their bellies with strange longings."*

Except Great-great-grandfather didn't really consider these longings strange, Clayton thinks. And neither does Clayton.

After all, you find yourself in a new land, surrounded by strangers. You need release, if only for seconds. The public women are, as Great-great-grandfather said, "raddled and unpalatable to a man of healthy parts."

Clayton allows himself to leave his body and embrace diversion for a few spasmic moments, watching the man.

Ah.

Afterward, the couple seems dirtier and only of professional interest. The woman is playing offended.

The man says, "You know I can't talk about work."

"Tell me something to make me feel less afraid."

And the man capitulates. A trained soldier, and a few licks in the right place and his brains become mush. Clayton picks up the pack of Chuckles, pops in a lime bit.

The man saying, "You can't tell anyone this . . ."

"I would never do that."

"Probably, what I'm about to say won't happen. I'm sure our scientists will kill Delta-3 first . . ."

"You may hear rumors," General Hauser tells his lover, "over the next weeks, about a plan to zone Washington."

The woman's eyes widen. "But you said the problem will be over soon."

"I said probably. But if it isn't, your apartment is south of the Calvert Street Bridge, so you'll be protected, no matter what happens outside your zone."

"What do you mean, zone? Protected?"

Clayton wipes himself clean with a tissue.

"Just don't move away from this street."

Clayton is delighted. *If they're going to zone the cities, they can't move their clean fuel. The situation is much worse than they're telling the public.*

Unfortunately the woman can't coax more details from the general right away. Timing is everything. Clayton has to sit through another round of sucking and kissing. He wants to puke.

After a while, General Hauser lights a cigar and blue smoke billows up. The recording equipment is so good it picks up the click of the lighter, the compressed bedsprings, the whispered breathing and exhalation of smoke.

The woman says, "If you won't tell me something important, tell me something unimportant. It makes me feel like part of your life. You never tell me things."

"Don't you enjoy what we do?"

"Tell me about that man again. The one you kicked off your committee, who bothers you. *He's* not important."

"Who? Gillette? He's still at Detrick. He actually thinks an American company started this thing. Can you believe it? Terrorists murder FBI families and Gillette wants to waste time sending agents to Nevada and Texas. Couple of Energy guys suggested the same thing. I stopped it," he brags.

Clayton shoots to his feet. He knows the home addresses of all members of the Rapid Response Committee. He's been checking out their home security systems in case he has to visit them. But he'd cut down spying on Gillette since the man's transfer.

"What's in Nevada and Texas?" the woman asks, dutifully.

"Nothing. That's my point."

Clayton hears himself breathing. Hauser is a pompous fool, but Gillette sounds diligent and dangerous. Will he stay away from Nevada?

After the man leaves, the woman talks directly into the camera, asking whoever picks up her tapes to assign her a different drop-off point. Not Rock Creek Park. Police patrols have been cut there. The park is deserted. Visitors to the park have been attacked. Can she be assigned a new drop?

Clayton destroys the tape, as always.

I need to learn more about Gillette and his family, just in case, he thinks.

NOVEMBER 10TH. 13 DAYS AFTER OUTBREAK.

"Sorry, all phone lines are out temporarily."

Gillette dials the number in Texas again.

"This is a recording. Due to the energy shortage . . ."

"Shit!"

He hangs up. He's been trying to call Halliburton for hours, but the problem is the same all over the west. Phone service goes in and out every day.

From the next computer over, Raines says excitedly, "You had a genius idea, boss."

On their screens are the corporate Web sites of Tangier and Cougar Energy Services. Gillette's idea—as about 12 million other investigators are tracking oil shipments too—was to hit the problem sideways. Yesterday he'd told Raines, "We start from zero. We take the top fluids and bactericide suppliers and just learn things about them. When they were founded. Who's on the board. PR releases. Hard news."

"Public relations releases?" Raines had asked, puzzled.

"Well, when a company wants to brag about something, they issue a release, right? Maybe the news isn't in the official records yet. Maybe it's just been announced. Why should we assume that Energy Department stats are the only relevant criteria? Let's fish, get the company spin on things."

"You mean, like, learn about new policies?"

"Deals. Plans. I know this stuff is easy to find. But easy-to-find information can be as valuable as hard."

Now, on his screen, he's looking at news of a labor strike that closed Tangier's fluids division this past August, in France.

And on Raines's screen is a glowing announcement on the Cougar site.

Due to the strike, Cougar supplied lots of Tangier contracts that month. "Plus," Raines says, "that month they sent out free samples of a new bactericide. So, doing the math"—he punches in numbers and gets more excited as results come up—"in August, we get a 42 percent match between fields that later became infected and Cougar products, and a 67 percent matchup with pipelines. Even Hauser can't ignore that, boss! It's *huge!*"

Ten minutes later Gillette is upstairs in Theresa's waiting room. Since their argument the daily briefings from her have stopped. It's harder to reach her. She treats him like a pariah. *But she better hear me out now,* he thinks.

On her anteroom TV, as he waits, mobs sack food stores in Tampa, Omaha, Dallas. RIOTS SPREAD TO TWENTY CITIES. FEDERAL GOVERNMENT TO TAKE OVER FOOD DELIVERY, reads the caption on-screen as Theresa's sullen secretary waves him in.

She listens to his theory, expressionless. Then she says she'll "pass it along" to Hauser.

"Colonel, someone needs to screen those factories for Delta-3 residue. It's basic. Logical. At least find out if the FBI checked. Or send me to Hauser to plead my own case."

"Are you deaf, Commander? You stay here."

When he gets downstairs, furious, he finds Raines jubilant.

"It's a girl! A girl!" Raines laughs, pumping Gillette's hand. "We'll call her Emily, after my mom. Six pounds seven ounces! And the nurse says Lizzie can stay with her awhile. It's safer. *You* did it, sir. *You* helped us. Because of *you* they'll have food and . . ."

He shuts up, realizing that Gillette is thinking about his own family. The big man's face fills with concern.

"You need to get into the field, boss. I can do the computer work here. We can coordinate by phone."

Gillette calls home and hears Annie sobbing in the background. A friend was raped in Rock Creek Park by two men who got away. Paulo says that a friend's dad stationed in Germany called and said that American troops there were rioting, demanding to be sent home. Marisa says that on the way to church she saw Gordon Dubbs and other men moving boxes of supplies into the Oasis. "One of them said something about rifles. Those men looked mean. I didn't recognize them, Greg. I'm scared."

NOVEMBER 13TH. 16 DAYS AFTER OUTBREAK.

"The sequencing is over. Final results are in."

Gillette stands in the lab, amid a circle of despondent scientists. Through the window he can see the deserted roads in the fort. Fuel is running low, even here.

"We have no idea what the dominant organism is. We've hit the wall," the chief researcher admits. "Like everyone else."

The scientists go silent, envisioning their families, and homes. Dread replaces hope.

"We'll go gene to gene, by hand, and hope we find the bastard responsible," the researcher says, rousing himself. "We won't quit."

Gillette goes upstairs and barges into Theresa's office, but she's out, the secretary says. The scene on the waiting room TV stuns him. It's Atlanta, where he grew up. Lullwater Road, near Emory University. A leafy, suburban street, a park on one side, expensive homes on the other. But one home is burning. Food robberies are rampant in Atlanta, the announcer says. Gillette casts back, sees himself in the backseat of a Chevy Impala, twenty-three years ago, sitting between gang buddies as they eyed the home on-screen now, a brick colonial with pillars in front.

"We'll be in and out in five minutes," one kid had said.

"I don't want to rob houses anymore," Gillette had said.

"Chicken. Faggot. The family's on vacation."

"I want to be a doctor."

They'd kicked him out of the car. They'd broken in while he'd walked off, and he later learned that his friends had surprised the home owner, and knifed him. They'd gone to prison, convicted of murder as adults.

But that night, Gillette had shown up at Dr. Larch's house on foot, and the man had asked no questions, given him a glass of orange juice, talked to him about the need for doctors, stood watching the boy walk off.

Now Theresa enters the office and halts. Gillette snaps back to the present.

"What are you doing here?" she demands.

"The sequencing failed."

"I know. And I ought to tell you also that the FBI cleared all fluids companies," she says stiffly, uncomfortable with the news. "They found no connection to the bug. Hauser's shut down that end of the investigation."

"After so little time?"

She gazes out the window, arms folded.

"Someone has to convince Hauser to keep looking. Are you going to let our fight stand in the way?"

"It wasn't a fight. It wasn't worth that."

"I know I was an asshole. But I'm right about this."

She sighs. Her hair is down, her posture magnificent. She seems to be trying to get her anger under control.

"All right," she says. "Take a leave. See your family. Keep your ID but don't abuse it, understand?"

What she really means is, *You can try to get into Hauser's office, but I didn't send you. Understand?*

When he gets downstairs, Raines promises to keep in contact daily, and follow Gillette's instructions while he's gone.

Gillette says, "I'll go door to door, try to convince people to reopen the fluids investigation. And I'll drop off some food for your wife on the way in, from the cafeteria."

"Feed Elizabeth and Emily and I'm your slave, boss."

That night, a van returns Gillette to Marion Street.

At least I can help my family while I'm here, he thinks, grateful to be home as Marisa opens the front door.

TEN

NOVEMBER 15TH. 18 DAYS AFTER OUTBREAK.

Gillette's up at five and out of bed before the alarm rings. He must hurry if he wants to secure a good place on line for food distribution, scheduled to start at eight, at the elementary school.

Because we're out of food if this doesn't work.

"The army held off yesterday, because of bad weather," Marisa says in the dark, over the steady assault of rain on the roof. She sounds exhausted. She tossed all night again. Water streams down windows and the house creaks in anticipation of winter.

"The rain's worse today, Greg."

Temperatures dropped below forty last night, he sees, eyeing the thermometer outside the pane.

Below, on the street, trudges a solitary protective figure in a rain slicker, a rubber hood over Bob Cantoni's head. The former Marine's on patrol again, moving between the Cantonis' Honda blocking one end of Marion, and the Gillettes' Subaru at the other end.

We blocked the street to keep strangers away. With police patrols curtailed, we'll watch out for ourselves.

Gillette's patrol shift had ended at midnight.

"Any more people show up?" Marisa asks from bed, not wanting to use the word "refugees."

He flashes to a man, woman and two small children wandering around in the rain last night. He'd heard the couple arguing. The woman snapping, "I *told* you we should have rationed!" The man hissing, "Rationed *what*? You bought takeout every night. We never had goddamn food in the house."

Nothing special about the fight or family normally. It had been the kind of argument many couples might have, the kind of family he'd see at a mall.

He tells Marisa, "They were from Reston. They'd heard there were beds at RFK Stadium, took the Metro in, didn't know that it's full. The guy offered me a thousand bucks to let them stay on Marion Street. I sent them to St. Paul's."

"You gave them your dinner, didn't you, Greg."

"I ate that peanut butter sandwich myself," he lies.

Every night more people come in from the suburbs. They're dead zones. Nothing moves in suburbs without a car.

He dresses in darkness to save energy, in corduroy trousers, waterproof hiking boots, a thick Irish sweater. The thermostat is set at fifty-eight degrees to save heating oil. Marisa wraps a terry cloth robe over her red flannel pj's. Her feet are encased in thick woolen ski socks. Everyone is losing weight.

No morning showers, to save fuel.

She says, "I'm sorry Hauser refused to see you again."

At least Gillette is back with his family. Grabbing cash for the food distribution, he calculates numbers and needs. They've got half a tank of heating oil left from last year, plus thirty gallons from Alice Lee's house. Alice has moved in.

Everyone except Gail doubled up to save oil.

"Greg? Remember to bring your driver's license. Make sure the neighbors filled out their proxies right."

He kisses her on the mouth. She knows he's fully prepared, but going over things is her way of handling tension.

Because anyone who can't prove they live within a mile from a distribution center won't receive supplies today.

He still smells sex on himself from last night. It's the only good part of not showering. From downstairs comes the soft hum of the living room TV, and when he gets there, Alice Lee is sitting under a quilt on the pile carpet, thin arms wrapping her legs as she stares at the news.

"You used to hate watching the news, Alice."

"Now I can't stop. I just hate the news."

On TV, soldiers unload A&P trucks outside a Los Angeles high school. An announcer says, "An experimental food distribution program expands today to Washington, Tampa and New Orleans. The program is designed to protect dwindling shipments, guarantee supply and save fuel."

Paulo snores lightly on the sofa, his new sleeping place since Alice moved in. Clutched to his chest is one of Gillette's books, *Germs That Changed the World.* Gillette feels a stab of failure. He knows his return from Detrick disappointed the boy. Paulo's lost faith in Gillette's power to solve problems, and that has lessened his confidence in problems being solved at all.

"CBS will be going off the air now," says the announcer. "Switch to ABC for continued coverage. CNN will take over at noon. We'll be back at six as part of the rotating news-coverage agreement, and in response to the president's plea for cooperative corporate energy saving. Stay safe today, especially viewers in the storm-ravaged Rockies."

The pantry hasn't looked this empty since before they bought the house. Nine foil-wrapped PowerBars remain on shelves, along with a box of ziti and half a dozen cans of string beans, thanks to Alice.

Alice is almost out of her heart medicine.

"Daddy?" Annie stands in the dining room doorway, wrapped in a quilt. "Will you come with me to the zoo today?"

"Sure, baby. After school."

"It's not even real school. It's like some old farmhouse school, with all the kids in the same room at church. Mom's teaching long division to third-graders."

"You calm the little kids, honey. They respect you."

Annie snuggles on his lap as he melts his vanilla PowerBar breakfast in boiled water. If Paulo's grown more distant and Marisa detail oriented, Annie's become more childish and needy since the unsolved rape of her friend.

"One of the baby cheetahs got sick yesterday, Daddy."

Marisa puts the last of the Quaker Oats honey raisin cereal on the table before Annie, mixed with tap water.

"Eat or you'll get sick too."

Annie's eyes tear up. "What if the cub dies?"

She never worries out loud about the oil, food, friend's rape. Just the baby cheetahs.

"We'll help 'em out today," Gillette says.

Is it only sixteen days since the plane fell on Ingomar Place? On the base, he recalls, you could walk around at night. You ate three meals a day. There were candy machines, and medicine in the dispensary, and Hollywood films to watch.

I guess that's why guys in third world countries join the army. To stay alive.

The bell rings, the front door slams. Bob Cantoni and Les Higuera tromp into the kitchen, dripping rain, arguing.

Les telling Bob, "Government control of food is a recipe for takeover!"

He's grown more nervous since he was laid off at ABC.

"You have a better idea? You want more riots?" snaps Bob, who loses his temper more easily now.

In their foul-weather slickers, Gillette's friends resemble Massachusetts lobstermen, not D.C. execs.

Marisa says, too brightly, "Have a PowerBar, guys."

"I ate already," Les lies as his stomach growls.

Marisa gives them each a bar anyway.

"By tonight we'll have real food. You know, we ought to plan a Thanksgiving dinner for the whole block this year!" she says.

Gillette makes his daily morning call to Raines, at Fort Detrick. His lone staff member sounds depressed.

"The last try to disinfect the pipelines failed," Raines says.

Gillette and his friends roll two-wheeled shopping carts up Nebraska Avenue. Rain overflows gutters. People pour from houses and side streets, bringing empty knapsacks, bags, pull-carts. During a Philadelphia distribution two days ago, so many people packed lines—tens of thousands—that Washington authorities decided to allow designated neighbors to pick up food for friends.

Les says, "I must have produced a dozen shows on oil dependence over the years. But I never really understood it. Every damn system we've built in this country since World War One has required oil."

Gillette hears sirens in the distance, and spots a black spiral of smoke smearing the sky.

"Another portable heating unit went up, I bet," Les says. "Every night. They oughtta be illegal."

Bob snaps, "Then how would old people stay warm?"

As they near Connecticut Avenue, they hear a man's voice repeating over a bullhorn. "Keep the line tight, folks. Stay calm. We have food for everyone, folks."

But when they reach Connecticut Avenue, Gillette sees that the line is already four deep, snaking around corners. People seem too quiet. Gillette hears a clip-clop sound and sees a mounted cop as they get on line.

Bob nudges Gillette. "Want to take a trip with me later?"

"Where?"

Bob whispers, "I knew a guy, in the Marines—"

"We voted *not* to buy guns."

"You voted with me. You know we need to be more prepared. You and I are the ones who've been in trouble spots overseas. Come on. I'll buy."

The conversation is interrupted as someone knocks into Gillette from behind. It's not a hard blow. But it's enough to move him half a step forward.

"Sorry," a familiar voice drawls. "I tripped."

Gordon Dubbs stands there, smiling and accompanied by two men Gillette does not recognize, who convey the same look of half-shaven toughness as they appraise Gillette. He flashes back to an alley in Atlanta. He sees himself giving the look to a bunch of gang boys from Chamblee. He knows that one day he'll fight the boys. He thinks now, not again.

"This is my neighbor, Joe Gillette," Dubbs tells the guys, intentionally getting the name wrong. "Joe, these guys just moved into my building. They're your new neighbors. Say hi, neighbor."

Distribution actually begins on time. The line inches forward to reach the school's parking lot. Gillette sees three Safeway trucks parked by the gym entrance. Soldiers carry boxes into the school, watched by stern-looking guards.

"What if they run out of food?" frets Les.

People exiting the school after being served hurry off with half-filled bags, light rations. They avoid making eye contact with people on line. The mounted cops move back and forth, wary, attentive, silent, wet.

Gillette hears Gordon Dubbs tell his friends, "Don't worry about the proxies. People in our building do what I say. Now, what did you think about 5105 Connecticut?"

"Connecticut's too open," comes a reply.

"Irvington Street?"

"Half the houses were deserted."

"What about that Tudor, behind the church?"

Looting, thinks Gillette.

The gymnasium is large, the food piled by the stands where Gillette used to sit watching Paulo play basketball on Friday nights. Clerks—soldiers—man computers at long tables. Hand-scrawled signs on walls proclaim, "Have ID ready. This food is for sale. It is not a donation. Pay promptly, in cash. Prices are fixed."

Gillette hands a bored-looking female clerk his driver's license and electric bill, confirming his address, along with forms Marisa printed off the Internet, signed by neighbors, authorizing him to purchase food for eight households on Marion Street.

The clerk studies his ID and types something into her computer as Les—always the journalist—explains to a woman on line, "She's comparing census data and real estate rolls to see how much food goes to each house."

Suddenly the clerk stops typing.

"Your block already got food," she accuses.

Gillette feels his pulse quicken. He tries to see the clerk's screen but it's impossible from his side of the table. "That's a mistake," he says, keeping his tone calm, friendly, logical.

"It says your food was allocated," she says, louder.

"To hell with what it says," Bob flares.

"You're holding up the line," says Gordon Dubbs from behind. "Back for seconds? Shame on you!"

Gillette squeezes Bob's shoulder to shut him up. He tries to convince the woman that the computer is wrong. Perhaps a different typist recorded incorrect information earlier, by accident, he says, feeling his stomach cramp.

"All I know is what I read. Come back next week. Sergeant! I have a repeater over here!"

It's Les who loses it, not Bob Cantoni. White-faced, Les grabs at the computer, tries to swing it around so they can see the screen. Gillette pulls him away. Once things get physical, they'll have no chance here, he knows.

"It isn't right!" Les wails.

"This food is for our *kids,*" Bob snaps as an older soldier—gray-haired and tough-looking—saunters over, clearly in no mood for problems. He's probably been dealing all morning with people lying to get extra food, begging, cajoling, misrepresenting themselves as locals.

Please, sir. There's a mix-up. A man stole my ID. I just moved here and don't have it yet. I lost it.

Hands on hips, the sergeant looks between the clerk and Gillette as Gillette tries to explain. The sergeant waves more soldiers over. "Sir, please leave. Call the 800 number on the wall, if you have a complaint."

Gillette makes a snap decision, reaches for his wallet and hands over the ID from Fort Detrick, which identifies him as a vital part of the oil relief effort. Beneath his laminated photo, the card—signed by the secretary of defense—orders all military personnel to fully cooperate with "the bearer of this ID."

As the sergeant reads, he looks angry and disgusted.

"This doesn't say to give you extra food."

"*Other* people's food," the clerk snaps.

If Hauser hears about this, I'll be arrested. But I'm not leaving without food for my family.

Gillette feels sweat on his back. "Make a call. Check the ID. Go ahead. Waste everyone's time."

Someone in back cries out, "What's holding things up?"

Someone else says, "They're running out of food!"

An electric current seems to shoot down the line. The sergeant instantly shoves the ID back and orders the clerk to give Gillette a second ration, and record that Marion Street received *two* deliveries. Both picked up by Gillette.

"You make me sick," the sergeant says, and waves Gillette toward more clerks who hand him eight blue coupons, one for each household on Marion Street, and point toward the back of the gym. There, other clerks give receipts for $300 cash received for each coupon. They count out cans of Dinty Moore stew, Bumble Bee tuna, bread, milk, butter, frozen peas. Also ramen noodles, Ball Park Franks, raisin bran, dill pickles, three eggs per recipient, Heinz baked beans and eight jars of maraschino cherries.

Bob stares at the cherries. "This isn't food."

"Everyone gets fruit," an exhausted clerk says. "Cherries are fruit. You don't have to take it."

"You call these fucking cancer balls fruit?"

"I don't have time to play favorites."

The three men keep the cherries.

"Some system. It won't last," Les says as they leave.

The old rusty Pontiac comes out of nowhere, gliding to the curb in a swish of water along Nebraska Avenue, its tinted windows humming, going down. Gillette hears vintage jazz—Nat Cole—from the speakers. Two men in front smile out at him. One's white. One's Chinese. The car's fuel must be clean, Gillette realizes. Are these two cops?

Are we about to be arrested because of what I did?

"Gentlemen," beams the driver, the friendliness meaning they're not cops. Both guys are in their early twenties, sharply dressed and wearing sunglasses on a rainy day. Approaching the window, Gillette notices thick sandwiches wrapped in cellophane lying on the backseat.

The engine keeps running. So the men are not worried about wasting gas.

The driver says, "It's wet out, fellas. Care to step in and talk? You hungry? Want a sandwich?"

And Gillette understands in a flash who they are.

He casts back to other scenes like this, except in Vietnam, during the SARS outbreak, the men who'd approached him had driven up in a Toyota four-by-four. In Rio Branco, Brazil, they'd hailed him from a Land Rover. Want to sell medicine? Bed space? Supplies?

And sure enough, the driver starts his spiel when he realizes Gillette is not getting in. "Man, that sergeant was a real asshole."

"I didn't see you in the gym," Gillette says.

"Those fucking clerks, man, act like kings," the guy says.

Gillette's at the window, staying close. His fist wraps a can of stew, below the driver's line of sight. He sees no weapons in the car, but they'll be there. Bob Cantoni's got a can in hand too.

The driver says, "Yep, that ID must really be special, to get you extra food."

"We didn't take extra."

"But they *thought* you did and gave it to you anyway."

"They were mistaken," says Gillette, sensing unsureness beneath the men's smiles. These two are trying to figure out how important he is. That's his temporary margin of safety.

"Gotta get home," he says. "Come on, guys." He straightens and heads off into the rain, going the wrong way on purpose while the men look on.

Les starts to say something about the direction. Bob stops him. They all walk off.

But the car glides forward and halts and the driver gets out, holding up both hands to make Gillette stop. He's still smiling. "Hey, everyone's on a hair trigger. I didn't mean to offend you. I was trying to explain that I have a friend who might like an ID like that, gets extra food."

"No," Gillette says.

"I don't mean give it to us. Sell it. Or trade. My friend has things. Steaks. Guns. Name it. We can copy an ID, change a name. Hell, you keep the original, if it's the right ID, that is. Can I see it?"

"It's the wrong ID," Gillette says.

"Don't give me that," the man snaps. "You know you weren't supposed to use it for food."

Gillette steps close to the guy. Sees the pores in his face. It's like he's twenty-four years back, in an alley in Atlanta, blood pounding in his head. He says, "You don't have a fucking clue what I'm supposed to do. Show me *your* ID. I'm off duty but I'll make an exception for you."

They eye each other. The rain falls hard.

If he moves fast, I'll hit him.

But the man throws up his hands, smiles and gives Gillette a white card with a scrawled phone number on it.

"Call if you change your mind."

Gillette motions his friends to wait right here until the car rounds the corner, and disappears.

"Well done," says Bob in admiration.

"Great story for the network," Les beams. "The black market! Starting already!"

Then he seems to remember he's been laid off.

"Everything's falling apart," he says.

Gillette watches the corner where the car disappeared. "They're new at this," he says. "Or they would have followed us home before approaching." A few weeks from now the vehicle will be better. The food too. The guys will wear better clothes. A bluff won't work.

Everything in transition.

Or did they already get our addresses from the clerks?

His left hand is shaking. He nods at Bob Cantoni, feeling air come back into the world, become normal.

He means, *I'll go with you to buy the guns.*

The rain lets up. Gillette and Bob ride mountain bikes along Grant Road in Rock Creek Park, near the stables. The air is cold, the park deserted. Gillette sees oaks cut down along paths, or just stumps where oaks had been.

"Wood poachers," says Bob. "For fuel."

Bob is carrying his Sig Sauer P226 under his jacket. Gillette is unarmed.

Gillette says, "Why did your friend want to meet here? Aren't people being attacked in the park?"

"Oh, that's Leon. He wants a fight. In the corps, he used to watch *Death Wish* nine times a month."

They reach a baseball field and a man pedals toward them on a mountain bike, from out of the bare trees along the right-field line. The guy looks young from a distance, but older, in his thirties, as he gets close. He's wearing a Redskins jacket and black stocking hat, and a knapsack on his back. His hard-looking face lights up with pleasure when he sees Bob.

"No problems the whole way," he says, disappointed.

"Couldn't we have done this at your house," Bob asks.

"Nah. My neighbor thinks I'm selling guns."

"Aren't you?" says Gillette.

"If I were selling, I'd charge twenty thousand dollars for these. I'm giving each to you for a fourth of that."

"Five thousand dollars?" gasps Gillette.

"That's half of what an apple will cost in a few weeks, bro." Grinning, Leon opens the rucksack.

Bob Cantoni and Gillette pay for a Ruger 9mm, a Browning .38 and boxes of ammo. Then Bob and Leon do the semper fi brotherhood hand slap routine, and Leon rides off toward downtown, the lone ranger, hoping to be ambushed, to find someone to fight.

Gillette says, "How will we even practice? The cops will still come if neighbors report shots."

"That's the least of our problems and you know it. You and me. We're like guys in old times. Food, shelter and weapons. *That's* our job. Making sure people in our cave survive."

"You sound happy about it. Like Leon."

"No. But I won't ignore the hard parts, like other people on the block. Judge Holmes thinks courts solve problems. Alice trusts the president to improve things. But you and I know how fast things fall apart."

"I'm not ready to start shooting, Bob."

"But you understand. We're all going to have to make choices about what's important. There's no place to go and no way to run. The distribution program won't last. There are too many people, too little supply. So you better start thinking about using that ID of yours to care for our own."

"You mean *really* steal food?" Gillette flares.

"Did you see the way Dubbs was looking at us, and hear what those guys were talking about? Breaking into abandoned houses. We should start thinking about that. And, Greg, you do know that we may have to fight them. Right?"

Gillette sighs. It's like he's back in juvenile detention, and the Aryan gang is gathering, and Gillette and his friends are stealing tools from the woodshop to protect themselves. He picks up a hammer in memory. He tells Bob, "We need to seal off the alley and backyards. I wonder if we could find security cameras, you know, set 'em up somehow to better watch the street at night."

"Thattaboy!" says Bob.

Gillette breaks out laughing, does a phony commercial. "He's the modern dad. He fends off mobsters. He buys guns. He plans battles with neighbors but still has time to take his daughter to the zoo."

"Just be back before dark," says Bob. "Or we'll come looking for you. You shouldn't let her go there anymore. It's not safe. The zoo?" he says, sweeping his arms wide to encompass the city, park, country.

Gillette eyes bare trees lining the walkway. They resemble bars of a cage. He finishes his friend's thought. "We're the animals. *This* is the zoo."

That night on TV he watches a panel of oil company executives meet with the president and agree that Islamic terrorists are probably behind the attack on the oil system. There's the chairmen of Exxon and Halliburton. Arco. Texaco. Cougar Energy Services.

"I wouldn't be surprised if Tehran ordered the terrorism," one exec says.

Gillette calls Raines.

"Widen the search. Check out the fluids companies in terms of legal action," he tells him. "Divorces. Civil suits. Land purchases. Cast a net. Look at individual employees. It's not that I discount terrorists. It's that no one seems to believe it might be anyone else."

"Get to Hauser, boss. We need a real staff."

ELEVEN

NOVEMBER 24TH. MORNING. 27 DAYS AFTER OUTBREAK.

The Energy Department is a madhouse. Half the staff is laid off, and the building barred to anyone lacking "special access." But hundreds of civilians, it seems, have managed to get in. They carry personal requests for favors from senators, cabinet secretaries, governors, generals, CEOs, mayors, White House friends and politicians' relatives.

"I've been trying to meet with someone here for three days," Gillette tells an assistant under secretary of energy.

"I know it's difficult," the man says sympathetically over the commotion from the hall outside.

Gillette tells the overworked official, "Based on our findings at Detrick, we believe it's crucial to send chemists to check the fluids plants for Delta-3 residue."

"Yes! Absolutely right!" The man glances at his calendar, sets his thin face, reminds Gillette of an undertaker. "The problem is, all our people are out testing wells and pipes. Plus, they've had difficulty traveling. Limited staff, you know. Limited travel. And so many senators," he whispers, "demand special consideration for their states, for testing."

Gillette says, "Send me, then. I'll do it."

"Oh, we'll get around to it, Dr. Gillette, once we complete our inventory of military equipment."

"I thought the Department of Defense handled that."

"There is some overlap, I know." The man dabs his sweaty forehead with a hankie. "If you don't mind a delicate personal question, sir, are you sure that you've not let a, uh, personal difference with a superior cloud your judgment?"

Gillette turns red. "Who did you talk to?"

"Well, we did propose a program similar to the one you're suggesting. It seems General Hauser already told you no."

The assistant under secretary rises and extends his pale hand. "Good luck, Doctor. So many appointments. So little time."

St. Paul's Church was built in 1852. The walls are gray stone and the doorways arches. The basement has been converted by Pastor Van Horne into a refugee shelter for families who sleep on purple velvet pew cushions laid in rows on the carpeted floor.

When I'm finished treating people here today, Gillette thinks, bending over a patient, *I'm going back to stealing. I haven't stolen anything in twenty-five years.*

"I'm fine, Doc," says the coughing man to Gillette. But he's wracked with the flu: shivering, shaking, dehydrated.

"You're running a fever of a hundred and two."

"It's just a winter cold."

The refugees, mostly suburbanites, have a docile, glazed look. They veer back and forth emotionally from shock to appreciation. The shelter's presence is not advertised. If it were, space would be overwhelmed.

Once a day, Gillette makes rounds at St. Paul's, checks for pain or fevers here, and dispenses the last of the over-the-counter medicines that came in a distribution last week. Today he also brought some donated pasta from Marion Street private stocks.

"You're not eating," Gillette tells the sick man, eyeing the red eyes, receding gums, saggy neck skin.

"Ain't that a laugh. My ex-wife used to say I ate too much. Anyway, my kids need the food."

"What was your job, before Delta-3?"

The man looks surprised. "State Department. Sub-Sahel affairs," he says with pride of memory, having had a job.

Gillette smells sweat and slept-in clothes and odors of thirty-six people who shower with cold water once a week.

He tells the guy, "I've worked African droughts. We both know what happens to the kids when the parents die."

The man's lips compress into a hard, thin line.

"Washington isn't Africa."

"It isn't Washington either anymore."

The guy replies loyally, "You heard the president. Hold on a few more days. Someone will figure out how to beat Delta-3!"

Outside, as Gillette waits to go stealing, the sky is blue. The Tuesday barter mart is starting up in the grassy lawn between church and firehouse, a public area in plain view of many apartment buildings, so it is safe. People bring goods to be traded or sold. They unroll blankets and lay out wares while Marisa teaches school inside the church.

Lawyers shout like fishmongers.

"Chickpeas! Vegetable soup! I also have toilet paper!"

"Lightbulbs! Flashlights! Candles! Batteries!"

Normally a scene like this might be fun, a bargain hunter's paradise. But there's no disguising the growing air of desperation, of as yet controlled need.

Gillette strolls among the offerings. Everyone except black marketeers are short of food, he knows, but many people—pack rats or bulk shoppers or those with special access—have a surplus of specific kinds of supplies.

People gambling that the crisis is temporary offer essentials at fantastic markups, trading food for jewelry, camping supplies for artwork, medicine for stereos, silverware, leather couches. Pessimists pay anything for survival supplies. Gillette spots Julie the flight attendant unloading a shopping cart of bulk Costco goods: Tylenol, maxi-pads, Slim Jims, Kirkland Cola and two huge jars of mayonnaise. Nearby, Les Higuera trades his 42-inch flat-screen plasma TV for heavy winter gloves and a small cooler filled with frozen hamburger meat.

"At least I hope it's hamburgers," he tells Gillette.

Gillette watches a set of brand-new Michelin racing tires go for twenty-five dollars to a well-dressed man and his burly assistant, who carries the tires to a waiting Ford SUV. Boxes of Marlboros—a nicotine addict's fix—go for a wedding ring. An antique coin collection changes hands for insulin and aspirin, ibuprofen, nasal spray.

Gillette spots Gail Hansen trading one of her smaller art gallery canvases for two cartons of vodka. She's refused to pool her goods with the block's trade-supply.

"I'll do what I want with my own stuff," she'd said.

He spots Gordon Dubbs and his two friends strolling among the sellers. They're not buying or selling today but seem to pay extra attention to people who walk off with valuables. They show none of the signs of hunger that Gillette detects among other people here: slow movements, baggy clothing, glazed looks, symptoms of oncoming disease.

So they have food and plenty of it. They're watching to see where people go.

The sight of Dubbs produces a hard, taut knowledge in Gillette's belly, the kind he used to get when he was a kid, when he'd spot guys from a rival gang.

"Anyone selling clean gasoline?" The well-dressed man strolls among sellers, looking prosperous, buoyant, happy, and followed by his human guard dog. "Anyone have heating oil? Heating oil will get top dollar! Name your price for heating oil!"

The residents of Marion Street have voted to divide up jobs. Bob Cantoni is in charge of defense. Les the TV producer will head work rotation, scrounging and supply planning. Marisa will liaison with neighborhood teachers and concentrate on kids, splitting her time between the block and the church, where Chris Van Horne plans to expand the shelter. Alice will handle meal prep, Gillette medical needs. Judge Holmes will head the "justice committee" to hopefully resolve disputes between neighbors. The Klines will monitor news on TV and blogs, to make sure that the street receives a daily roundup. And Joe Holmes will figure out ways to save on heat in homes by caulking, zoning off rooms, increasing fuel efficiency.

By two, Gillette, Bob and Joe are wedged into the front seat of Bob's Suburban for the five-mile theft run to Takoma Park. The sun is high. The temperature a balmy, unseasonable fifty. Garbage spills off curbs. Traffic lights are off. Each day fewer private vehicles venture out, and tend to do it midday, when travel is more safe.

"I can't believe you didn't tell us before this that you had heating oil," Bob snaps at Joe.

"I wanted to be sure we'd need it. I don't generally steal from my own company. Technically this is looting."

"Well, I just hope someone else from your company didn't take it already."

"Your wife's not a judge, Bob."

"Every guy's wife is a judge."

Gillette reassures Joe. "When the emergency's over, we'll give back the oil."

"Eleanor says all thieves say that."

Since the only purpose of car trips is to gather supplies, anyone on the road is a target for ambush. Drivers avoid eye contact. They run stop signs, as it's safer not to slow down. Vehicles are occupied by several people, usually armed one way or another, with guns, bats, golf clubs, mace.

"Greg, when you were in that gang, you must have stolen lots of stuff, right?" Joe asks, flicking his eyes to four empty black oil drums in back, taken from another of Joe's construction sites nearby.

"I loved stealing motorcycles. Tell Paulo that and I will kill you, my friend."

"Gables Glen"—Joe's project—is an assisted-living home that had been scheduled to open in late October. The two-story brick building sits on a wooded acre off Piney Branch Road. Beyond a razor-wire fence, a long driveway snakes past a guardhouse to the entrance. The guardhouse is empty. Windows are smashed in the main building. Looters have spray-painted walls, Gillette sees as they park. Even Washington's vandals make political accusations.

"Delta-3 was a secret weapon that got out."

"We filled up the boilers back in September," Joe says, eyeing the damage. "We got a great price from Sav-Mor."

They park by the smashed-in rear door and stand listening to hear if anyone is inside. Power is off. With mixed hope and fear they advance into the dark basement by flashlight. Rats skitter along the floor. Gillette feels his pulse rise when he spots the looming oil tank in a corner, curved like an elephant's belly.

Joe shines his flashlight beam on the meter. "Thank God! Still full!"

"Let's do this fast," Gillette says, not liking the confined space, knowing that rare police patrols are still out, and under looting laws, they'd be locked up for this.

Joe and Bob bring in the first empty drum on a dolly. They run a hose from fuel tank to drum. Dark oil runs into the drum. Twenty minutes later all drums are full, and the men drive out of the compound. The Suburban smells of oil. Joe grins as if they've stolen the Hope Diamond. Bob keeps glancing back as if expecting cops to appear, and Gillette tells him to drive slow. At Marion Street they roll the Suburban into Les Higuera's backyard, where neighbors await to unload. The oil will top off boilers, and the extra is to be stored in the flight attendants' unoccupied corner house.

By seven, they've transferred over 2,000 gallons of home heating oil to Marion Street.

"Now we can move back into our own homes," Gail Hansen proposes happily at the block meeting that night, in Gillette's house.

"It's smarter to conserve and keep doubling up."

Gail's drunk. "I'm sleeping in my bed. In my house. With my things. You're not Congress. You're just people."

They vote to donate 20 percent of the oil to the church.

"Now we'll stay warm for the whole winter," Paulo says. "And that jerk Teddie Dubbs will freeze!"

The kids get a lecture. They must never, ever let anyone else know about the extra oil, a warning that is lost within minutes on Gail.

"If we have enough to donate, we have enough to swap," she says, holding up her highball glass. "You people need to relax. Let's throw a party, a goddamn big party. We can trade oil for vodka. We have plenty to spare!"

When the meeting breaks up, the Cantonis and Higueras stay to play Pictionary and watch the news. The Russian government has fallen. An Iranian mullah has called for religious war, saying, "God has sickened oil until we make the world right."

"Gail's going to be a problem," says Bob.

"She's just excitable," says Marisa.

"She's an addict."

At least the thermostat is up, at a balmy 65 degrees, which is celebration enough for Gillette.

"Heat," Marisa says deliciously, later, as she and Gillette make love, and forget, for an hour, about oil.

I took heat for granted, Gillette thinks as he falls asleep. But the next morning the frustration returns when he visits the American Petroleum Institute to ask the oil industry lobbyists there to help request from Hauser renewed screening of fluids and bactericide companies.

"Does General Hauser know you're here, Dr. Gillette? We were warned you might show up."

TWELVE

NOVEMBER 26TH. 29 DAYS AFTER OUTBREAK.

His days are nightmares. His nightmares are hell.

Hassan el Kader, former executive producer of Al Jazeera's *Faces and Places*, CIA spy and new arrival in the U.S., sits back in a leather armchair in the agency safe house (what a laugh—it's not safe, not even a house) four hours into one more debriefing session with the man who originally recruited him back at U. Kuwait. And assured him that the agency would protect him were he ever in need.

"Have another fig, Hassan."

"I'd rather go back to the hotel."

The overworked heart in his fat man's chest is galloping like one of the emir's thoroughbreds. He'd thought his fear was bad before the oil emergency started, when it was the dry-mouthed day-to-day kind. A few bad dreams. The occasional glance over the shoulder. A man could live with that, pretend it wasn't there sometimes.

But ever since the Al Jazeera broadcast, he barely sleeps. His chest aches all the time.

I've endangered my family, he thinks.

Within minutes of his "source" informing him that his cover was blown, he'd fled with his wife and daughters to the American Embassy, and demanded asylum, practically doubled over with stomach pain. And *that* fear had been nothing compared to the terror that engulfed him upon boarding a leased 757—"safe," he was assured—to be flown to the empire itself. The shining city by the Potomac and birthplace of modern democracy. Hassan's new home.

"Hassan, perhaps you've had some new thoughts on who among your staff stole the videos of the caves."

Hassan clutches his intestines. The room is done in dark wood. His old "professor" has thinned over the years, but still dresses in British suits, still has that quirky shock of white electroshock hair, like Albert Einstein.

"Hassan, have you ever heard of a piece of oil equipment called a 'sucker rod beam pump'?"

He had thought that if he actually reached America, his terror would dissipate. But it had escalated into throat-clutching breathlessness during landing, as the ground rushed close, and the seatback video monitors showed newsreels of planes crashing in Paris from Delta-3.

"You said you'd give me a new home and identity," he says. "And a new job in TV. But you've got my family and me alone in a hotel."

His "professor" looks sympathetic. He never mentions his own problems, and he must have plenty because of the oil. He just says, "Do you know how dear hotel space is? Everything has changed for a while. We're not going back on our word. We just don't have extra agents now. They're out hunting terrorists. And frankly, if people wanted to kill you, they wouldn't have phoned you. You'd be dead."

There will never be a safe place for my family again.

"Hassan, help us with the immediate problem, tracking down the people who did this."

"How? On foot?"

"Don't be dramatic. We still have resources."

"Then why," Hassan demands, "are we sitting in a bank building instead of the CIA building? Why did we move downtown? You're evacuating headquarters, aren't you?"

"A few agencies are relocating, that's all."

Hassan stands and walks to the window. He can see the White House a few blocks away. Below, work crews carry boxes in and out of office buildings. Trucks load up. Whatever is going on is important enough so there's plenty of gasoline for it. Normal businesses seem to be vacating. Government types seem to be moving in.

They're pulling in the wagons, like in American movies.

"Hassan, did you hear anything about a new refinery in Ak-Darya, Uzbekistan, shutting down due to problems, *before* it opened, six months ago?"

"Why? Did Delta-3 infect it?"

"Hassan, did the Imam or caller mention anything about a gas station blowing up in Jakarta, in May?"

"Are you saying somebody tested the bug then?"

They never answer *his* questions. And he doesn't know the answer to theirs.

Hassan tells his old professor, "I'm done for today."

Twenty minutes later, walking toward his hotel, he sees a man jump off the World Bank building. Suicides are occurring daily now. Men can't feed their families. Hassan stands in the crowd and gapes down at the smashed head, limp limbs, pool of blood on the sidewalk. The man, his gray suit ripped, seems Hassan's age, seems like he was a lawyer or white-collar official. Hassan's chest pains start up again. He makes it to the Mayflower Hotel, slips into the elevator, opens the door of his room and thinks, *What's that smell?*

"Hello? I'm home!"

Nobody here. The girls' cots are made. So is the bed.

They must be out walking. They're bored. Or maybe they left because of the smell. It smells like something died in the wall. I'll call the front desk for help.

He doesn't, though. He can't.

Because something rubbery presses over his mouth.

He can see in the mirror opposite that the man behind him is lean, handsome, white.

The pain comes then, extreme, and getting worse.

"You'll tell me what they ask you about," the man says. "You'll tell me exactly what they know."

* * *

Gillette's been turned away at the FBI. He's been turned away from the *Post.* He never even reached the head of the Senate Energy Committee. He's been ignored politely at the barely staffed, overwhelmed institutes, think tanks, advisers' offices, and by an assistant deputy secretary of Homeland Security, when he finally cornered her in an elevator.

"Who sent you to see me, exactly?" she had asked.

Now he listens to the news over headphones on his iPod, while in the Metro, taking Annie to the zoo.

"Looters are being shot in Los Angeles," he hears. "In Los Angeles, martial law is in effect."

"The cheetahs seemed better yesterday," Annie says.

"That's because caring people like you look after them," he says. They're crazy to keep coming here, but the zoo is the only thing that calms Annie. The trips are her reward for working with the small kids at St. Paul's. Reading to them. Playing with them.

At least I can make my daughter happy, Gillette thinks.

"This just in," he hears. "CIA officials confirmed that the man murdered this morning with his wife and daughters in the Mayflower Hotel had a key role in the Delta investigation."

Christ. The whole family, Gillette thinks.

"Baby Hatari is the cutest," Annie says.

Gillette's "leave" is almost over. He'll have to go back to Detrick soon, that useless place where he accomplishes nothing, and can't even help his loved ones. Temperatures have plummeted, even in the train. He and Annie wear parkas, stocking hats, gloves. The announcer goes on to explain that the Mayflower killings had matched exactly the murder of the FBI man's family in early November. The bodies had been found with another note quoting Mohammed: "Allah changeth not the condition of a folk until they changeth what is in their hearts."

There goes my last chance of diverting agents. I'll never convince anyone to pay attention to the fluids plants now.

The Metro is packed and in recent days has become filthy, smelling of sweat, garbage, desperation. People clutching suitcases flood toward the city. But airports are closed. Travelers are alone. So where are they going? It doesn't make sense.

Gillette asks a man with a knapsack where he's headed.

"To visit my son," the man replies too quickly. "Uh, in New York. Uh, thank the Lord for the electric trains."

Gillette tries a woman, who turns red.

"Shopping. My suitcase is empty. I'll fill it as I go."

"How can you shop? Stores are closed," Annie asks.

The woman flares, "It's rude to be inquisitive."

Annie looks her squarely in the eye.

"So is lying," she says as Gillette hides a smile.

The train glides to a halt at the Cleveland Park station, formerly popular with zoo goers and tourists, but today, Gillette and Annie are the only two getting off in midafternoon. The long platform is empty. Their steps echo off the arched roof.

"Several of Pakistan's nuclear missiles have been reported missing since last week's Fundamentalist coup," the radio announcer says. "It is feared that the missiles will end up in the hands of terrorists."

As they near the mezzanine, he hears a collective murmur of voices, radios, a man singing "Fool on the Hill." The ticket booth comes into view and with it a scene out of the London Blitz: people lining the walls, sitting, sleeping, eating whatever meager food authorities provided.

Annie says, "They cut rations for the cheetahs again."

How do you feed animals when people can't eat?

A little girl calls out, "Lady? Have any spare food?"

"No one ever called me lady before," Annie says.

At that moment, a woman screams on the platform.

"A wolf! A wolf!"

People start leaping to their feet.

Gillette whirls. He can't believe it. The animal is on the opposite end of the mezzanine. It's too big to be a dog, gray, frothing, shivering, tail tucked, backed against a wall. Its head swings back and forth as it looks between the escalators and the crowd. It's as scared of people as they are of it.

It's bleeding. Limping. Has it been shot?

A loudspeaker from up top calls, "Hold your fire!"

That does it. The crowd surges for the exit. Gillette grabs Annie and feels them swept along in the panicked rush. They fight their way up the escalator, toward the glow of sun. They spill topside as black-clad SWAT cops push past, heading toward the wolf.

"Keep down," the man on the bullhorn is repeating.

Gillette pulls Annie to the cold street. They're in a bigger crowd now, outside the zoo, facing the lion statues and American flag and row of police cars and armored officers.

"We're going home," he tells Annie.

"But the cheetahs, Daddy!"

I've been crazy to humor her.

And as if to confirm this, shooting erupts from the north, along Connecticut Avenue, the direction he had been prepared to walk. He recognizes the rapid bursts of automatic weapons. He's heard them often enough in the Third World.

How many wolves got out? And how?

Breathless voices in the crowd fill him in, the snatches of conversation like more headlines.

"People broke into the zoo last night, from the back."

"They stole animals, to eat, or sell for food, I bet."

"Stay where you are!" the bullhorn blares.

And then, a long anguished howl, not from the Metro or zoo, but another direction, beyond the shuttered shops lining Connecticut Avenue. The hills that climb toward Wisconsin Avenue. An area of private homes.

I'm going mad, he thinks as Annie jerks away and stands. He grabs her before she can run for the zoo. She's crying hysterically about the cheetahs. He tells her she's forbidden to go to the zoo anymore. And that's final.

As he drags her off, they hear two cops talking loudly.

"The park rangers were fired yesterday," one says as Gillette catches sight of a monkey running along the roof of a dry cleaning shop. "Want to bet one of them did it?"

"I'd skin the panda," says the other. "Little garlic. Little salt. Probably tastes like chicken. Now I understand poachers in Africa. Their kids need the food."

Gillette's cell phone rings during the nightly after-dinner sermon that Pastor Van Horne delivers in St. Paul's.

"There is a lot to be thankful for," Van Horne says. "Family. Friends. Home."

"It's Jim Raines, sir. I have good news and bad news."

Marisa elbows Gillette and mouths, "Take it outside." He makes his way to the church steps. It's cold here, with clouds coming in. The sermon blasts through an open window.

"We hope that when the president speaks later tonight, he will announce progress," Van Horne says.

"Cougar again," Raines says as Gillette's interest flares. "This is from juvenile court records, Elk Valley, Nevada, 2001. Two sixteen-year-olds were arrested for breaking into the labs three times that month, stealing equipment. Painting graffiti on walls. Genitals. Swastikas."

"You think the kids are involved in Delta-3?" Gillette asks.

"Nah. One's in Yale now. The other one died in a car crash. I reached the Yale guy at his parents' home. He said the security system was a joke. Anyone could get in."

Gillette draws a breath. Three people on horses clip-clop past on Connecticut Avenue. They're not cops, but the first civilian riders he's seen inside the city.

Is that a rifle over one man's shoulder? I can't tell.

He also spots a woman's silhouette moving across Connecticut Avenue from the direction of Marion Street toward 5110 Connecticut. Could it be Gail? he thinks. But it couldn't be. Why would she go there?

Then he forgets about the woman because Raines says, "On to Nevada employee relations records, you know, lawsuits from disgruntled employees? Comes to Cougar, we have complaints up the wazoo. Fired without cause. Sex discrimination. Reneged on bonuses."

Gillette feels his pulse speed up.

He says, "So we have an unsecured lab, and workers who hate the company. Did the lab do genetic research?"

"No, but they were looking at ways to kill bacteria in oil wells, keep pipes clean. Hmmmm?"

He can feel Raines grinning on the phone.

"Didn't the FBI check this?"

"Yeah. They never make mistakes. I have complete confidence in them, especially when they only spend a few days on an investigation before being pulled off."

"Okay, what you need to do next is—"

"Actually, sir, I haven't told you the bad news. We're ordered to stay away from Cougar. Someone high up complained. I got a call from upstairs."

Gillette lets out a long breath. "Cover-up?"

"Or we just bothered them. Either way, Cougar gets a pass. We've been told to stop harassing them."

Gillette feels fury rising. "Harassing?" he says. "Every time there's an outbreak I hear the same bullshit. 'We have things under control.' 'We don't need help.' 'Stop harassing our people.' Then afterwards they never said it. Raines, someone needs to get out there. Time's running out."

"You don't have to convince me, boss. But who can go?"

The president looks unwell on television, pale, hunched, dwarfed by his desk. Gillette's basement is filled with silent neighbors. Marisa has put out her best water glasses, filled from the tap.

The president says, "In order to guarantee smoothly functioning government, we must zone our major cities until the situation improves."

Gillette, feeling sick, understands why so many people had been moving into Washington. They were the chosen ones, who will be living now in a fully secure zone.

"I had hoped we would never reach this point. But we must guarantee order," the president says. "Essential personnel will live and work in areas where fuel and safety are guaranteed."

The president is normally well liked in this room, but at the moment no one seems to be breathing. Little Grace Kline takes Paulo's hand.

"Our scientists continue to unravel the problem. We must not allow order to collapse."

Inserts on TV show the plan for major cities, starting tomorrow. In Washington, full fuel will go to an egg-shaped area around the White House: to the Naval Observatory in the north and the Navy Yard in the south. It will include Bolling Air Force Base and run along the west shore of the Potomac into Roslyn, to serve the Pentagon and Fort Myer.

Beyond that a second zone—to receive slightly less fuel and military protection; a buffer zone, so to speak—will extend from the Calvert Street Bridge to Marbury Point, east to D.C. Hospital, west to Arlington Heights.

"We're outside both zones," Alice Lee gasps. "They're abandoning us, saving themselves!"

Marion Street, a mere four and a half miles from the White House, will as of tomorrow receive less protection.

Gillette walks outside. His head throbs. For once he hears no sirens, sees no glow of fire, as if even nature is out of fuel. Joe Holmes, on patrol, trudges past a gap between houses, through which is visible wreckage from the air crash on Ingomar Place.

Triage, Gillette thinks, horrified. *Like in an outbreak. They've decided who can be saved, who might be expendable for the greater good. My family has been black-tagged this time.*

At that moment, he has the idea.

Someone has come up beside him. Marisa drapes a coat over his shoulders. He puts his arm around her. Zoning changes everything, he knows, pulling out his wallet.

In the moonlight, the Detrick ID—giving him special power, special access, special influence—is unreadable. To misuse it is to risk arrest, at best.

He tells her what he'd like to do—and how he must lie to the neighbors in order to protect them—and she kisses him and says she agrees. Back inside, the speech is over. The commentators look shocked and useless. As important journalists, they'll probably get to live in zone A.

Gillette shuts the set. No one is talking. Grace Kline is crying. Paulo's got his arm around her, unsure who to direct his fury at: the president, the microbe, or all adults.

If I'm going to leave my family, I need these people to help them. Because things will fall apart fast now.

He's been sharing information here all along but now he says he's been ordered to travel. If he tells the truth, the neighbors could be liable for not turning him in. He'll be gone for a few days, he says, during which time he'll have a small chance of making progress against the oil bug. Will they watch out for Marisa and the kids while he's gone?

Judge Holmes says, "Of course. We're on our own now."

Les agrees. "We can't stave off disaster forever. One canceled food distribution, one bad riot and we're sunk."

"Don't go," snaps Gail Hansen. "My gallery was sacked. I don't want to lose my house too. We need all the men to protect the block. Why should *you* go? You're just one man. You won't make a difference. This zoning . . . people will be like animals. Tell them you're sick."

Chris Van Horne cups a hand to his ear. "Perhaps if we discussed this tomorrow . . ."

"I've been ordered to leave right away," Gillette lies.

The vote goes Gillette's way. They also decide to double patrols from now on, cut rations by 10 percent, and to start learning firearm safety from Bob Cantoni.

"I'm proud of you, Dad," Paulo says when the meeting is over. "I'll protect the family. Nobody gets through me."

He sounds brave. He looks like a tough-talking kid. Gillette wonders if he's any less deluded than Paulo. Maybe Cougar is a wild-goose chase. The odds are it's clean. But Larch trained him a long time ago to ignore odds. Larch trained him to pay attention to intuition, to listen to the small inner voice.

Gillette lays out his uniform, about to go AWOL.

What if Gail was right and I'm wrong, he thinks. *What I didn't tell Marisa was, if they catch me and tell Hauser, I won't get back home at all.*

THIRTEEN

NOVEMBER 27TH. MORNING. 30 DAYS AFTER OUTBREAK.

Gillette looks like he belongs. But he doesn't.

Surprise will be the key, he thinks.

He's back in the Metro at eight, in pressed uniform, carrying an overnight bag and attache case, heading at thirty miles an hour into the barred zone of the capital. In minutes he'll encounter whatever security barrier will block access. If he doesn't get through, his plan will fail before it begins.

I overheard at the FBI that some flights are going out. That's how their investigators get around.

The car is packed, despite the new zoning law. But the mood is tense. Many riders are resolved to challenge the rules, he gathers from overheard conversations. The car reeks of human need. People are trying to reach zone A to plead for food, jobs, protection.

The best chance I have is to act like exactly what I am, a CDC doctor on an important mission. If I can get to Cougar, I'll say we're worried that the bug has jumped to humans. I'll ask medical questions. Then I'll check facilities for Delta-3.

In his attache case he's packed epi-aid supplies: syringes for drawing blood, intravenous tubing, ten pounds of bogus medical questionnaires that he'd printed off the Internet last night before power failed at two.

The Metro halts at Cleveland Park, last stop before the buffer zone. Riders get on, but no one gets off. The train sits at the platform longer than usual. Suddenly soldiers come into the car, polite but firm, and wielding sidearms. Loudspeakers order everyone lacking proper ID to exit the train.

Here we go, Gillette thinks, crossing his legs casually, pulling out an article on sporulating bacteria. He wills his heart to beat slowly, like he used to do before a gang fight, car theft, arrest.

Most passengers grumblingly leave the train. They'd expected to be challenged. But others argue with soldiers.

"I am the mayor of Fargo, North Dakota, and *demand* to see my senator! It took three weeks to get to Washington!"

"Sorry, sir."

"Fargo has no food!"

"You'll be arrested if you don't leave, sir."

A pretty young woman in a business suit smiles at the soldier. "I was supposed to get a zone A card. There must be a mistake . . ."

Gillette hears, "I have an *appointment* with the congressman . . ."

And, "I am an ACLU attorney!"

A hand is thrust at Gillette's face. "ID," snaps a pimply kid wielding a rifle, already self-important, able to order around people who yesterday paid him no attention.

Gillette lazily produces his laminated card from his breast pocket, and goes back to reading about symbiotic partnerships between deep-sea vent microbes. He's now officially misused his special ID, a military crime.

He reads, *"Some microbes actually prevent others from doing damage. They function as a natural check, limiting or altering dangerous behavior."*

"I never saw an ID like this before, sir," the soldier says, awed by the secretary of defense's signature. Then the troops are gone. The Metro starts up. Passengers remaining—the privileged few—relax, having breached the barrier. They grow talkative, belonging to a new exclusive club.

"That worked rather well, I thought," a man—a lawyer—in a gray suit and raincoat tells Gillette smugly.

The man works for the Department of Housing and Urban Development, he says, adding that Gillette has "no idea" of how "nightmarish" the logistics had been, to set up security zones. Journalists had agreed to sit on the story, had been threatened. People already living in key zones will receive extra food for sharing apartments. Lawyers had worked night and day to produce documents authorizing temporary seizure of offices, shops, autos, horses, even Smithsonian museum wagons.

"May I ask about *your* job?" the man inquires.

"Disease prevention."

"Ah, headed for the hantavirus outbreak in Iowa?"

Gillette masks his surprise. He's not heard of any outbreak. The man gets off and Gillette rides to the stop for Bolling Air Force Base, D.C.'s only functioning airport. As he gets off, it is clear that full electrical power is being supplied to zone A. All lights are on. Escalators work. The sense of safety is physical, and people around him seem to move with more assurance. Reaching the surface he is surprised to see a fleet of electric golf carts at the curb, marked TAXICAB. There are more cars on the streets, small ones. Priuses. Coopers. Motorcycles. The world works.

He has to show ID again at Bolling's front gate, and before being admitted into terminal A, which resembles a civilian version in military green. It's a madhouse. Anyone with clearance to fly is here. Fighting his way onto a line, Gillette sees politicians he recognizes and generals he doesn't, traveling with staffers smart enough to stay close.

Announcements blare. A combo flight serving Boston, Albany, Cincinnati, Minneapolis and Seattle is boarding. A flight from L.A., Phoenix, Salt Lake City, Dallas, Tallahassee and Atlanta has just landed at gate B.

The arguments at the counter are the same ones he heard in the Metro. "Do you know who I am?" a man shouts.

Gillette produces the ID and tells an Air Force sergeant—a ticket clerk—that he needs a ticket for Nevada. The woman's eyes widen. The sergeant excuses herself, carries the card to her CO and gets into a heated conversation, waving the ID as Gillette holds his breath.

If they call Hauser I'll be arrested.

But she comes back and prints out a ticket, to be charged to Gillette's American Express card.

"We had to bump a colonel on your flight. Usually people charge tickets to their department, sir."

"I'll get reimbursed," Gillette lies, glances at the cost and almost falls over. The price of fuel has gone up, all right. Round trip just cost $7,900 of his savings.

It would be funny, after all this, if Cougar has nothing to do with the outbreak.

Thankfully, there's actually food for sale on the way to the gate, half sandwiches and Oreo cookies in foil. He wishes he could bring some home. The gate area is packed. Outside the window, dozens of mechanics go over the jet. A delay is announced due to "mechanical difficulty."

There's no other plane to use.

The delay stretches to two hours. Four. Dusk comes on. Overhead monitors—TV news—show Navy blockade ships on the Potomac turning away civilian boats. A boat turns over. People are drowning. Then suddenly the news changes. The screens start showing good news only. A corporal in Florida has donated a kidney to his nephew. A rock band is shown playing charity events.

At 8 P.M. he hears, finally, "Now boarding flight two!"

The engines cough, warming up. The jet rolls down the runway. Gillette, at a window, realizes he's grabbed the seat rest. The plane vibrates, takes to the air. Below, he sees no lights south of the Potomac. Northern Virginia has lost power. He sees only four pairs of headlights on roads. The shaking stops. They climb steadily.

He's filled with triumph. He's made it this far.

The country looks darker than normal, he thinks, thirty minutes later, looking down during a dinner of reheated chicken. But airline food never tasted so good.

The converted 737, a high-level puddle jumper, spends the night landing and taking off, hopping west as Gillette's hope grows. The flight reminds him of trips he's taken around third world countries, except this plane is filled with some of the most powerful people in the United States. A different person sits next to him on each leg, each traveling on different business, each a source of news.

Out of Washington, he chats with a woman from the Nuclear Regulatory Commission, who's on her way to inspect nuclear power plants near Chicago.

"The accident was never on the news," she says. "The staff was exhausted. Someone fell asleep. At least no one died, but the core almost melted. Half the Midwest would have gotten the cloud."

After Chicago, where the West Side was engulfed in flames from race riots, he finds himself beside a man from the Commerce Department. "The lubrication oil is infected in food factories. Machinery's breaking down," the man says.

They land in St. Louis. Kansas City. Dallas, on the early morning of the 28th. The woman who joins him after that—a tough-looking, attractive brunette—turns out to be a lawyer working with the judge advocate general's office, on her way to Camp Pendleton in California.

"That's where the high-level trials will be," she says.

"Trials?"

"Don't you know? Then why are you going to Nevada?" She orders a third gin and tonic and touches him on the wrist.

"Fill me in," he says.

"This is going to sound forward," she says, changing subjects. "But I'm drunk. And you're handsome."

"Married too."

"I've always wanted to join the Mile High Club."

"I recommend it. But not with me. What went wrong in Nevada?"

She orders more gin, grows morose. "Everyone where I live is fooling around. I never even slept with Bill . . ."

"We'll be landing in Denver in fifteen minutes," the captain says. "All Nevada passengers must deplane. We will be skipping Nevada . . . No flights are stopping in Nevada . . ."

The woman says, "All my friends made the Mile High Club. I always wait too long. I never do anything except work. How much time do we have left?"

The woman starts crying. Gillette takes her hand.

He says, fiercely, "There's a lot of time left."

NOVEMBER 30TH. AFTERNOON. 33 DAYS AFTER OUTBREAK.

No one will tell him what's wrong in Nevada. Not at the airport, or the zone A hotel where he's wasted two days waiting for a flight, worried that he'll be discovered, poring over more articles on heat-resistant bacteria, deep-sea bacteria, oil well bacteria, spores. News shows have become worthless. He calls home and asks Les Higuera if he's heard anything about Nevada yet from his ABC friends.

"No. It's been decided by the White House that *real* news upsets people. Real news causes riots and depression. They got a court order . . . We're not allowed to . . ."

"Les, are you all right?"

"It's not journalism anymore. It's public relations. The finest minds turned into shills for incompetents. Try the blogs for real news, but you never know which is true."

He has more luck learning tidbits in the hotel bar, since everyone stuck here has been interrupted while traveling on important missions. He's surprised to run into the lawyer from HUD whom he'd met on the Metro, the once smug bureaucrat who had helped set up the new zoning laws.

"It's been a disaster," the man moans, looking ashen. "I don't understand it. Riots. Killings. Average people took it the wrong way. They think we did it to hurt them, not help them. Can't they see we need to keep government running?"

He signals for another double Manhattan. "I could lose my job. My boss asked me, in Washington, who started the fighting?" he whispers into his glass, talking to himself as much as Gillette, as a TV news show over the bar broadcasts a cheery piece about a boy's choir. "I told him, the police blame rioters. The rioters blame police, Jews, Catholics, neighbors they don't like, family members they used to like, politicians. Oh God! Bartender!"

"Fighting? What fighting?" Gillette asks.

"In New York, it started at the bridges. In San Francisco, those idiots looted distribution supplies. In Tampa, the National Guard was ordered to fire on the mob, but they *let the people through.* I knew zoning was a bad idea," says the man who, three days earlier, had bragged about it. "I told my boss, people won't go along! No one listened to me."

Gillette goes back to the hotel computer room and checks the blogs. Russia reports wholesale mutiny on its ships. Britain reports Russian subs attacking working oil platforms. The pope has called an alchemist cult—people trying to turn gold into oil—heretics.

Then hotel power fails.

His cell phone chimes in the dark.

Gillette grabs it, hoping that the airport is calling. But it's Colonel Novak's secretary, thinking he's still home, ordering him to stand by to be picked up by the Fort Detrick van, tomorrow, and come back to work.

Gillette starts coughing. "I . . . have . . . a terrible flu."

"Be ready at five, Commander."

Shit. When I don't show up, they'll know I'm AWOL. I haven't even reached Nevada yet.

The phone chimes again.

Now what? he thinks.

"Can you come to the airport right now, sir?"

DECEMBER 1ST. 6 A.M. 34 DAYS AFTER OUTBREAK.

The big C-130 finally starts backloading troops, tough, heavily armed men heading for Las Vegas. The loadmaster eyes Gillette's ID, then waves him aboard, in the falling snow.

"First a mutiny," he says. "Now disease, eh? I were you, I'd take a gun. Vegas still ain't safe."

They climb through a storm, the plane shaking and heaving, but finally they get high enough so Gillette sees stars above. Beside him sits the troops' commander, a major general named Winston, a black man from South Carolina, about fifty years old, gray, bespectacled, sad.

"Who would have believed it," Gillette says, knowing that the best way to learn information is to pretend you already know it. "Mutiny."

"All they wanted was what was theirs," Winston says.

"I know."

"What kind of officer sells his own men's food?"

"The worst kind," Gillette says.

"He turns off the AC on his base. It's over a hundred degrees out there. He sells the fuel for the generators, and fires on his own troops when they protest."

"And *you* had to put it down," Gillette sympathizes.

"They wanted to seize a plane, to get out of the country. We drove them out of the airport and hunted them room to room on the strip. Two hundred and fifty Americans," sighs the general. "Arrested. Killed. Buzzards feeding on our own. Martial law, Commander. We're trying looters and executing them. How's that for a day in America's fun vacationland? Huh?"

Black smoke rises over the city as the wheels touch down. Gillette smells chemicals burning over the unseasonably hot desert when the rear door yawns open. He's been in McCarran International once before, during a weekend convention of tuberculosis experts. He and Marisa had played blackjack at the Luxor, seen a Cirque du Soleil show at the Mirage.

Now he's shocked to see bullet holes marring walls, ads and restaurants in the terminal. Half the slot machines have been destroyed. Crusted blood smears floors and walls.

"The mutineers stole vehicles on base, and made their way to the airport," says the captain who has been assigned by Major General Winston to help Gillette.

"Where did they want to go?"

"Anywhere safe. There's no such place."

They pass a large group of soldiers waiting for a flight: sullen men and women, some bandaged, sitting together. But then Gillette realizes that these soldiers are not armed. He sees the handcuffs, shackles, guards.

"Don't worry, Commander. We'll get you to Cougar."

People always want to help the CDC.

Outside, in the bright sun, odd-looking cars are lined up by the taxi stand. Gillette sees an old blue Fairlane with the trunk lid off and the interior filled with bunched-up batteries. There's a Dodge sporting a sign, "Vegetable oil fuel." And a Prius with a window sticker reading, SET AMERICA FREE. HYBRID CAR.

The driver is a civilian. The captain waves over four soldiers, and orders two of them to follow Gillette in a second car, to protect him, and two to ride with Gillette. Gillette gives the driver Cougar's address, outside the city. The kid frowns.

"We can't use I-15," he says, nodding in the direction of the black smoke. "There's still some fighting. But Las Vegas Boulevard should be safe if we keep moving."

The car starts up silently, and glides out of the airport at twenty miles an hour, new legal limit. The driver explains that he's a grad student in engineering. All the modified cars at the airport—impounded by the military—had gathered here before Delta-3 hit, for an annual "Fuel Saver Race."

"I've got extra batteries and a power cord in back. I can plug into any electric outlet. I get ninety miles a gallon with a fifty-fifty mix of gas and electricity."

Gillette eyes the hotels coming up, the pockmarked facades, smashed windows, flash of binoculars on roofs. *Army snipers,* he thinks. The boulevard is empty except for burned-out cars. A barrier of smoking tires has been pushed aside. Half a dozen soldiers on patrol disappear into the Tropicana. The fences that normally keep pedestrians off the street have been knocked down, smashed through.

Gillette sees an upturned helmet outside the Aladdin. A roulette wheel without a table, in lightly blowing sand.

It's funny, he thinks. Usually in cities, even late at night, you get a sense that buildings are occupied. Maybe it's just logic, knowing people are there, or perhaps it's a psychic connection with others in the species. But these glittery hotels feel as dead as an Incan monument.

There's something about these empty hotels . . . something useful they remind me of, but what?

"Pull over," he says as they approach a stalled bus.

"They warned us not to stop, sir. It's dangerous."

"Do it," he says. The driver halts in front of Bally's casino. Gillette recalls words of his dead mentor, Dr. Larch.

"Sometimes on an outbreak you'll get an urge to do something that even you don't understand. It's your unconscious working, making connections. Good investigators work from knowledge. Great ones use instinct too."

"Sir?" a soldier urges. "We ought to go."

What am I looking for?

Suddenly the windshield shatters. Gillette feels something wet and warm blow onto his face.

The world goes red as his escort grabs him.

Mutineers firing M-16s pour from the stalled bus.

The driver's exploded head has been blown sideways against his window. Gillette's guards pull him safely to the hot street, where he wipes the kid's blood from his eyes. He's uninjured, but his escorts wedge him in between them, firing when they see attackers.

"Give me a gun," Gillette says.

"Sir, I'm not supposed to do that."

From overhead, and a pedestrian walkway, two mutineers shoot down at them. Bullets whine past and riddle the car. The half dozen raggedy-looking soldiers who had poured from the bus have fanned out, using stalled cars for protection, trying to circle behind Gillette.

Gillette reaches out and takes the corporal's hefty .45 from his web belt as an amplified voice cries from the overhead walkway, "Give us your cars! We won't hurt you! Walk away from your cars!"

"Shit, let's give it to them," says Gillette's other soldier.

"No way in hell," the corporal snaps, and calls for help over his helmet communicator. But the sound of shots will draw aid too, Gillette knows.

In the distance, he sees a copter, smoke, mountains. Gillette sees the glittering pyramid casino, the Luxor.

Brrt . . . Brrt . . .

He fires. Men above dive for cover.

Brrt . . . Brrt . . . Brrt . . .

The mutineer in charge starts screaming over a hailer at his men. "Not the engine! Not the tires!"

Of course, Gillette thinks. *The safest place for us is right here, pressed against the car.*

Clearly the attackers had hoped to kill everyone in the cars with the first wave of firing. Now they must be enraged at their failure and terrified that regular troops will arrive. Gillette ducks as a fusillade shakes the chassis. Whatever little restraint the mutineers are exercising in choosing targets won't last.

They don't have time to wait.

They look like my guys. Exactly the same.

For weeks now the third world that Gillette regularly visits, and the world of home and relative order, have been coming to resemble each other. Now they converge as the mutineers charge, going for broke as their own leader screams for them to take the cars.

Bam!

Attackers fall as Gillette shoots, feels recoil, hears something buzz past his ear. He thinks, *I left the microbial testing equipment on the backseat.* He thinks, *It better not get damaged.* He thinks, *That soldier is running toward me. Stop thinking about the equipment.*

Things moving so fast.

Gillette sees his own hands extended, feels more recoil and watches the attacker stumble as his weapon clatters onto the street. And now a duck has appeared in the street! It must have flown over from the Hotel Venice "canal." It's flapping around, squawking amid the soldiers. A duck. A damn duck.

Gillette's finger pulling the trigger, but the .45 is empty.

The firing falters. Both of Gillette's escorts from the second car are down. Bodies lie amid stalled cars.

I made the driver stop. I'm responsible for this, he thinks, standing, looking out at the carnage as Humvees arrive, disgorging troops.

He starts to check the killed and wounded, but if he stays to help the Army doctor here, he risks being detained. He's vaguely aware of the corporal explaining to a furious officer why they stopped in the first place. It's because Gillette wanted to look at the hotels. Gillette feels angry stares directed at him. His left hand has started shaking. An officer is shouting into his face, demanding to know why he ordered the cars to stop. The officer takes Gillette's ID back to his Humvee.

The officer writes down all information on the ID.

"I asked," sneers Gillette's corporal, "was our stop useful for you, sir."

"Take me to Cougar," he says.

Twenty silent minutes later, on the outskirts of the city, the car pulls up before a barbed-wire fence. Beyond it is a horseshoe-shaped building of brick and dark glass.

Gillette's pulse races. He's still fighting off nausea from the fight. Guarding the fence he sees soldiers, and more troops inside the entrance booth. The dark windows make it impossible to see into the building, but clearly researchers are hard at work in there. This place hasn't shut down.

Now to look for the needle in the haystack. To hope the trip wasn't a waste of life.

I'm due at Detrick tomorrow. And that officer back at Bally's is probably checking my ID right now. Will Hauser find out I'm here?

Gillette knows he is running out of time.

FOURTEEN

DECEMBER 1ST. LATE AFTERNOON, EST. 34 DAYS AFTER OUTBREAK.

The convicts and food convoy come together in a freak ice storm, on I-95 south of Petersburg, Virginia, at a roadblock of cut pine trees that the escapees laid across the road. It's a perfect spot for an ambush. Communication towers are out here. SOS calls can't be sent.

For four hundred starving felons, a cornucopia of supplies is coming. Deserted by their guards days ago, the starving men had broken out of Sussex Prison this morning. They've been rampaging on foot through the countryside, looting and murdering after breaking into the abandoned National Guard armory at Garderville. Some of the convicts are veterans, and know how to use the M-16s and rocket-propelled grenades they took from the arsenal there.

The approaching convoy is the largest relief shipment sent to Washington—and Marion Street—since the oil crisis began. The ninety-nine tractor trailers are ten hours out of Wilmington, North Carolina, loaded with clean fuel and canned meat from Raleigh packing plants. Convoy routes and destinations vary nightly. Humvees guard the trucks, inching forward behind a plow.

Drivers peer through view slits. Ice coats the close-packed pine forest lining both sides of the road. Refugees had been out earlier, walking the interstate, begging for food in the wake of the new zoning law. But with the worsening storm they've dropped away, looking for shelter.

Now the plow halts before what looks like a multicar accident, and its crew reports back to the convoy commander that the cars were actually stopped by logs blocking the road. Then the report grows strident. Passengers stripped of clothing lay shot or clubbed to death in cars. Trunks and glove compartments have been rifled. A dead state trooper is sprawled in the median strip. Ice covers pools of dark frozen blood.

"Go around," the convoy commander orders. "Now."

But at that moment the first rocket-propelled grenade hits the plow, blowing the side door off and killing the driver. Convicts in looted coats rush from the forest on both sides, and rise out of the median strip, shooting as they rush forward, screaming for courage, attacking in a wave.

The Humvee gunners pour murderous fire back, driving off the convicts at first, but the storm is blinding, and the guards used up too much ammo this morning, repelling snipers on the Michael Jordan Highway in North Carolina. Now attackers crawl forward, and call out to each other in the woods. They spray the Humvees with M-16 fire. For each convict that falls, two more seem to appear.

Molotov cocktails sail into truck windshields.

With the plow disabled, the trucks can't back up.

Within the hour, the soldiers are dead, the lead trucks on fire. Dead convicts litter the road, but healthy or wounded ones loot the semis, ripping open boxes, wolfing down food.

The convicts get the plow running. They drive off in the functioning trucks and exit I-95 in rural Virginia.

One imaginative convict has even taken photos with a looted Nikon. Thirty minutes later, in a farmhouse that has electricity, he sends them out over the Internet, on the Web site of Riseup.com, a new oil emergency blog.

Millions of people check Riseup.com hourly—when electricity works—for photos, news, rumors of battles and oil cults (whose adherents worship derricks), gossip from inside the White House. Alongside the photos of the smashed convoy, Riseup reports the president praying in the Oval Office, weeping, asking the Lord for guidance, unsure what to do.

At 5:48 P.M., on upper Connecticut Avenue, Gordon Dubbs stands in his new top-floor apartment, eyeing the convoy photos. His pulse races happily at the sound of explosions coming from the southeast and the distant glow of fires burning on Rhode Island Avenue, despite the freezing rain. Riseup.com reports pitched battles there, between mobs and soldiers, at the border separating zones B and C.

Washington's poor are trying to break into the zones that have food, Gordon reads, and believes.

Like an animal sensing an oncoming season, Dubbs is waking to opportunities, experiencing a sense of power that was not there before.

I have food. Medicine. Liquor. I've turned this building into a fortress. We have everything we need to live well except heating oil.

Gordon Dubbs, reborn amid fire. The ex-cop dreams of becoming neighborhood warlord if the emergency becomes permanent. The homes nearby in flames in his mind. The police pulled back into zone A. The Oasis a feudal fief with Dubbs as lord, as Darwin's natural law replaces civil service in Washington.

The key, he knows, is to know how far to go each day, to wait for police to collapse completely and then make a big move.

He's always felt special, even back at age five when he saw that people treated him differently because of his fantastic smile, his film star–quality face. Gordo at ten talked his way out of detention for cheating on tests. Gordo the high school football tackle talked his way out of charges that he and his friends used brass knuckles on an opposing running back the night before a game. Gordo the U.S. Army sergeant testified that *he* never tortured prisoners, and turned into the cop who laughed away rumors that he sold cocaine from the evidence vault, and shot a suspect after the man surrendered. He sweet-talked his way into marriage. And after that, dozens of affairs.

Because of that great smile.

And then at the ripe age of thirty-five came the shocking revelation: there are some accusations you can't talk your way out of, especially when tapes prove you lied.

"The kid pulled a gun," he told the review board.

"You beat him to a pulp. You planted that .38. The boy didn't even break the speed limit. You're a disgrace."

"Tapes made without consent are inadmissible in court."

"Which makes you lucky, Dubbs. The deal is, quit and we don't press charges. The family will accept a settlement. But you will never work as a cop again."

Just thinking about it makes him want to break things, especially when he remembers the humiliation after that. His wife divorcing him. The begging for jobs. The come-down of becoming security chief for Three Faith Charities, which sends donated supplies to disaster areas overseas.

Gordo remembering driving out to an industrial park near Baltimore-Washington Airport every night, listening to the rap of his shoes in the warehouse, shining a flashlight on tons of canned meat and vegetables, vitamin drinks, bicycles and pumps, antibiotics, Land Rovers, hand-crank radios, propane fuel, antibacterial soap, tarpaulins, sledgehammers, Levi's, aspirin, Betty Crocker cake mix, dental drills, tires and about twelve million plaster statues of Jesus Christ, bound for people who probably never worked a day in their lazy, overpopulated lives.

Gordon hating every minute until he found opportunity! He realized that the home office back in Minneapolis didn't have a clue about what happens to all their booty once it left the warehouse. Those Midwest do-gooders were no mathematicians. Manifests were signed by morons. The volunteer "accountant" was more interested in watching prostitutes strip to Donnie Osmond music in the office.

And then things got even better, because Gordon finally got into the caged area in back one night, and couldn't believe what those idiots had stored there, ready to ship to Sudan, Indonesia, Zimbabwe.

Bribes! That's what. For local officials. Watches. Perfumes. Cigars. Champagne. Even a few Russian AK-47s in the mix. Can you believe it? A charity!!! Or maybe even not a real charity. Maybe a cover to ship arms.

Time to hire the proper staff, he'd thought.

Next thing he knew, the two honest guards who were working for him were gone, and new guys, whom Gordon knew in the street, were his staff. Lester Gish and Basil Prue were failures who believed him when he told them how they got screwed by life and deserved better. They were guys who would kiss *his* ass for a change, and wholeheartedly throw themselves into pilfering.

Now, as police protection in Washington has diminished, Gordo's proclivity for violence has returned. The first death had been an accident, he tells himself. The old man had returned home from the swap mart to find Gordo and friends looting it. The guy had started screaming so Gordon whacked him with a poker. The man had seen their faces.

It was his fault, Gordon thinks. For coming home early.

The second time had occurred when a gang of public-housing kids—Metro Ferals, the blogs call them—hit an Ibar's Electronics store at the precise moment Gordon's bunch were relieving it of closed-circuit cameras. The shooting had been over in minutes. The three men still relive the thrill when drunk. "Whooo-eeee! Did you see the tall one go down?"

Tonight there's a fire warming the hearth. The Oasis was once luxury, and all D line apartments feature working fireplaces. This apartment had a whole library of books too, and the glow of volumes that Teddie is burning—like *The Three Musketeers* and *The Plague*—extends to mounds of boxed supplies against the walls. It's a cornucopia of warehouse goods stolen after the oil bug hit, before the Three Faiths Charities Land Rover ran out of gas.

If only we had heating oil. My people won't stay loyal unless I give them heat. I must find heating oil!

He's distracted as his cell phone rings. At least it's working at the moment. Basil, guarding the lobby, reports that Gail Hansen of Marion Street wants to see Gordon right away. She's one of several women who occasionally sneak over, to ask for food, or flecainide acetate for a heart patient husband, or baby formula, or some personal treat.

She's old. But I could use a blow job.

Gordon's dressed in his postapocalyptic best. Looted donated cords—last year's models, tax write-offs—from Barneys. Donated flannel shirt by Calvin Klein. Looted Timberland boots and thick sweater, that had been bound only weeks ago for Turkish earthquake victims.

On the way downstairs, he stops at apartment 6C to make sure that the elderly Roths have enough food. Mike Roth is a retired cop, and has always treated Gordon with respect. The college brats in 5B are gone—fled back to Mommy and Daddy. Basil and Lester share that apartment now.

He inspects as he descends, using the stairs, not the elevator, in case power fails again. Let Gail wait and grow desperate. Apartment 3-D houses Dr. Raskovitch, and the building always needs doctors. Gordon drops a bottle of pinot noir over. Todd and Jeffrey, the two gay interior designers in 3C, have been scared off. Their apartment went to Jovina and Ike Mimeaux, both former chauffeurs for the state department, and ferocious fighters in looting situations. Both were ejected from St. Paul's for fighting with other refugees after snatching their food.

Tomorrow I'll have them steal some metal traffic signs, to block off windows on the lower floors. And I'll ask Frank Gerard, the ex-bouncer from Clyde's, if he wants to join up.

Gail is already in 1F when he arrives, let in by Basil, who knows the drill when women show up. The studio has been set up so he can receive visitors: potential new residents, attractive women who need favors, Army clerks who give tips about food distribution, and soon, hopefully, ambassadors from other fiefs.

Ah, he thinks walking in, appreciating the looted leather furniture. A terrific wet bar and silky double bed. Weird expensive wall art traded for vodka by Gail Hansen. Class A Nakamichi CD player and a thousand CDs: jazz, rock, even classical for the snooty intellectual types who show up.

But it's freezing in here.

"Hi, Gordon. I have something special to trade today."

He smells the booze on her from four feet off. Her attractiveness is plummeting by the hour. The old haughtiness in Gail's face—once an invitation to knock her down a peg—is growing into pinched and permanent fear. The clothing is rumpled, since she has no dry cleaner anymore. The cosmetics are gone. The veneers of the professional woman—BlackBerry, hair stylist, even shoe polish—recede into the past.

"I brought you a superb new oil by Savilliari," she says, flicking her hand at some ridiculous-looking canvas, black on solid black. Not even a picture. Just a mass of paint.

"I don't need more art," he says.

"Au contraire, Gordon. It's an incredible investment."

"Then why don't you keep it?"

"Why don't we discuss this over a cocktail," she says, crossing her legs, which aren't bad-looking, he has to admit.

"Why don't you shut up and suck a cock," he says.

She's shocked and furious and he couldn't care less. He wonders with excitement how much humiliation she'll take. She forces her rage into bright shallow laughter. "What a crude sense of humor you have, Gordon," she says.

"I'm Mr. Dubbs."

He tells her that he's sorry, he's almost out of vodka. He needs all his remaining liquor for "important" trades, not "favors for neighbors." He tells her that if all she has to offer him is artwork, he's sure that some other building—that has liquor—will be glad to take her little oils to adorn their walls. He tells her, enjoying the raw need growing in her eyes, that "unless you have something really valuable for me," his dwindling supply of Smirnoffs, Chivas, Icelandic and Dewar's must be carefully allocated.

Then he gets up to leave and she grabs his shirt.

After the blow job, the last one he'll accept from her, he decides, he offers her a quarter-filled bottle of Three Czars vodka, watered down, for the shitty painting, and watches her swallow her pride and accept the trade. He's half distracted, and thinking that he ought to get Teddie some girls too, when he hears her say, "I *told* them to trade our extra heating oil. But they wouldn't listen!"

"Eh?"

"Gillette and the men managed to find two thousand gallons of home heating oil. We stored it in a basement. It's more than enough for us. Maybe I can convince them to trade some to you. Do you need heating oil, Gordon?"

"We have more than enough of that," he says, pulse soaring.

After she leaves, told that she can come back again, he goes upstairs, thinking furiously. He can see thick smoke rising from every second chimney on Marion Street. The furnaces are roaring over there, for sure. But was Gail exaggerating? *Two thousand gallons?* And where exactly *is* the extra oil?

He needs to go scouting.

From the storage area he selects as a "peace offering" a few cans of Starkist tuna, a few packages of black bean soup.

I'll say that I noticed how the kids on Marion Street look hungry.

His unlicensed Glock 9mm pistol goes into the belt holster, in case another gang is out tonight.

He makes sure that Lester is monitoring the building's security cameras, making sure no one gets to the loot upstairs. That done, the future leader of upper D.C., General Gordo and El Supremo of the weak-minded—trailed by safety escort Basil—trots through falling rain across Connecticut Avenue toward Marion Street.

When he gets within thirty feet of it, he sends Prue back, tells Prue to wait for a call, stay sober and return when phoned to escort him.

Prue says, "How about bringing me that cute black girl. The Gillette girl."

"What did I tell you about young girls?" Gordon snaps.

Prue hangs his small head.

"It wasn't my fault. She started to scream."

Gordon leaves Basil looking like a dog in heat, walks the last quarter block and turns onto Marion Street. He passes the Subaru blocking the north end, steps onto the porch of the Gillette home and reaches for the bell.

I'll ask them if they want to team up for mutual protection. I'll casually mention heating oil. I'll see how their faces look when I do.

But it turns out he doesn't need to do it. Because as he reaches for the bell, he hears, through the open window, Gail Hansen arguing with Marisa Gillette.

"Why can't we trade the extra oil?"

"Nobody can know we have it, Gail! Don't you understand?"

"Hey!" a man's voice snaps, behind Gordon. "What are you doing?"

Gordo spins. Materializing through the rain is a stranger in a hooded parka who, close up, turns out to be the ex-Marine prick, one of Gillette's goody-goody friends.

"Gordon Dubbs! What's in that bag?" the guy demands.

I don't believe this! It's a free country. I can go where I want.

Now a second man appears at the porch, holding a gun!

They're armed, Gordon thinks, startled.

Gordon backpedals. "I'm here on a social call."

His bag is snatched away at the precise moment that the door opens, and Mrs. Gillette looks out.

Gordon tries to avoid a scene. He goes charming. He flashes the old winning movie-star smile.

"There's been a mistake."

"There's *food* in here," snaps the Marine, holding up a Starkist can like it's exhibit A in a courtroom, like Gordon stole it, which he did, of course, but from *other* people, not these people. These people are supposed to *get* the food.

"Whose house did you rob, Dubbs?"

Gordon drops his hand slowly, seeking the cell phone. "I noticed that the children on this block look hungry."

"Gun!" shouts the former Marine, mistaking the glint.

The blow catches Gordon in the midsection, drives him onto the ground. He can't breathe. His rage is titanic. They're groping at his belt, taking away his Glock pistol. But they also see now that he had reached only for the phone. They help him up, return the bag, and propel him off the block at gunpoint, without even apologizing. They even give him back his gun but keep the magazine.

Gail told the truth, he now knows, through the pain in his gut, the sick throbbing that intensifies his fury.

Those fuckers have two thousand gallons of home heating oil.

FIFTEEN

DECEMBER 1ST. DUSK, NEVADA TIME. 34 DAYS AFTER OUTBREAK.

Dr. Veejay Varunisakera looks more worried by the second.

"Are you telling me that Delta-3 has jumped to humans, Dr. Gillette?" he gasps.

"That would be premature. Consider my visit a symptoms survey. Nothing more."

Gillette smiles with patently false reassurance—as if there really *is* a danger of an outbreak at Cougar Energy—and concentrates on the man across the desk. He must push away images of the fighting in town, the dead men on the boulevard, the soldiers shooting. He must improvise strategy as he goes along.

I'll take samples of blood, tissue, urine. Then I'll sample oil fluids they make, and their tanks, and eyeball the labs and factory. But I need time to finish.

The fastidiously groomed Dr. Varunisakera is small, fortyish, soft-spoken and Southeast Asian, judging from photos framed in silver across the rear of his oversized wooden desk. Gillette sees an old man wearing a conical hat, leading a water buffalo by a rope. Then a younger man with Dr. Varunisakera's gray eyes and sharp cheekbones, dressed in medical whites. And lots of big happy family shots with the wife and two daughters, from the Grand Canyon, Disneyland, Glacier National Park, the Everglades.

The more scared he is, the more he will cooperate.

"The rashes that CDC found on lab workers may be coincidental.

We're surveying all oil and fluids companies," Gillette lies. "I wouldn't panic just yet."

He smiles in a practiced bureaucratic way, guaranteed to heighten a listener's anxiety.

"I'll help any way I can," Varunisakera replies, and adds something that makes Gillette feel guilty. "I'm living in an A zone. Unfortunately, my wife and children are not."

Reach a certain point in carrying out a deception and it flows naturally or self-destructs. Gillette's ID has worked magic at Cougar. He sailed through security but then had to wait two hours for Dr. Varunisakera to return from town. The head of Cougar's Fluids Research Division dropped all work to speak to Gillette.

"You'll want to interview my staff," he says.

"Certainly."

"The fight to stop the ODB is our top priority. It's all we've been working on."

Which means the microbe is already here. So how do I determine if they created it or not?

Gillette says, "I'll start with some standard questions. The FBI microbiologists who came here may have asked these. I hope you won't mind. I've found it's better to go back over things sometimes. Answers can vary."

Varunisakera understands. "Frankly, Dr. Gillette, I was surprised their visit was so short. They seemed more interested in trying to find terrorist connections. I don't recall any microbiologists among them. Just investigators."

Figures, thinks Gillette in disgust. He asks, "Give me some general background on your work on fluids and bactericides."

"As I said, our regular projects are on hold. But normally we look for less expensive ways to kill bacteria that cause degradation to oil machinery. We work on chemicals and bioremediation, microorganisms that decompose organic compounds with enzymes . . . Is this what you want?"

Gillette nods, taking notes to study later.

"We work with recombinant DNA," Varunisakera adds, "to improve on organisms that show potential against corrosive bacteria. Using plasmids as vectors we microinject genetic material into target cells and see if the 'bad guys' die."

"Sounds fascinating."

"In the old days, before I came here, the company also experimented with contact-dependent inhibition. As you may know, Escherichia coli, the same bacteria that causes urinary tract infections in humans, contains genes that block growth in other E. coli they touch. They don't kill the other bacteria. But they halt growth. No one knows why. Cougar hoped to find natural inhibitors in nature."

Gillette looks up sharply.

"You mean that Cougar collected new organisms?"

Varunisakera nods. "In caves. Meteorites. We sampled soil and sands near wells, and water around hot vents. But we never tried to find anything that attacks refined oil. Just crude oil spills."

"You refer to the work in the past tense. Why?"

"It was costly. There were no good finds, or leads."

Gillette asks hopefully, "Are the samples still here?"

I can get them sequenced.

Dr. Varunisakera spreads his hands. "Destroyed, sadly. When the ODB first appeared I checked our sequence bank, tried to match the genomes with the bug. Nothing came close. We passed this information to the FBI. Would you like copies?"

"Thank you. Maybe someone who worked on the project might remember something."

"The FBI asked that too. Contact-dependent work was popular with many companies years ago. But the dream of finding a useful superbug in nature never came true."

Gillette swallows his disappointment. He'll examine the data bank later. But time is passing. He asks, "Tell me more about the bacteria you're trying to kill in wells."

"Well, you always find some RDMs—reservoir-damaging microbes—at wells. They're in unflushed reservoirs, injection water, even cores. Some people believe they get into fields from the surface during drilling.

Others think they live down there. Either way, oil companies are very interested in finding cheaper bactericides. Several joint research projects are under way. But cooperation is problematic. Companies are secret about bugs in *their* wells, but want to know everything about what their competitors found."

Gillette nods.

"So trusted third parties like Cougar act as go-betweens. We take samples at thousands of wells. We maintain common libraries of microbes. We manufacture biocides, test them and provide results to all clients simultaneously. A strictly supervised process. Guards at collection points. Independent oversight. Step-by-step reports. Our clients are very suspicious of each other."

Gillette tries to envision the vastness of such a project. *He's saying that companies like Cougar serve as clearinghouses for samples going in and out of fields around the world. A perfect vector point.*

"Are the biocides you test made here?"

"Across the compound, in the plant."

"During tests, are they injected directly into wells?"

Varunisakera shakes his head. "No. Mud samples are extracted at different depths at wells around the world, tested for bacteria and shipped here if they're infected. None matched the ODB."

Dr. Varunisakera frowns as if a new thought has come to him. "I thought you were investigating ODB jumping to humans? May I ask why you're asking these questions?"

Gillette taps his pen on his pad impatiently, flashes a look of bureaucratic annoyance.

"We're trying to establish baseline information, Doctor. To decide which facilities to test for contamination, which to pass up. Of course that decision depends on what goes on at these places."

"I wasn't trying to challenge you."

Gillette smiles in a way that says, *Sure you were.*

"Please explain more about your cleanup work."

"Bacteria create sulfates that sour wells and corrode pipes. Long-chain paraffins in crude oil form deposits that impede flow. We're trying to design bacteria that can survive extreme conditions, emulsify the paraffins."

Like Dr. Frankenstein at the microbial level. Did one of these microbes mutate? Or have a bad side effect?

Dr. Varunisakera says, "By applying heat-tolerant anaerobic bacteria to fields, we've increased crude oil recovery rates 24 percent in Nigeria. We've sequenced the genomes of over 253 bacterial strains which may help reduce contamination."

My God! Delta-3 is heat-resistant!

"Finally, we use microbes against spills and oily sludge generated at refineries. Cougar's CLEANUP 3 has helped eliminate hundreds of thousands of tons of waste."

Dr. Varunisakera says proudly, "I'll show you the labs while we wait for that printout you asked for—staffers' names and addresses. Talk to my people. Perhaps they will remember something useful that I do not."

The "war screen," as Varunisakera calls it, is a gigantic monitor dominating the back wall of the lab. The Mercator world map on it is like the one Gillette saw at the Pentagon, except for a key difference. It only shows Cougar operations.

Gillette takes in the capillary-like global network of crimson-colored oil pipelines, emerald gas pipelines, purple derricks signifying fields in which Cougar works. Dark-blue barrel symbols show refineries.

From California to Kazakhstan, Louisiana to Liberia, Cougar's contaminated operations are lit with red lights.

Gillette starts analyzing. *When Raines checked there was a 42 percent correlation with infected Cougar fields. Now the number has jumped to at least 80 percent. Did the FBI miss that or ignore it? Or are they still investigating behind the scenes?*

Dr. Varunisakera brings him back to the present. "Just about everyone here has loved ones in trouble spots."

Gillette takes in the banks of fluorescent lights above workstations. Researchers in white work at microscopes, grinders, incubators. The high-speed computers and automated sequencers are the kind Gillette would see at any large university.

But the people here look more intense. The urgency is obvious in the way they consult in low voices, and show complete lack of curiosity about a visitor. Photos tacked to corkboards show children, spouses, pets, homes, parents.

"Samples are over there," Varunisakera says, indicating floor-to-ceiling glass cabinets in which lie rows of petri dishes filled with amber or black liquids. "Those come from infected operations, although it's growing impossible to transport samples here. Sequencing gave us some information, but not enough."

They must have a thousand samples.

At the closest workstation, a man wearing a Boston Red Sox cap transfers liquid by glass pipette from a dish to a test tube as Varunisakera explains.

"In site-directed mutagenesis, we know where a gene is in a sequence but not what it does. So we inactivate it chemically. Gene by gene. Then we see, with each gene inactivated, if the microorganism still destroys fuel. If yes, we move on to the next gene. If no, we've found our gene."

"No luck yet, I take it."

"But hope."

At the next station a woman peers at a computer screen on which Gillette sees magnified shots of frothing bacteria. Fixed beneath the lens of her microscope is a squarish plate of clear plastic, lined with what looks like tiny bits of greenish dust.

"Microarray," Gillette says, recognizing it. "It's how researchers identified the anthrax strain found in Congress."

"Yes. We put a whole array of genes on a chip, fragmented DNA. We subject it to heat and refining chemicals. If a spore forms, we've isolated the trait."

Dr. Varunisakera sighs. "Sporulating bacteria tend to live in soil, not oil. And none that I've heard of degrades oil, until now."

"You seem to be doing a good job," Gillette says.

"Dr. Gillette, my fear is that we may miss the link even if it appears in front of our eyes. Why? Because what if the trait is caused not by a single gene but several, acting in concert? What if we miss a *mix*? Genes often don't work alone. Bacteria in nature don't grow in monocultures either, but in complex food webs. So many questions, they drive one to madness. Once the ODB survives refining, how does it continue to live when the other microbes in its community have been killed?"

Varunisakera stops work so Gillette can take blood samples and give the staff a talk on the "possibility" that Delta-3 has spread to humans. Has anyone contracted bullseye rashes over the last few months, he asks, or high fevers and joint pain? Dr. Varunisakera hands out copies of Gillette's bogus symptom questionnaire. He asks people to return the completed forms by tomorrow morning.

But instantly the staffers ignore work and start studying the forms. There's nothing like the fear of contagion to distract people. Gillette feels a stab of guilt at halting real work on Delta-3.

Someone needed to check Cougar better, he thinks.

But he also thinks, *The likelihood of someone here coming up with answers is greater than me doing it. It's funny. These people are like normal bacteria, functioning in a interacting community. I'm like Delta-3. Working alone.*

A tall, bearded scientist raises a hand. "These symptoms sound like Lyme disease. My sister has it, in Connecticut."

That's because it is *Lyme disease,* Gillette thinks. *When I made up symptoms, those came to mind.*

Another scientist, a young woman with long, prematurely gray-streaked black hair, asks, "Where exactly have people gotten sick?"

He makes up answers based on red lights on the map, at Cougar locations. "So far, Sri Lanka . . . Indonesia . . ."

Damn General Hauser. I wish a microbiologist had been included in the original investigation here.

Suddenly Gillette can't help but flash back to the battle in town, the sight of his driver bleeding onto the hot street. Then the scene switches and he's staring at the empty hotels, trying to understand why he asked the driver to stop.

A wave of discouragement washes over him.

What if Hauser was right, and the problem is terrorists? What if I'm what Hauser said, a self-important bastard who should have stayed home to protect his family and friends?

"It's late. Let's all go home and continue tomorrow," says Dr. Varunisakera.

"Oh, I'm not tired," says Gillette.

"I've learned, Dr. Gillette, that a tired staff is an inefficient one. Besides, no rush you said. Right?"

"Right," says Gillette, eyeing the clock.

DECEMBER 2ND. MORNING. 35 DAYS AFTER OUTBREAK.

He wants to get to the factory but the researchers have become so fearful that they keep him in the lab, after a sleepless night spent in the site dorm. Gillette must treat their questions seriously. After all, he's supposed to be concerned with human disease.

"Dr. Gillette, I had a jagged rash on my arm in April. Are you sure you only want to know about bullseye ones?"

"Dr. Gillette, I didn't get a *high* fever in Peru, but a *low*-grade one that lasted three months."

"We're only interested in high," he says.

Varunisakera hovers close, listening to Gillette's reassurances, wasting more time. Every time his phone chimes, Gillette wonders if Washington is calling. Every time Varunisakera gets a call, Gillette wonders if he's been found out.

Now one scientist asks the others, "Hey! Anyone remember that virus that went around the lab a few months ago, that no one could identify?"

Five more minutes and I'll make an excuse and head for the fluids plant. I cannot afford to make Varunisakera suspicious, but at this rate I'll need to stay here for several days.

The scientists and technicians start telling war stories, bragging about illnesses contracted in the field.

"I took daily shots of arsenic against leishmaniasis, in Sudan. Those shots *hurt.*"

"Ever get blowflies? Those little buggers erupted right out of my head!"

"My favorite one," Gillette hears, "was Lyle."

They laugh. Most of them already know this story.

"Good old Lyle. Twice he comes down with the bends and *still* he wants the assignments, for Christ sakes."

Wait a minute, Gillette thinks, a ticking starting in his head. *Deep-sea divers get the bends when they surface too fast, and nitrogen bubbles in their blood expand. Why was a Cougar scientist getting assignments in the ocean?*

"Lyle's a diver, I take it?" Gillette asks.

The speaker is the man who'd asked about Lyme disease. "Dr. Samuelson worked here six, seven years ago, on recombinant DNA. The company must have wasted millions, sending him around the world, checking out deep-sea thermal vents, looking for new life."

Gillette's heart starts to pound.

Thermal vents contain the most heat-resistant bacteria on earth.

"Did Lyle find anything useful?"

"Oh, Lyle had theories about worlds of undiscovered microbes. Deep rich biosphere and all that. He was sure ten billion new species are waiting to be discovered down there. He just wanted an excuse to go diving."

A few researchers chuckle.

"I did not work here then," Dr. Varunisakera says.

Gillette persists. "He went down himself? Aren't those vents too deep to reach by diving? And too hot?"

"Sometimes he went in minisubs, and collected samples in cooler water further from the vents. Sometimes he'd dive alone and wait for the robots to come back. Any excuse to get into the water."

"Where is Dr. Samuelson now?" Gillette asks.

"He took the buyout and moved back to Massachusetts. He custom-makes kayaks, last I heard, and lives alone. He never returns e-mails. We stopped trying to reach him."

Gillette checks the printout of current and former lab employees and spots the name and social security number. In the men's room he calls Raines at Fort Detrick.

"Where are you, boss?"

"Home," he lies, because anyone who abets his trip knowingly has broken the law too, and can be arrested.

"Well, you won't believe what's coming over the wire. Cannibalism, boss. In the *Bronx*."

"Find out everything about Dr. Lyle Samuelson. Bank accounts. Travel. Credit. Affiliations. Criminal records. Education. Anything you can think of that might relate to planning or carrying out genetic experiments, either at Cougar or at home."

"By the way, boss: the van busted. So it won't be out to Marion Street to pick you up until tomorrow."

"Whatever you find, keep it between you and me."

Dr. Varunisakera finally starts to get suspicious when Gillette asks the wrong question.

"Ever work on oil-related biological weapons programs here?"

They've finally left the lab, and are driving in an electric golf cart down a private road flanked by fenced-off sagebrush desert. The fluids plant is idle, ahead.

The question startles Dr. Varunisakera, not in a way that makes him look guilty, but like it made him wonder something unpleasant. Varunisakera holds Gillette's gaze. His first thought has led to a second.

"That sort of project is illegal, Doctor, under the Biological and Toxin Weapons Convention!" he says as they walk inside.

Gillette says, "I know, but you haven't the slightest idea of the kinds of things we've found out at other plants."

Varunisakera's jaw drops, and his eyes widen.

"Which other company violated the law?" he asks.

"That's confidential, sir," Gillette snaps.

Varunisakera seems to be fooled again.

Gillette opens his small travel/sampling bag.

"I'll walk around a bit, and scrape some surfaces," he says. "The same procedure is in effect at several companies."

"You're looking for spores in *here?*"

"If a mutation occurred, it could have started anywhere. And the spores may have traveled to other spots in the compound, through the air."

The two men stand in a vast high plant, where the conveyor belt, Varunisakera explains, normally carries barium powder in from outside, mixes it with chemicals, and incubates the mix in four large round vats that take up most of the floor space. The five-foot-tall vats are made of copper, and covered with glass domes. An automated system of valves and thick pipes runs the mix to a second building, where it is bottled and shipped out around the world.

Gillette carries his kit to the first vat as Varunisakera opens the dome. He reaches down and with a spatula scrapes off a beige residue left over from the mixing stage. He deposits the residue into a plastic vial and stoppers it. He does the same thing in all the vats. The dome interiors are sparkling clean. No dust. Probably no spores.

I have to sample the belt too, and the barium outside.

"Now I remember! The FBI did take some samples in here," Varunisakera says, watching in fascination. "Perhaps they did have a microbiologist on their team."

Which makes the likelihood of finding anything even less likely, Gillette thinks, but keeps going.

He needs twenty minutes to finish the sampling, then tells Varunisakera to take him to the bottling plant. But at the door, Gillette is seized by the same feeling of missing something that had bothered him in town.

He stops in his tracks.

"Is something wrong?" asks Dr. Varunisakera.

"I don't know. Can I have a few minutes to think?"

He returns to the main floor, walks, looks, lets his eyes stray to walls, floors, a glassed-in manager's office. He goes into the office and scrapes the desk surface and file cabinets. The feeling gets stronger, as if he is missing the answer to a problem instead of figuring it out.

What could connect a factory and hotels?

Think, he tells himself, screaming with frustration inside.

Okay, envision the trip here. Try to recall the exact moment when the odd feeling hit.

What's the connection between empty casino hotels and a fluids plant? That they're built by the same people? That workers from here may have stayed there? No. Start again. The buildings look completely different . . .

Shit.

Dr. Varunisakera asks, "May I help you?"

"It was because they were empty," Gillette says out loud.

"Excuse me?"

"They'd been evacuated."

"What are you talking about?" Dr. Varunisakera says.

"Evacuated hotels and microbes," he says, seeing it, his heartbeat picking up, his eyes now looking for something specific here, that he had not sought before.

"What kind of air-conditioning system do you use?" he asks.

"Are you hot? I can turn it down."

"What *kind*? Central? Or individual units?"

Gillette spots vents now, on the far side of the huge room, away from the closed vats. They're on the ceiling, grated and small and *the plant is equipped with central air-conditioning, like the last evacuated hotel I visited during an outbreak.*

That hotel had been in Seattle. It had been a hotel where Legionnaires' disease had killed six.

Legionnaires' disease doesn't sporulate, but it travels through AC units. Shit! Just before Delta-3 hit, that's what I was doing in Washington! Running drills on how to handle spore contamination from ventilation systems.

"I need a ladder. Now."

They find one in the maintenance closet. From the top rung, Gillette unscrews the air-conditioning grate and peers inside. It looks clean but he takes samples anyway.

"Now the bottling plant," he says.

It's next door. Inside, rows of open plastic bottles sit on idle conveyor belts. Dr. Varunisakera explains that normally the bottles flow under nozzles. The nozzles spew bactericide or drilling chemicals or fluids inside.

Which means that the bottles are open on the belt.

Gillette spots an air-conditioning vent directly above the bottling area. When he unscrews it and looks inside with a flashlight, he sees greenish film lying in the beam.

Could it be?

Excited, he takes samples and scrambles down just as the door opens, and three silhouettes appear in the light. They're men, from the size. Strangers, when they step out of the bright sun. And, unfortunately, Gillette sees with a sinking feeling, tough-looking, no-nonsense soldiers.

"Dr. Gillette?" demands an officer, drawing a .45 from his web belt, looking menacing.

"Yes?"

"You're under arrest. Give those vials to me, please. Slowly. Then, sir, put your hands behind your back."

SIXTEEN

DECEMBER 2ND. 35 DAYS AFTER OUTBREAK.

Pedaling home after scouting out Marion Street's defenses, Clayton Cox stops on Reno Road to give a spare package of soup to a starving beggar boy. The boy—in rags—grabs the soup and scrambles off without looking back. The ones Clayton helps are always about twelve. Back on the bike, he is transported. He sees himself at that age following the man who will become his mentor through the twisting lanes of Amman's old city. The foreigner—browsing in stalls—is about to be robbed by other boys.

He looks like me, young Clayton thinks, fascinated.

The man looks to be in his late twenties, lithe, blond, tanned. He ducks into the stall belonging to Abu, the seller of stolen antiquities. Amman's bazaar is a maze of thousand-year-old stone-walled paths. Abu, or "Uncle Camel" to tourists, beams, bows, waves the stranger toward a stuffed easy chair. He pours sweet tea into a cracked china cup. He rummages through his wares for Iraqi rings he's bought from refugees, Bedouin necklaces, pendants and silver bracelets that have graced female wrists for decades before landing up in his shop.

"I make you a good deal," he says in English.

The beggar boy knows he should leave. What happens to this European is none of his business. What had caught his attention at first was the man's regal aspect. He seems at home in the alien world of the bazaar. The man in cool white against the heat; confident in loose-fitting slacks and a cotton knit shirt. The expensive watch a lure to thieves.

Clayton's counted four bedraggled boys trailing the man for the last ten minutes. They drop back when he goes into a shop, come closer as he walks, to test his alertness, flash hand signals to one another.

Meaning, not yet. At the next corner, okay?

Did my father look like him, Clayton wonders. And at that instant knows he will not leave the man.

Other boys who work the bazaar call Clayton "Feranj" (foreign), or the abomination, sand-face, bastard, *beni naji,* which means son of a dog. But even the strong ones avoid him, as they did in the camp. He's sent too many of them to doctors with stab wounds. He's left attackers bleeding and moaning, staring into his devil blue eyes as he whispered that next time they came he would take revenge on their sisters and parents.

An eye for ten eyes. A tooth for ten teeth.

Since escaping from the sprawling camp two months ago he's been sleeping in alleys and supporting himself by foraging the packed bazaar each day. Snatching a wallet from a tourist, a fig from a stand; a new cotton shirt or pair of Levi's from a rack. The gay tourists prefer clean-looking boys, and invite them to the Intercontinental for a few hours of imagining you are somewhere else, and being rewarded with a hamburger, shower, and dollars, francs, yen.

You are so handsome, the men tell him. So white.

"I am not of you," he says, remembering that his mother had told him often that he carried the blood of a famous man, a revered warrior. Not pigs like these.

"Who was my father?" he'd ask her.

But she'd been killed before he found out.

Lives change in seconds. A boy makes his choice. From inside the stand, he hears the man tell Uncle Camel, "I *told* you I want an emerald."

Uncle Camel holds a ruby. "This *is* emerald! I swear!"

The man pushes himself out of the stall, out into the bazaar.

"Idiot," he mutters, but he is smiling.

I wonder if my father shopped here.

The man wanders past the shoemaker, fig seller, poultry and hookah stands. The air smells of dung and spices. The bookseller is situated beside the musical instrument repairman. The shoemaker hammers. Voices call to shoppers under the ancient vaulted roof.

The ambush will happen at the next corner, near the stands belonging to White Jihad members.

Clayton hears a low hissing from the left, a curse. The boys are warning him to leave.

Instead, Clayton catches up to the man.

"Hello," he says.

"Go away."

"The boys will rob you," Clayton says in stilted English. He's talented at languages; picks them up from missionaries, men at hotels, a stolen tape recorder, the English-language shows that foreign camp doctors used to watch.

"Four boys follow you. I will help."

The man breaks out laughing. His teeth are white and straight. His hair is durum-colored, brushed in twin wings back from his narrow skull. His posture is perfect, brows fine, eyes a bright, intelligent blue. He holds out money.

Clayton shakes his head, no.

The man frowns, meaning, *Why help me? Why care?*

Clayton says, "You look like me, Feranj."

The man starts walking again, but this time the eyes flicker back. The altered carriage—the tenseness in the shoulders—says he's more aware now, more alert.

"You speak English well, strange boy."

"The doctors spoke it, in camp."

"Where are your parents?"

"Men beheaded my mother," he says, anticipating the usual shock this brings. But the man merely peers down with more intent appraisal, as if Clayton just passed a test.

"It will happen after the next corner," Clayton says, feeling his own senses ratchet up before a fight.

The lane narrows when they make the turn. The walls press close. Here there is room only for foot traffic. Light dims. The sixty-watt bulb hanging from the roof is broken, or unscrewed. Most shops are shuttered for the midday meal. The owner of the carpet stand—also White Jihad—stops pulling down his steel door. His glance at the tourist is filled with anticipation. But it changes to befuddlement when he spots the beggar boy.

The shopkeeper makes a quick, cutting hand motion. He means, Get away.

The boys behind us must be his relatives. He must get a cut of what they steal.

What is supposed to happen, Clayton knows, is a quick knockdown, a few kicks and curses. A rapid extraction from the pockets and a scattering of boys.

"My friend! He is hurt!"

Ahead, an older boy—a market bully—lies on the ground, writhing as if injured. A second boy, crying for help, leans on a crutch.

"Please help him, sir."

Clayton whispers, "Now."

He spins and draws from beneath his checkered shirt a spring-knife that he clicks open. The sight of it stops the boys materializing from the shadows: one wielding a banana knife, three hefting cudgels, thickly nailed sticks.

They really want to hurt this man, not just rob him.

Clayton hears the tourist tell the boy on the ground with surprising calmness, in perfect Arabic, "The prophet said, the thief will be punished a thousand times."

Clayton breaks out laughing. Then comes another surprise. He has no idea that the shop owner was leading the attack until the man throws himself at the stranger.

Then the boys charge.

As always, in battle, Clayton reads intent, and sees speed as if it is slow motion. The lead boy holds his knife too high. Clayton's blade whips out. He avoids the cudgel that flies past, close enough to disturb the air by his ear.

The lead boy is screaming, "My face!"

The "cripple" lifts his crutch, starts to swing it. Clayton kicks the boy's good knee and hears a snap.

The bully runs away, not looking back.

When Clayton turns, the blond man stands over the shop owner, who lies unmoving on the cobblestones, head twisted back limply, so he looks like a chicken with a broken neck.

The blond man has done this damage with his hands.

"Are you all right, my new friend?" he asks.

"The police will come, Feranj. We must hide."

The man seems amused.

"Police are no problem. But those people were hired to kill me. And that is what will happen to you if you stay in this place. We can't have that."

The boy is surprised that someone cares about him. He feels a lightness in his chest.

The man considers him. "I should have listened to my bodyguard and not gone off alone during negotiations. Come to my hotel."

The boy starts to smile.

"That's not what I mean. Tell me your name?"

"Call me whatever you want."

But the usual answer seems to irritate the man. "Never say that," he admonishes as they hurry away. Outside the bazaar he hails a taxi. "Names are important. They say who you are. My name connects me to a thousand years of responsibility. Choose a new name if you like. A British name. You look British."

"I am not of you or of anyone I have met."

"Everyone knows who they are not. The trick is knowing who you are."

"Then I will be your new bodyguard," the boy says as they head for the hotel.

The man laughs. "It could happen. I can always use a good yeoman. But first we test you. First we see what you have to give."

The school in which the man enrolls him is in Bahrain. He boards there with the sons of people in the oil business: middle managers, accountants, lawyers, engineers. He has to wear a tie and jacket, and when the boys try to haze him he sends three to the infirmary. He learns languages, science, politics, geography. The mentor never visits him, but the boy reads about the man in newspapers. He always remembers that the mentor had said this school would be a test.

His roommates go on at night about dreams of fame and fortune. "I want to be a rock star." "I will be a great writer." "I'll discover a cure for cancer."

"I want to be a soldier," the boy says. "And fight."

"Fight who?"

The boy can't answer exactly. He can't stand being shut into classrooms. His muscles need violent exercise, more than merely soccer and wrestling. But he stays because the man from the bazaar wants it, and he wants to please the man. He has never wanted to please another person before.

Even pariahs dream of being special. At night he lies alone and, when no one can see, gives in to loneliness, floats through space, unattached. He imagines being the warrior his father had been. He sees himself slaughtering the men who killed his mother. *He* is one of the men in the books, whom others feared, loved, served, respected. Men who lived inside themselves. *He* is Tamerlane. Suleiman.

"Hey! I figured out who you are!"

It starts one night jokingly, in the second school the man sends him to, in Austria, in a dorm room he shares with two other boys. The sketch his bookish roommate holds up shows a white man wearing Bedouin robes. Head raised proudly. Dagger in hand. Eyes seemingly fixed on Clayton.

"He was an Englishman who lived among Arabs. He led them in a great revolt, long ago."

Clayton edges closer.

"Who is he?"

"He was sent to Arabia during World War One. He fought the Turks. He blew up trains," the boy says, pointing at the chiseled jaw and sad, sensitive eyes, the delicate brow, that does, indeed, resemble Clayton's.

Other boys in the room laugh. It's a joke. That's all.

Clayton thinks about it. "People feared this man?"

"Greatly."

"Arabs followed him?"

"He swept the desert. He met generals and kings. He was revered yet never belonged anywhere."

"I am not of that man," Clayton says, meaning it, but that night dreams of conquest and vengeance; sees himself, sword held high, charging across the dunes.

And he remembers that night now, years later, pedaling a bicycle onto Macomb Street toward his rented house, seeing looters breaking into homes on the block. His front door is open. This had to happen sooner or later, he knows.

Time to become Bartholomew Young, Clayton Cox thinks, eyeing the men and women streaming in and out of smashed-in doors, loading up stolen shopping carts. Bigger men guard the carts. Residents have fled. No one calls police anymore since zoning started. What's the point?

"Hey! You!"

One of the looters has spotted him staring. Two more appear at the front door of his house, bats in hand, thinking that the odds make them safe.

"What are you looking at," one guy snarls, as dangerous to Clayton as a yapping dog. "Give us that bike!"

Clayton Cox ceases to exist at that moment, and Father Bartholomew Young takes his place. He leaves the bike and walks off. It attracts too much attention anyway at this point. You always keep dual apartments in cities, and Clayton moved anything worth keeping to zone B a week ago. He morphs step by step into the man on his new license. He goes gray. He acquires spectacles. He removes the mustache and slows down.

Names matter a great deal to me now.

DECEMBER 5TH. 38 DAYS AFTER OUTBREAK.

"Dr. Gregory Gillette will identify himself immediately! Attention. Gregory Gillette! Stand up at once!"

Prisoners are housed in the U-Nevada sports arena. Military trials go on around the clock. Gillette's been sitting on the basketball court for three days, beneath the big monitor, in a sea of thieves, looters and rapists; legs crossed, muscles aching from inactivity. Standing makes him dizzy. He's had nothing to eat. The sound system is turned up too high. The booming announcements never stop.

The soldiers took my samples.

"You Gillette?" a trooper asks.

The woman gestures with her M-16 toward the exit sign beyond the court, and hundreds of cuffed prisoners. The stands are filled with more captives, with new ones arriving by the hour.

Loudspeakers blare, "Prisoners may not speak to each other. Remain in place. If you need to go to the bathroom, signal a guard. You are now in the custody of the United States Army. Disobedience will be punished."

A man starts screaming up in the stands. Guards converge on him. Rifle butts rise and fall up there.

Where are my samples? Gillette thinks.

Prodded forward, he passes below the basket, into a cinderblock corridor, past locker rooms. The guard stops him before an elevator. Other prisoners wait here too. Overhead lights flicker. The arena, prisoners have been cautioned, has backup generators so even if power fails, do not move.

Gillette has gathered from whispers that fellow captives also include motorcycle gang members who refused to turn over impounded Harleys, convicts from the city jail and snipers who'd fired on soldiers. Apparently northwest Vegas has become a no-man's-land. Soldiers are in no mood to baby prisoners. They act as if all prisoners fired at them.

"Out of the elevator, scumbag. To the right!"

He shuffles across a beige carpet spotted with blood, in the administrative office hallway. Walls are hung with university basketball team photos. But the glory images don't go with the pleading voice coming from down the hall. "It was just a sandwich! I was hungry! A lousy sandwich! That's all!"

Gillette sees his reflection in a trophy case as his escort jabs him, stops him outside an office.

"Go in and meet your lawyer, asshole. I'll wait outside, but try something and I'll be there."

Inside, an exhausted-looking black man behind the desk—scanning printouts—wears the rumpled uniform of a U.S. Army captain. EVANDER, reads the tag. The room reeks of tobacco. The ashtray is filled with butts. Outside the window, night has fallen. Gillette sees the glow of fires. Smoke spirals against the moon. Blue / white searchlights crisscross the sky.

"Gregory Gillette?" Evander asks, sounding like he has a cold, not looking up.

"What happened to my samples?"

Gillette estimates the haggard face that looks up to be in the midtwenties. The eyes behind the wire-framed glasses are veined red. Evander shakes a filter cigarette from a pack.

"Someone needs to test my samples," Gillette says.

"What are you talking about?"

Gillette must explain carefully, so the importance comes through. "I'm a CDC doctor out of Fort Detrick. I'm working on Delta. I found dust in vents at Cougar and—"

Evander's eyes are glossing over.

"The samples may be evidence. The soldiers have them."

"Did they give you a receipt? They're supposed to give a receipt for anything they take."

Evander holds up the printout. His accent places him originally from somewhere in the southeast United States.

"I don't know about any samples. You're going on trial in a few minutes. I'm your lawyer. We need to prepare, if you want to stay alive."

Gillette gasps. "A few *minutes*?"

"You're charged with four counts of misusing ID," Evander reads, "reckless endangerment of troops, resulting in nine deaths, and sabotage of the antiterrorism effort. You caused key personnel to miss flights. Criminal misuse of equipment. Directly disobeying orders. And let's not forget looting. How will you plead?"

"Looting?! Are you crazy?!"

Evander clicks a pen open. "You admit the rest?"

He scratches something on a yellow legal pad.

"What are you writing? I don't admit anything!"

"Take my advice. Contest the looting. Looters are being shot. The court will be more inclined to believe you if you admit something. After the ambush General Winston called Washington. Apparently you looted food at a distribution there."

"That's not true."

Gillette feels faint. Evander's cell phone rings. For an exhausted-looking man he grabs at it quickly. His panicked whisper is audible. "Shawna, how *is* he, Shawna?" he says.

Evander looks relieved. "Nicky's fever's down? Thank God."

Gillette thinks, *How do I reach this man?*

Evander wipes his eyes as Gillette realizes that this office belonged to a PR man. A wall plaque proclaims, TOP BULLSHITTER OF THE YEAR.

Evander is saying into the phone, "You traded the ring for antibiotics? I'll buy you another wedding ring."

Gillette flashes to his own family, whom he last spoke to three days ago. Marisa had told him that Gail Hansen had locked herself in her house, and was drinking heavily. Where does she get booze? Paulo had announced another halving of rations. Annie had pleaded to be able to visit the zoo.

"No."

Evander's hand is trembling when he hangs up. He fixes his tie, tries with small physical routines to switch from distraught father to professional lawyer. Gillette needs the man's full attention.

But roles are breaking down everywhere. People are pulled in different directions. Duty versus family. Country against home.

"I lied to get to Cougar," Gillette says finally, "so I could track Delta-3. It may have been introduced into the oil supply at the factory, through air-conditioning units."

"Uh huh. Air-conditioning."

"Look. It's technical. You see, sporulating bacteria . . ."

"Commander, we don't have much time and I've been listening to lies for days. You can't believe the crap that comes out of people's mouths in here. You tell one story. Witnesses report you misusing ID at a D.C. food distribution. Stealing food is looting. Let's talk about that."

"*The report is wrong.* The clerk thought we'd already received supplies. We hadn't. My kids were hungry. My neighbors too. I couldn't walk away. So I used the ID, but is it looting to take food that's yours anyway? I only took our allowance. If I were stealing, why not take more?"

Evander stares at him wearily.

The guard sticks her head in. "Time."

Evander tells him as they walk, "There *have* been glitches with distributions. But I must tell you, Commander, the court is looking for excuses. There are three thousand prisoners here and not enough food. So when we get in there, if you did something small, admit it. Act respectful, no matter what happens. My job is to try to keep you alive. And shut up about those samples. If they think you're jerking them around, you're done."

"You believe me?"

Evander avoids answering. All he says is, "I know I look like shit, but I'm good at my job. It's just that with no prep time, I can't check facts or call witnesses."

They enter the conference room where three stern-looking officers sit beneath more sports photos.

Evander tells Gillette, "This isn't exactly the way things were in law school, in mock trials. Do you believe in God, Commander? If you do, pray."

The major on the right looks sympathetic, the captain in the middle antagonistic, and the other major has a twitch in one eye that makes his expression hard to read.

The trial has lasted forty minutes so far, and has gone the way Evander predicted. Gillette's told his story, and answered questions. He's admitted misusing the ID in Vegas.

He's brought up the samples as Evander shook his head.

For a guy who had no time to prepare, Evander did a good job questioning the looting allegation, Gillette thinks.

"The court will retire to reach a decision."

Gillette finds himself sitting out in the hallway, sipping a welcome cup of tepid water while Evander returns to his office to prepare for another case while he waits.

A voice in Gillette's head—Dr. Larch—says, *"There's never an excuse for avoiding the field."*

He feels as if Larch is beside him. He even smells the man, the combination of wet wool and Borkum Riff tobacco.

"If they test my samples and the result is positive, we can start tracking the bug," he tells Larch.

"Remember the names I taught you? The heroes? Doug Cruise went into the field and found the vector for cryptosporidium, the waterborne parasite. DeVries isolated E. coli 0157:H7, saving kids' lives. All the researchers who went into remote villages and put their lives in danger, to find the particular microbe that was coming out of hiding that day. Take Ruth Berkelman."

"You always tell this story."

"I always will. Postoperative infections were killing patients in a Michigan hospital. Doctors had taken throat and nasal cultures from everyone present during operations and found nothing, so hospital officials assumed staffers were clean. But Ruth cultured them again more thoroughly, and found a nurse infected in her anus and vagina."

"She was the vector," Greg recalls.

"The hospital insisted the nurse had never come in contact with patients or the surgical instruments. They said it was impossible she was the vector."

"Ruth proved it was possible."

"She asked the nurse to put on sneakers and surrounded her with petri dishes, filled with sterile culture. She had the nurse run in place. She incubated the dishes. Four days later they were infected . . ."

"Because the bacteria traveled through the air."

"Always visit the field. You did nothing wrong."

And now Gillette snaps out of it, as Evander's voice says from beside him, "They've reached a decision. And just so you know, I found out about your samples. The vials broke. That's why you never got a receipt."

Throat dry, Gillette takes his seat in the conference room. The officers look stern. But maybe they always look that way, no matter what verdict they announce.

"Commander Gillette, we reached General Hauser and were told there is absolutely no substance to your story about samples."

He feels sick. "Hauser wouldn't know."

"We've spent quite a bit of time on you. We find that your direct disobeyance of orders resulted in nine deaths. It diverted key resources from the war on terror. It stopped work at Cougar. We can't have every Tom and Dick throwing monkey wrenches because they have a theory."

Gillette's head is swimming. *This can't be happening.*

"Normally you would have been imprisoned for these infractions. Your motives and record inclined members of this court toward leniency. But these are unprecedented times. The tribunal gives you the benefit of the doubt on looting. We find you guilty on all other counts."

Gillette hears the words as if from far away.

"We believe that you did not act for your personal benefit. But order must be maintained. You are sentenced to death under the New Emergency Powers Act. This sentence will be carried out tomorrow. Perhaps you meant well, Commander. God bless your soul."

SEVENTEEN

DECEMBER 6TH. 39 DAYS AFTER OUTBREAK.

The troops come for Gillette at 7:30 A.M. He is squeezed with thirty other men and women into a covered truck.

Great use for fuel, he thinks as the vehicle rumbles out of the city and into the desert. Prisoners stare into space, or sob or clutch one another. Last night in the darkness he'd heard their whispered stories and excuses. Some people denied charges against them. Others boasted that they were true.

"Yeah, I shot a soldier. They broke into my house."

"The damn hospital was empty, with food inside."

The flap is down, the heat intense. When the road turns bumpy, Gillette imagines them driving on dirt. And when the truck halts ten minutes later, and the flap is raised, he sees that he is right.

One by one prisoners' names are called. If you don't answer, the soldiers get rough. Gillette sees lavender-colored mountains in the distance, and smells boot leather, sweat, alkaline, sage.

"Gregory Gillette?"

He climbs down. His phone and ID have been confiscated.

It's a beautiful morning.

In his head he tells Marisa that he is sorry.

I failed you, he thinks.

The prisoners march toward the mountains beneath wheeling buzzards and small fluffy clouds.

"December shouldn't be this hot," a guard says.

* * *

An hour later, in Washington, Major General A. L. Hauser steps purposefully down the Pentagon corridor toward the suite occupied by his boss, Under Secretary of Defense Dennis Ames. He's on his way to give Ames the daily early-morning progress report, filled with facts and plans.

Hauser thinks, *I don't wish Delta-3 on anyone, but these last few weeks have been an opportunity for me to show my talents.*

Hauser the White House consultant, the high-level desk jockey grown powerful, due to his Rapid Response Team job.

"Coffee?" Ames asks when he walks in.

"No thank you, sir."

Hauser launches into the usual edited, upbeat report, despite the growing riots, power and phone outages, and mutinies. FBI agents in Sacramento have in custody several militant Muslims who've provided "promising leads" as to the identity of terrorists responsible for Delta-3.

Ames nods.

Homeland Security liaison work with the Europeans is keeping "close track" of militants overseas.

"Uh huh."

Cooperation with Swiss and Grand Cayman governments vis-à-vis tracking terrorist monies is "excellent," thanks to the worldwide suspension of national banking secrecy laws.

"In fact, sir," Hauser says, "yesterday's roundup is sure to produce more leads, now that we can intensify interrogation procedures, under the new rules."

Ames lifts a finger, halting the flow.

He says, "I understand there have been some problems with one of your people in Nevada."

Hauser's good mood instantly evaporates, to be replaced by rage. He'd gotten the call about Gillette yesterday from the judge advocate general's office. He'd told them coldly that the man was AWOL, that he should be punished for the deaths he'd caused and his disgusting theft of food. Hauser had felt a little sorry for Gillette, but there was no way he would jeopardize his career for a subordinate.

"Gillette is no longer on the team, sir," Hauser says. "I removed him from the Pentagon weeks ago. He was ordered to stay away from Cougar. He'll cause no further problems, I promise."

"Then you didn't send him to Nevada?"

"On the contrary, I've done everything in my power to keep him out of the way. I would have removed him from the staff entirely, but Colonel Novak insisted he might be of use to her."

Ames hits a buzzer on his desk.

"Send her in."

To Hauser's surprise, Colonel Novak enters, holding a report, nods at Ames and takes the stuffed seat into which Ames waves her. The desk clock chimes 6 A.M. EST, which means, Hauser knows, that Gillette has been shot in Nevada. Yet still the pest is making trouble, even from the grave.

"Tell General Hauser what you told me," Ames says.

"I ordered Commander Gillette to go to Las Vegas."

Hauser flares. "I specifically told you to keep him away from there. I ordered his samples destroyed! We're not wasting any more time on him. He's delusional."

Novak says stiffly, "I offer you my resignation, sir."

"Accepted, by God!"

"By the way, Colonel," Ames asks, "what's in your hand?"

"A report that came in from Nevada last night, Mr. Secretary. After Commander Gillette's samples were destroyed, a Cougar scientist named Varunisakera ordered his researchers to duplicate his work."

"This is preposterous," Hauser flares.

"They found Delta spores in the AC compression system. The vents spewed the stuff into shipping bottles. The bug spread from there, it seems, in a new bactericide, of all things."

Hauser turns white and feels his balls start to shrink. "But . . . the FBI insisted . . . But that means . . ."

"It means you're off the team," Ames says. "And that Lieutenant

Colonel Novak is now—I understand—a full colonel. We stopped the execution. Colonel Novak will give Commander Gillette whatever help he needs from now on."

DECEMBER 6TH. AFTERNOON. 39 DAYS AFTER OUTBREAK.

The new apartment is quiet, tidy, furnished in maple, filled with Kennedy Center posters of operatic and symphony performances. The sad clown in *Pagliacci*. The wild-haired guest conductor Kurt Masur, signaling the national symphony with a baton. The view out the fourth-floor window is a sun-drenched park that Clayton Cox—now Bartholomew Young—sees patrolled by police regularly. It's a haven for moms pushing strollers. His new buffer zone neighborhood is, in a word, safe.

This morning he actually bought coffee from a small shop at the corner of Q and 23rd Streets. People are always outside; walking, greeting each other, relishing the fact that they've been blessed to live in zone B.

"Darling. You're *thinking* too much. Relax," the woman on tape says.

On this latest tape, made hours ago at midday, Major General A. L. Hauser gets out of bed and walks naked to the window. He seems unhappy. Not his usual horny self.

He usually stays at work until night, Young thinks, enjoying the man's anger but knowing from previous viewings that the woman will have to work on him awhile now, before he talks more.

The tart coos and coaxes, rubs and kisses. Young flashes back as he waits, sees craggy mountains beneath dark clouds, rain shrouding green moors, plateaus draped with heather.

Gigantic logs crackle in a hearth. The mentor's library chandelier once held torches instead of electric lights. The tapestry—a wolf hunt scene—dates from the time of the Crusades. The mentor looks like the ancestors whose portraits hang on the walls: his face long, narrow, balding and mustached. Only hairstyles and collars on the paintings date them in a time line. From breastplate to lace, to starched cotton, to a turtleneck on the man pouring them both single malt scotch.

"You had quite a record in the SAS," the mentor says. "Before you went private."

"I still want to work for you."

"Afghanistan. Iraq. You don't seem surprised that I have your file."

The mentor sips his drink. The soldier has been here for two days now, having turned up on his own and been made an honored guest. The cooks downstairs have served the family since the time of the Picots, he's learned. The groundskeepers' ancestors lived here before the Spanish Armada sailed. The villagers in The White Knight tell tales about the Crusades, when their ancestors followed their liege lord to Jerusalem. They've raised sheep for the mentor's family and fished for it through the centuries, mined coal for it and brewed whiskey for it. These days many men here work on oil platforms that the mentor owns.

"Name the most powerful narcotic in the world," he asks the former beggar boy.

He still stands straight, still has that look of amused appraisal in his bright blue eyes.

I know he's not my father. But he's the closest thing to it I will ever have, the former soldier thinks.

"*Oil,*" the mentor says. "More than opium, more than heroin. The pipelines are syringes. The addicts pay anything for their supply, kill for it, steal for it, topple governments for it. Do you want to commit yourself to selling a drug?"

"I'm committing myself to you."

"I wondered if you would come here one day. You never told anyone what happened in the bazaar. You passed all the tests. I have a gift for you, my yeoman. But first, meet some friends. People who understand that our world must change."

There has never been a night like this one for the soldier. He meets a neoconservative member of Parliament; the CEO of a Japanese oil company whose wells have just been nationalized in South America; a Texas oilman whose wife was kidnapped in Iran; a French oilman whose pipelines have been blown up by saboteurs.

Meet my most valuable friend, who saved my life, the mentor tells these people. *Meet a treasured man.*

It is as if the man saw into the boy's heart years ago. He knows exactly what to say.

"He is like my son."

And the soldier has killed men and women on battlefields, in hotel rooms, parking lots, in deserts. He feels as if a door is opening inside him. He has controlled his essential loneliness and locked it away for years. He had not known it was there. He can defend himself against violence, but not against respect.

And later, after the guests leave, the mentor brings him back into the library. There has been a certain ceremony in this house for hundreds of years, he says. It holds no legal meaning. It's just words. But would the former beggar boy like to recite these words?

The former beggar boy fights back tears.

Then the mentor goes to his desk, and produces from the top drawer a leather file wrapped with blue ribbon.

"I had your blood tested. In case you ever came."

"Blood?" This is baffling and unexpected.

"DNA," the mentor says, "is God's fingerprint. Irrefutable proof that the ancestor matches the man."

The soldier's heart starts pounding. "No one knows my father," he says. But he's also anxious, because the mentor would not bring up a subject if there were nothing to tell.

"You're correct. I didn't locate your father. I found your great-great-*grandfather.* I admit it was a hunch. A long shot that occurred to me after the boys at school teased you about it. You probably don't remember. It was years ago."

Of course he remembers.

"You do look amazingly like him, you know. The photographs are clear. And the sketches! It was a guess, but," he says, offering the file, "it's no wonder you turned out the way you did. It's in the blood."

This cannot be happening.

"It was hard to obtain his DNA. Like you, he was a hero."

My mother told the truth, the boy thinks, cursing the tears forming at the edge of his eyes.

"He wrote a book that I would like to give you. You must read about him. His trials. His loneliness. His subordination of personal goals for a greater good."

This is the most magnificent gift I could imagine.

"It's an old book. Not expensive, but beautiful."

"It's expensive to me."

And now, years later, Bartholomew Young returns to the present, where, on tape, Major General A. L. Hauser is finally ready to talk.

"Fucking Gillette," Hauser says.

Young stirs, leans forward.

"He went to Nevada by himself. He found traces of the bug at Cougar. He was lucky."

Father Young stands up.

"They're giving him carte blanche in Nevada. I'm out."

Ten minutes later Young is on the encrypted phone, to Europe. The mentor says nothing when he hears the news. Young hears only breathing over the set.

Then the mentor says, "Now it is a race, my friend. Will they collapse before they find us, or find us before they collapse?"

"You need another way to find out what they are thinking, now that Hauser is gone. Gillette will know. His family is in Washington, even if he's still in Nevada. If I can get to the family, I can get to the man."

"It's too dangerous for you, my friend."

"Great-great-grandfather said that nothing is written, except that which you write yourself."

The mentor does not reply for several heartbeats.

Then he says, "My most valued yeoman. Soon we will bring you home."

EIGHTEEN

DECEMBER 6TH. 1 P.M. WCT. 39 DAYS AFTER OUTBREAK.

I can't wait to hear how the murder of a part-time security guard relates to Delta-3, Dr. Gillette."

"That's what I'm trying to figure out."

Las Vegas detective Duane C. Hardy is a stocky man with thinning wheat-colored hair and a permanent cynical sneer on his lips. But the tough effect is offset by his clothing: baggy checkered shorts that show pale legs, a guayabera swelling out with the belly, shiny loafers without socks. Unshaven, he reeks of cologne.

"Cuts the smell inside," he says.

He's met Gillette outside city hall, near the fallen police officers memorial, where he'd been having a smoke when the convoy arrived. The sun is bright, the ten-story building made of brick, steel and glass. Hardy glances out at Gillette's ragtag fleet of jury-rigged armored vehicles, an apocalyptic fleet parked by a half dozen impounded mules slurping water from an improvised trough, an old bathtub.

Gillette is getting full cooperation from authorities now. It's amazing what one phone call from Washington can do. He's still reeling from the last-minute reprieve from Ames's office; the revelation that the spores had been found; the horrified apologies and fawning cooperation from officials at the sports arena; the subsequent return to Cougar Energy Services and interrogation of the security chief, a thirty-year company veteran who wore cowboy boots and a string tie and found it hard to believe that anyone he actually knew could be involved in terrorism.

"Which employees had access to the air-conditioning system this summer?" Gillette had demanded. "That is when the bug was introduced, based on rate of spread."

The security chief had turned defensive. "The whole staff was checked for terrorist connections, even Bob Grady, the university student murdered on the night of the first crashes. It was in the papers. Remember?"

"It wasn't in the papers in Washington."

"Well, it was a big deal around here. The chairman's nephew, killed in a hotel. He was a part-time summer security guard. A goof-off. But I had to hire him."

"If he died the night of the crashes," Gillette had said, flabbergasted, "didn't the FBI follow up on this?"

"Of course they did! That's my point. Bob owed money. He gambled. You're barking up the wrong tree asking about him. Trust me. It was coincidence."

"Then the police caught the killer?"

"No. But they have videos. The guy wasn't Middle Eastern. And Bob wasn't even working for Cougar when he died."

"He was working there when the bug was introduced into the air-conditioning system," Gillette had snapped.

Now Gillette follows Detective Hardy into the building and stops dead. The blast of noise is tremendous, but it's not as bad as the stink.

Holy Christ! It's a refugee camp in here!

The lobby has been partitioned into cubicles, each housing a family. People lounge on cots in their underwear, play Scrabble, read books, give children spelling lessons or, Gillette sees with envy, eat half sandwiches. He smells urine, Lysol, dirty diapers, soiled perfume, garlic, feet. Televisions blare. A man plays a guitar. Someone shouts, "Who took my shoes?"

"You sure you don't want some cologne, Doctor?" Hardy asks.

Gillette splashes on Paco Rabanne.

Hardy explains as they weave toward the elevator, "We were losing officers every day. Their families weren't safe so the sheriff and city officials moved the families here. We're packed. Hallways. Cells. Even the sheriff's office. But Vegas has the lowest percentage of AWOL cops in the U.S. Hi, Leon," he tells a kid wearing underpants and socks in the elevator.

"Tommy stole my Frisbee. Arrest him."

Gillette asks, "Did you identify the killer?"

"No. But I set up the tape machine." The doors open. The roar seems louder on the second floor. They squeeze past a poker game and a Bible study group. The noise abates slightly when the detective closes his office door. Inside is the illusion of normalcy. File cabinets. A police league softball trophy. Hardy puts a tape in the machine.

Gillette seats himself to watch.

"I spent a lot of time on this case. Watch."

On the set, an overhead hotel security camera has recorded a college-age boy checking into a hotel.

"That's Grady. He gives the bellboy his bag. He turns to go gamble."

Suddenly, everyone around the boy spins to look in the same direction. Gillette sees lights flashing on a slot machine. But instead of waiting for his payoff, the winner stands and walks off behind Bobby Grady.

"He just left $20,000 in winnings," Hardy says, popping gum in his mouth. "This picture ran on every local station. All the hotels reviewed their tapes. Who's the shy winner? Turns out—according to a clerk—that he was another guest at New York–New York. Named Lewis Stokes."

"The killer?"

"Ah." Hardy replaces the first tape in the machine with a second.

"This one is from the Monte Carlo casino, thirty minutes later," he says, pointing to figures captured from overhead. "Here's Grady, sitting to play blackjack. And here, going into the men's room at the same time, is Lewis Stokes. No cameras in the men's room. But does Lewis ever come out of the bathroom? Nope. Instead, this guy does, and finally . . ."

He inserts a third tape.

"Now we're back at New York–New York, just before the murder. Here's Grady, waiting to board the elevator. And look who gets in with him."

"Lewis Stokes," Gillette breathes.

"Yep. Plus, earlier that day Grady's girlfriend got a call from Lewis. He told her he's a college buddy. He gave some bullshit story about sock baseball games in the dorm."

Gillette sits up straighter, remembering Marisa's words about an "old college buddy" who had called the house and told the same quirky story. What had the name been? What the hell had it been? Lewis?

Clayton Cox, he recalls after a pause.

Hardy says, "Turned out Stokes had insisted on checking in to the room next to Grady's. He said it was lucky. We ran his credit card and home address. Both phony. Professional hit, we figured, but we thought it was about money. So why is it about oil?"

"Someone infected the fluids plant with Delta-3. Did the girlfriend tell you who Grady hung out with? Or borrowed money from?"

"I tried to reach her again when the Pentagon phoned. The school is closed. I tried the parents. Their phone's dead. You can't find people anymore. But if you want, I'll run you out to her place on one of our trusty mules."

There's no time. Gillette checks his watch. Another milk run flight is leaving for Albany—near western Massachusetts—in ninety minutes. If he misses it, the next one might not go off for days.

I need to talk to Lyle Samuelson about his work with microbes six years ago.

"Call me if you learn anything more," he tells Hardy, and asks the question with which he always ends interviews. "Anything else you remember, that I didn't ask about? Anything you noticed, even if it seemed small?"

Hardy dabs on more cologne. "Well, there *was* something. The name. The bellboy remembered that he called Lewis 'Lew' when he checked in. The guy gets mad, snapped at him, like, my name is Lewis, you idiot, not Lew."

Gillette pays close attention. With microbes every hint identifying behavior helps an ID.

"Same thing happened with Grady's girlfriend. She misheard the name, and called him Les on the phone. *Again* the guy explodes, even spells L-e-w-i-s."

"Why would he care, if it was a phony name?"

"I said it was a small thing."

"So is Delta," Gillette says. "Look at the damage it does."

The flight leaves on time, scheduled to again meander Gillette across the continent. Vegas to L.A., Tucson and Tulsa. Then Dallas, Mobile, Atlanta and more stops before Gillette's: Albany.

"Passengers may use the seatback phones," the pilot announces when the Boeing 737 reaches 10,000 feet.

Worried, he calls Marisa for the third time since his release from prison and reaches her during a block meeting. He tells her about Clayton and Lewis, and that if anyone using either name calls again she is to phone Theresa at Fort Detrick and demand soldiers on the block right away.

Then she tells him, "The last convoy never reached Washington, Greg. There's no food this week. Joe and Chris want to evacuate. They say if we wait any longer we'll be trapped here. Chris says he heard there's food in Richmond, lots of it at the churches. He wants a vote."

Gillette feels his heart freeze up. All he has to do is look down from the plane to see refugees on an Arizona highway, walking in both directions, amid hundreds of broken-down cars. The scope of the migration is awesome.

"There's no place to go," he says.

As if to answer him, he hears Joe Holmes over the receiver, trying to convince the neighbors to leave. "The cops can't protect us. Eleanor is sick. Ed is sick. We can't eat heating oil."

"Put me on the speakerphone," Gillette tells Marisa.

He tells them he's in a plane, and describes what he's seen. Refugees. Fires. Carnage. "So stay put."

Pastor Van Horne speaks up, "But the Internet says . . . "

"The Internet? Are you crazy? There are rumors all over the place! Clean fuel in Florida. Surplus grain in Ohio. Always someplace impossible to reach."

Gillette might as well be calling from Mars. He can feel the bonds stretching: of faith, friendship, neighborhood. The phone is a black hole for sound.

Joe Holmes says, "If we can get to Union Station, the electric train lines have been extended."

The voices all speak at the same time.

"We can bribe soldiers at the barricades."

"If you had kids you would vote to leave!"

Gillette must get through to them. The urge to move to solve problems is coded into the DNA of Americans. Go west young man. Or north. Or east.

"We're making progress on Delta-3," he insists. "We know where the bug entered the system. We may have identified people who created it. Hold on a little longer. If we can immobilize the genes, maybe we can reverse the effect."

"Maybe," snaps Gail Hansen.

"Maybe's the best we get."

"You zone A people always tell us to stay quiet. Be nice. Sit and starve like good stupid citizens."

Marisa's voice snaps, "Don't talk to my husband that way! He's risking his life for you, you drunk!"

Gillette hears Bob calling for order. The former Marine will be stepping into the middle of the room, hands raised for silence. Bob says that Greg is right, that there's nowhere to run. No one will save them. He says, "*Let's trade our oil.* Let's find people with extra food. Between fishing in the Potomac and maybe, for survival's sake, maybe even trying a little *borrowing* from abandoned homes—"

Judge Holmes's angry voice breaks in, "You mean looting!"

"Give me a break, Eleanor! You're not a damn judge here. You're a mother with hungry kids. You ate the meat I brought home Friday, didn't you? Well I stole it! From a house. You liked eating it."

Oh no, Gillette thinks.

Suddenly Paulo's voice cries, "Alice fell down!"

He hears a chair fall over, and a scream. He hears Gail Hansen shout, "That's what will happen to *all* of us!"

Gillette says, "Marisa? What happened?!"

The line goes dead.

He can't get through on the phone anymore. Gillette's seatmate explains why. "I work for Verizon. Lines are down. People are cutting up poles for firewood. Tower maintenance crews can't get to work. Keep trying."

Gillette calls Raines to keep his mind off Marion Street. At least he gets through there, and orders Raines to run checks on the names Clayton Cox and Lewis Stokes.

"I've got good news and bad, boss, on Lyle Samuelson."

"Good first," he says, praying that Alice Lee will be all right.

"*Follow the money,* Commander. Cars may not run. Food may not grow. Oil may not flow. But money still talks. Lyle retires four years ago, right? Takes a buyout, you said. Well, eight months later, he pays taxes on one million dollars he earned as a 'private consultant' in microbiotics after leaving Cougar. All legal."

"Who paid him?" Gillette asks.

"Some outfit called Applied Technology, registered in the Caymans. Dummy corporation, turns out. No building. No staff. Headquarters is a closet."

"That's the good news? The trail just ends?"

"Nope! Follow the money! The Caymans government is cooperating. Applied Technology's money was wired from Austria. The Austria funds came from Switzerland. Classic laundering. I've seen it a thousand times."

"So Switzerland is the original source?"

Raines sighs. "That's the bad news. All monies in the Swiss account were deposited by hand."

"I thought hand was better."

"No. You can't track hand deposits. A guy walks into a bank with a suitcase. Bing. Bong. He walks out. The security tapes were erased. But Interpol found the banker who opened the account, and described the client. Male. British. False name."

Gillette snaps, "Don't Swiss bankers require proper identification?"

"Yeah. It's called money. That's part one, bad news. Lyle Samuelson was murdered in Massachusetts, in September, before Delta hit. Killed in his house, in the woods."

Gillette squeezes his eyes shut.

"And no one knows who did it, right?"

"Mind reader," Raines says. "Go on TV."

The sky seems bluer without contrails in it. The air below is clearer without cars on the road.

He finally reaches Marisa, an hour later. She sounds shocked and exhausted.

"Alice had a stroke, Greg. She's alive, but barely. Dr. Neuman came over from Jennifer Street. But he's out of medicine. He doesn't think she'll last the night."

Marisa sounds close to tears.

Alice, he thinks, grieving over the sheer volume of goings-on, remembering the old lady when she babysat the kids. Alice making blueberry muffins at the annual July blockfest.

I'll call ahead and tell the FBI what to look for in Massachusetts. I've done enough. My family needs me.

"I'm coming home, Marisa."

"You better not! You know why? Because with all the experts, all the authorities running around in all the countries, I want you there," she says. "I love you. We all do. You found the vector. Don't you dare come home yet."

"Thank you," he says, blinking back tears. Night is falling outside.

"Kill the bastard, Greg," his wife tells him, giving the old war cry. "Kill it. And *then* come home."

Around him the mood is quieter than during his last flight. People seem lost in thought. They keep their eyes glued to their windows. The Verizon man, now on the phone, seems to be apologizing to his estranged son for missing the son's wedding. Everybody catching up, making up, setting things right.

And when he reaches Colonel Novak to thank her for saving him in Nevada, she sets things right too.

"You were right about why I came to your house, Greg."

"You don't have to say this."

"I do. I had feelings for you. I figured, if I met your family, that would put them to rest. I was mad because you knew."

"It's not like I didn't have feelings too. Look, I want to say this right, so you don't misunderstand. I love my family. They're the most important thing in my life. But if I were unattached . . ."

"Yeah, yeah . . . I know. Honest times. The Chinese curse."

He takes a deep breath, envisioning Lewis Stokes out there somewhere. "Theresa, I have a request. Can you get my family housing at Detrick, or in zone A?"

Silence.

"Not for me. Just them," he says, knowing that when he gets back to Washington, he'll go to Marion Street to help the neighbors. But he must keep Marisa and the kids safe.

She thinks I'm afraid for myself. She thinks all I want is the favor.

But Theresa just sighs. "I wish I could help. Anyway, the zones are going to be disbanded. They've been a disaster. They've caused more problems than they solved. The president will announce it tomorrow. We're all in the same boat now, and it's sinking fast."

DECEMBER 7TH. 2 A.M. 40 DAYS AFTER OUTBREAK.

He's jolted awake by turbulence as the pilot announces that he's forgoing landing in Albany—Gillette's stop. Conditions are too snowy, and deicing equipment isn't working below. The lube oil is infected, so even if the plane manages a landing, it may not again take to the air.

I need to get down there. We're running out of time.

Gillette unhooks his seat belt and lurches toward the cockpit, ignoring protests from the air hostesses. He has to grab seatbacks to stay up. When he starts banging on the cockpit door, other men—burly military types—get up to stop him.

Gillette produces the magic ID.

"Sit down, gentlemen," he tells them. "If you don't let go of me I'll have you arrested when we land."

The plane pitches as they bank east, toward Boston. When the cockpit door opens Gillette finds himself facing two angry pilots, who grow even more furious when he orders them to turn back.

"I don't care what your ID says, Commander Gillette. In this plane I'm the final authority," says the captain. "I'm not jeopardizing everyone for you."

The plane's floodlights come back at them, reflected off a wall of clouds. Doppler radar shows a scythe-shaped arc of red ahead. Beeping warnings pierce the cockpit. The captain's forehead is wet with sweat.

"Return to your seat or I'll put you in handcuffs."

Which is how I started off yesterday.

Gillette speaks firmly, reasonably. "The man responsible for the oil bug is down there. Do you understand? We think we found the guy who made it, *there,*" Gillette says, jabbing his finger earthward. "Not in Boston. You want to put me in cuffs? Go ahead."

The captain looks at the copilot. "Oh shit," he says.

The state capital of New York seems an odd combo of first and third world when they break from the clouds. The downtown area—zone A, certainly—blazes with electric power. The rest of the city is blacked out except for fires. Trash fires. Home fires. A huge blaze—a factory maybe—by the Hudson River.

"Hold on to your seats. This landing may be rough."

The wheels thump down. The plane skids, catches and straightens and Gillette sees swirling snow out the window, and runway lights flashing by. Wind blasts into the plane when the flight attendants open the door. "You better be right," the captain growls as Gillette exits onto portable steps. The hydraulic walkways don't work. The lube oil was infected. A lone figure in a hooded parka waits at the bottom of the steps, looking up through fierce snow.

"Commander Greg Gillette? I'm agent Patricia Saiko, from the FBI's Albany office."

She's a medium-sized, slim woman, whose dark slanted eyes measure him from beneath the hood. When they get inside the terminal and she takes off the hood, her face is narrow, intelligent; her hair is black and cut midlength. She's dressed for the woods. So are half a dozen academic types he sees wolfing down food at long folding tables. Sleeping bags are laid out on the terminal carpet.

"We'll be heading into Massachusetts at first light," she says, as if the state were wartorn Somalia, "with five agents and the best staff we could dig up for you on short notice. Microbiologist from EPA. Toxic cleanup. DNA expert from SUNY. We'll pull supplies on sleds, by snowmobile. Roads are unplowed once you leave the interstate. Hungry?"

"We'll leave now," he says.

Agent Saiko smiles in a way that tells him she likes his spirit, but that she's the boss. "Have some coffee."

"I don't need coffee."

"You do. I understand the importance of this trip, but we can't risk traveling at night. Roads aren't safe. I had to cut the number of agents coming along to make room for your staff. If something goes wrong, we're on our own, capisce? Go meet your scientists. Eat a tuna sandwich."

"So we just waste time and sit until dawn?"

"It's not a city out there. You don't know what to expect. We'll do this right so you don't get killed, Commander. I heard about the ambush in Nevada. I would have hoped by now that you've learned not to jump the gun."

Only one lane on each side of Interstate 90 is plowed, flanked by drifts that get higher the further east they go. Gillette sits in the lead Land Rover beside Saiko, who stays in radio contact with military stations along the road. Tollbooth areas are minicamps for soldiers. Drifts top five feet. The Rovers pull flatbed trailers loaded with snowmobiles and large-capacity deck sleds.

I remember the Christmas we took the kids to the Berkshires, Gillette thinks. *We stayed at the Revere Inn in Lenox. We cross-country skied on Mt. Greylock. We ate in fine restaurants in Great Barrington. Annie loved the Norman Rockwell Museum and Herman Melville home.*

He's briefed the scientists and told them that they're heading for a house in the woods, and that they're going to tear the place apart. And try to find any study, note, paper, photo, reference, video, disk or book that in any way, no matter how far-fetched, might relate to Delta-3.

We look like winter troops in our white camouflage parkas.

This is the way he used to go into outbreak areas in Central Africa, escorted by troops. Except in Africa, they didn't pass cheery billboards for the Butternut ski resort and the Boston Symphony at Tanglewood. Highway rest stops are mini ghost towns. The sky over the Berkshires is gray.

The Rovers stop at exit 2, which is unplowed. Gillette watches agents unload their two-stroke Yamahas and sleds off the trailers.

"Food deliveries to this part of Massachusetts stopped a month ago," Patricia Saiko says, checking her M-16. "It's been snowing for days. There have even been unconfirmed reports of ambushes and . . . shall we say . . . imaginative eating?"

Horrified, Gillette stares out at the quiet landscape. A moment ago it resembled a Currier & Ives engraving. Valleys. Quiet towns. New England forest. Snow.

"I said they were unconfirmed," Saiko says, and waves everyone onto their assigned spots. The convoy winds down the exit ramp, through the deserted tollbooths and into the small town of Lee, past the Kiwanis club sign.

Lee looks deserted. No smoke comes from rooftops. No birds sit in bare trees. It's like a winter diorama dominated by a New England church steeple.

Gillette wonders if hungry people are looking back at them. In the vast quiet, with no other motorized traffic out, the sound of even the muted engines will carry for miles, alerting anything living of their approach.

"I'll get you there, Commander," says Agent Saiko through Gillette's earpiece, grimly. "And I'll get you out."

NINETEEN

DECEMBER 7TH. AFTERNOON. 40 DAYS AFTER OUTBREAK.

Rural Massachusetts Route 20—normally two lanes wide—is buried under snow so deep that speed limit signs seem to float inches above powder. The travelers pass secondary-growth forest, granite outcrops, mountain bogs. The convoy floats back in time, according to town markers. Now leaving Lee, established 1777. Entering Becket, established 1765.

"Everybody listen up," says Agent Saiko in Gillette's earpiece, from the front of his sled. "The map puts George Carter Road about eight miles from Lee, so we're close."

He sees ice fishermen on a frozen lake with shotguns over their shoulders. A white clapboard home spurts sooty smoke from a chimney. A hooded man with binoculars watches from a swamp. Gillette realizes that lumps in the snow around a smashed-in barn are junked cars drawn into a defensive circle that clearly hadn't worked.

The voice in his earpiece says, "Look for the sign. Jacob's Pillow Dance Festival."

He starts, realizing that his family once visited the festival's campus—a national historic landmark—and stage in the woods. *The trees were leafy. We watched Israeli dancers on an outdoor stage.*

Saiko cautions, "Do not stand up in the sleds."

Gillette thinks back. He's a med school student, asleep on a Saturday night after drinking too many beers. Pounding on his door wakes him. He opens it to see Dr. Larch in the hall, wearing a dripping raincoat and rubbers over his shoes.

"I need help," Larch says.

Ten minutes later they're driving across the Key Bridge into Roslyn as Gillette fortifies himself with steaming Folger's. Larch hunches over the wheel, peering through creaky wipers and midnight spring rain.

"His name is Firoz Khashoggi," Larch says. "He trained at Iran's science and tech center, its bioweapons lab in Tehran. He rented a house in Reston. He's tried to obtain components for anthrax, and fungus to make T-2 mycotoxin. The FBI asked for CDC help on the search of his home."

"What exactly are we looking for?" Gillette asks, excited and fighting the headache. Larch has told him about the FBI assist calls, but never taken him on one.

"Notes. Components. Samples."

Gillette's enthusiasm flags when he learns that the FBI is unsure if the Reston Firoz is the one they're seeking. Two hundred other Firoz Khashoggis live in the U.S., Larch says.

"All their houses are being searched?"

"I guess so."

Gillette's mood worsens when he learns that over twenty biohazard experts from other federal agencies have been combing the Reston house for hours.

"Why do they need us then?"

"Protocol," Larch says, exiting Route 267. Suburbs here mark the capital's decades-long expansion toward Dulles Airport. They pass new housing developments, office parks and strip malls. McMansions precede cineplexes and follow chain restaurants in malls.

The cul de sac into which Larch pulls is filled with black Fords, and the Khashoggi house, lit up, is a one-story ranch. The suspect is elsewhere. Rain drenches Gillette before he gets inside. He finds himself fumbling around in a biohazard mask, bumping into other searchers, opening drawers and cabinets that have already been searched.

Larch yawns. "I'll sit awhile. I'm tired."

Gillette's itchy and hot. It's clear Larch brought him along to do the older man's work. Still, he spends an hour carefully going through

Firoz's home office, looking under picture frames, the carpet, looking through the oak desk.

"This might not even be the right guy," he tells Larch.

"Probably you're right."

Emergency has degenerated into bureaucracy. Soon the other investigators announce they're going home. Larch and Gillette can stay if they want.

"This is ridiculous," Gillette snaps.

Larch apologizes, takes him to the car, turns around and leads him back to the house, to the pantry. Gillette's already searched it. Larch pulls out a plastic bag.

"What's this, Greg?"

"Lima beans." Gillette reads the label sullenly, but he's starting to understand already. He thinks, oh, shit.

Larch opens the bag. "Do they *look* like lima beans?"

Gillette starts to feel hot. It's another of Larch's endless tests. Later he'll learn that the "agents" were CDC volunteers and the house belonged to one of Larch's epidemiologist buddies. CDC trainees have been failing the test here all day.

"They're castor beans," Gillette says, sniffing.

"Which produce what?"

"The deadly toxin ricin."

"So what did you learn tonight, Greg?"

"To force myself to be extra thorough if I get assigned a job and think there's no point to it."

"Anything else?" Larch yawns and stretches. It's dawn.

"Emergencies happen on their own schedule, not mine."

"Feel free to speak up if you think of more."

"You said we were looking for fungus, but you planted something else. I didn't look for anything else."

And now, years later, the convoy proceeds through a light, windless New England snowstorm. Flakes feel feathery against Gillette's face. No one talks.

JACOB'S PILLOW DANCE FESTIVAL, a sign says, ahead.

They turn onto a narrow forest road and ride for a quarter mile.

"Trees down ahead!" says Saiko's edgy voice in his ear.

They draw up before two large trunks laid across the road, one behind the other. "Ambush," an agent says. "Hurry. We need to move these logs."

The forest is too thick to allow passage around. The map shows no other route to the Samuelson house. As nervous agents stand guard, Gillette helps rifle through supplies and affix chains to trunks. Beyond the barrier are wooden bungalows of the dance festival. He remembers touring the dance studios. The archive building. An administration house. A summer barn theater in weathered wood.

"I smell meat cooking," Gillette says as the odor hits.

The smoke is tantalizing. They're almost ready to move the logs. He feels exposed. Then a voice on a loudspeaker booms from the woods. "Stop where you are! This is the police! Lay down your arms and put your hands above your heads."

The agents, pivoted toward the woods, remain ready, weapons up.

"Calm, everyone stay calm," Saiko orders. "But hold on to your arms." She raises her hands and shouts to the speaker. "FBI! We're trying to reach a house on this road to follow up an important lead on Delta-3."

Gillette peers into the forest but sees no one, holds his breath and admires her poise.

One of the researchers in the sled starts whimpering.

"What are they grilling? It's meat! Where'd they get meat? Oh God! The rumors are true!"

"If you're from the FBI," the voice in the woods booms, "prove it."

Agent Saiko steps up to the snowbank—ID raised—and disappears into the forest. Gillette hears the crunch of boots receding on snow. Then Saiko's voice comes over the hailer, ordering the agents to stand down.

Saiko reappears, accompanied by a large figure bundled into an orange hunter's parka, a shotgun over his shoulder. The guy helps Saiko through a drift. Gillette judges him to be about fifty. Closer up, Gillette sees a blue shoulder patch on the parka. "EMS."

"I'm Alvin Natkin, head of the Becket Ambulance Corps," the man says, turning talkative and helpful, as if embarrassed by his lie of being a cop. "All us neighbors live at the festival now. We take turns guarding the barrier. I'll guide you to Lyle's if you want. There's a path in the woods but you have to know it. Give me a minute to call for a replacement. You say Lyle's work might help you figure out Delta-3?"

Patricia Saiko seems unhappy about having to trust a stranger. The agents keep their arms ready as they start off. At least the whimpering scientist shuts up.

Twenty minutes later, Gillette stands in the wreckage of Lyle Samuelson's isolated A-frame home, two miles up the same road, but reached a back way. Drawers tip open. Windows are smashed. Cupboards are empty and couches ripped up. Snow blows inside the room.

The front door was open when we arrived. So much for not disturbing the crime scene.

"Did this happen during the murder?"

"No. Looters did it," says Alvin Natkin.

The elevation is higher here. So are the drifts. The skylight is snowed over and electricity turned off. Gillette's powerful flashlight beam illuminates a thermostat reading 31 degrees. Small animal tracks crisscross the snow-dusted living room floor.

"Fisher. Racoon. Mice," Natkin says.

Will I really find anything useful in this wreck?

Instantly Dr. Larch's voice comes into his head. "Maybe you'd rather give up and go home," it teases.

The living room—largest room in the house—adjoins a butcher block kitchen and master bedroom. Stairs rise to a railed balcony and two smaller rooms: a home office and a TV room/den. The kayak making went on in the basement. Down there, half-built frames lie on sawhorses near worktables filled with tools, chemicals, blueprints.

As agents wait outside to guard against attack, Gillette orders the scientists, "Search every inch."

"Can I help?" asks Natkin, who seems sincere.

"Tell me about Lyle."

"He was married to my sister before she left him. He was a good man who had bad breaks. He loved his work but had problems at Cougar Energy. He never made much money from the kayaks, but at least he felt appreciated at the end."

"Unhappy, huh?"

Natkin sighs. "Disillusioned. I guess all of us, when we're young, think if we work hard and follow rules we'll be rewarded. Then you get older and things don't turn out that way. Lyle never adjusted. He slaved away in school but the kid with connections got the scholarship. He was a good husband but Sis fell in love with someone else. When he got back from Nevada, he was bitter. He wouldn't talk about it. He threw himself into the kayaks. A detail-oriented techie all the way."

"Could you see him creating Delta-3?"

Natkin looks shocked. "*Create* it? I thought you said his work could help you *understand* it."

"He may not have intended for the microbe to be used this way," Gillette quickly says, needing cooperation.

Natkin gets angry. "How many people are you wasting on this? No wonder Homeland Security can't find their own ass. No wonder the country is in shambles."

Natkin snorts derisively.

He says, "I've known Lyle since he was six years old."

Time is running out. Gillette starts in the study, stands pivoting in the center of the room, papers crackling underfoot. The looters left the flat-screen TV and laptop computer here, but tore the rest of the place apart. Were they looking for food? Proof of Delta-3? The flashlight beam roves over a series of blow-up undersea photos hung on the pine walls. A man's bearded face peers from a bathysphere porthole. A huge albino fish with a glowing filament on its head drifts in the deep sea. A fantastic undersea bouquet of white tubes—ghostly and synchronized by current, and topped by blood red gills—sprouts from the ocean floor. Black smoke oozes up through water.

Is that oil?

He picks up the papers, goes over each one. They're kayak invoices, designs, catalogues, e-mails from other kayak makers. The level of detail confirms that Samuelson was neurotic in his dedication to work. Margins are filled with notes, cross-references, prices, descriptions of types of wood to use on frames.

Nothing about oil.

He wouldn't leave something important lying around in plain view, if it was here.

In the file cabinets Gillette finds bank statements, health records, a stock investment file, a renovate-the-bathroom file, all also covered with tiny notes.

Nothing about oil.

Samuelson's steam-obscured face peers back at him from the bathysphere. Gillette asks the photo, "Did you really know anything about Delta, or are you just some bad-luck innocent who dived too deep and got the bends?"

Gillette switches on the laptop and is gratified to see that the battery still works. A kayak swims up on-screen. Gillette accesses the hard drive easily and goes through files. Again, Samuelson seemed almost fanatic in his attention to detail about anything in which he was interested. Snowshoes. Studies on oil-degrading bacteria used to clean up Antarctic bases. A paper he'd written back at Cougar Energy, on designing bactericides to keep oil wells and pipes clean.

Why is there nothing here about bacteria living at the sea bottom, if it was such a great passion of his? Why nothing on diving? Did he lose interest?

Frustrated, they quit for the night at ten. The temperature inside Samuelson's house has dropped below 20.

"Come back to the campus," says Natkin, as if trying to convince them of Samuelson's innocence with kindness. "We shot a moose last night. There's food. You can sleep in the dancers' shacks. It'll be warmer than camping out."

* * *

The festival site has been transformed into a cooperative. The dining hall is toasty from a woodstove fire. People seem friendly, and Gillette gathers that they normally dwell in homes along a network of dirt roads here. The kitchen—built to feed over a hundred students and staff in summers—is hot with delicious smells of grilled moose meat.

"We ice fish in the lake," Alvin Natkin explains as they sit to eat. "We've shot some deer and wild turkey. Chuck James owns the egg farm down the road. He brought his poultry here. Then we lucked out when a Price Chopper truck broke down on Route 20. We've got about four thousand packs of snap peas frozen out in the snow. We've got propane. And electricity. If outsiders leave us be, we'll be okay."

The feeling of camaraderie reminds Gillette of his own street. Steaming food is brought out on platters, laid out on cafeteria-style tables. "We got lucky early on," Natkin says. "See, Ella—she runs the festival—opened the doors to neighbors. We voted to let in outsiders if they have skills." He points around the room, where men, women and children are laughing, telling stories, eating their late dinner slowly, a sign that they are well fed. No one grabs more than they need. They leave Gillette's group alone.

"Joe is a builder and hunter. He got us propane gas early on. Josh is a mechanic from Pittsfield. He knows how to butcher meat. Liz Neering is a computer expert. She rigged up bicycle power to recharge cell phones—when phones work, that is. We've got a cell tower on the road, luckily. And I know first aid."

"What if someone doesn't have a skill and wants to live here anyway?" Gillette asks.

Natkin looks unhappy. "That's why we need the barriers. You try to do the right thing but you have to make choices. If you don't have a skill, we turn you away. Some people try to come back. They don't get in."

DECEMBER 8TH. 41 DAYS AFTER OUTBREAK.

Gillette starts with the bookshelves today, reminding himself to go slowly. He finds several volumes on undersea life; dog-eared and tea-stained, highlighted in different colors. Clearly Samuelson went over these passages often.

In *Below,* Gillette eyes a passage in which the author claims that recently discovered microbes hint at a vast undiscovered world of life deep beneath earth.

In a collection *Best of Earth Science,* he reads by flashlight an essay called "Development Theory of Oil and Gas," by Russian professor N. A. Koudryavtsev. The Russian argues that oil and gas are *not* the crushed remains of prehistoric plants and animals, as is generally thought, but the byproduct of hydrocarbon-eating bacteria living beneath earth. "The planet's life started off below and moved upwards, not the other way around," he writes.

In the margin, Samuelson wrote, *Yes!!!*

Gillette scans David Karl's *The Microbiology of Deep Sea Vents.* And a looseleaf filled with articles like "Oil Prospects in the Cretaceous-Tertiary Basin of West Greenland"; "Connections between Martian and Earth Bacteria"; and "Thermophilic, Anaerobic Bacteria Isolated from a Deep Borehole of Granite in Sweden."

Samuelson wrote in the margin there too. *"Check vents near Iceland!!! Two thousand degrees necessary or no methane gas for oxidation!!! No food!"*

Gillette stares, his mind racing. Two thousand degrees is twice the heat generated during oil refining. Had Samuelson concluded that life could somehow exist at these temperatures? It was impossible. Out of the question. No thinking person could seriously entertain it.

And why the reference to methane gas? Why the link between "methane" and "food"?

Samuelson clearly once looked for new microbes in the deep, Gillette thinks. *And he was still interested in the subject because his notes look recent. So why is there nothing in his computer about it? Not a single file?*

The last essay he sees is, "Bacterial Gene Swapping in Nature: How New Organisms Combine DNA Outside the Lab."

Oh boy, Gillette thinks.

Still, all the work is by other scientists, not Samuelson. In fact, nothing in the room seems to represent any of his work relating to hot vents at all, even though he'd spent years studying it voraciously, kept detailed notes on everything that interested him, and obviously retained interest at his time of death.

By ten, when the group returns to the dance festival site to sleep, Gillette has found nothing more, and neither have other searchers. The residents leave them alone, displeased by the FBI's suspicion of Samuelson. Gillette's cell phone won't work tonight. At least the bungalows are heated by newly installed woodstoves, and propane, and Gillette is grateful for a toasty sleep.

They're back at Samuelson's house at sunup, December 9th, plowing through six fresh inches of snow. Two FBI agents have come down with the flu. The temperature in the house has dropped to 11 degrees. Fingers cramping, even in gloves, Gillette finishes up with the computer, desk, cabinets. Agents are helping search the house now, looking for hiding places, ripping apart furniture and wall beams and looking behind heating units and in toilet tanks.

Gillette keeps thinking, Os Preston said we'd hit the point of no return sometime around fifty days.

Did someone steal your work, Dr. Samuelson? Did you destroy your records? Are they at Cougar? Did you sell them? Or are you innocent and I'm just plain wrong?

The day ends as a half moon rises. Ice stalactites glitter inside bedroom windows. Agents outside start fires to try to stay warm. Agent Saiko has caught the flu.

Gillette calls home that night—gets through for a change—and learns that Alice Lee was buried in her backyard after Joe Holmes broke up the frozen ground with a "borrowed" electric jackhammer. Chris Van Horne read selections from the Bible. The block's residents will not leave Washington. Thank God for that, he thinks, and then the phone goes out again.

Back at Samuelson's house on December 10th, last day here, he starts in the basement instead of upstairs. It's already been searched, of course, but he does what he promised Larch years ago. He pretends it hasn't been searched at all.

There are fresh mouse tracks on the floor, and ice dust blows in through shattered glass doors. Small animals rustle in the chimney. With a screwdriver he pries open paint cans and frozen drawers. He checks the toolbox, fuse box, water pump, water filter. He's going crazy with frustration. One agent has come down with frostbite. Once again, Gillette thinks, my doggedness caused people to suffer.

There just isn't any evidence here, he thinks as his phone chimes. Marisa has gotten through for once, after trying for hours, but her faint voice goes in and out. She's shouting to be heard.

"Talk to Annie," she says. "She won't come out of her room unless I let her go to the zoo. It's insane! Annie? Come out! Your father is on the line!"

Gillette looks around the basement one last time as he tries to soothe his daughter. She feels as far away as Mars. "Be sensible," he says. "The zoo is dangerous."

Annie is crying. "But they're going to *shoot* the animals left, to feed other animals. That's cannibalism."

"I want to hear you promise that you'll stop this behavior, and stop locking yourself in your room."

"Just because they're animals doesn't mean they don't have a right to live. God made cheetahs too."

There's a long pause. He hates how long.

"I promise. Happy, Daddy?" But her words sound more like an accusation than an endearment.

She hangs up as more rustling—the small-animal sound—comes from the chimney. No, he realizes, not the chimney but inside a kayak-in-progress on a sawhorse.

He frowns. *I already checked inside but saw no animal.*

He runs his gloved hands over the shiny wooden surface of the half-made kayak. He rechecks the area behind the seat, opens the baggage compartment and peers inside. He sees nothing. So where did the rustling come from?

Frowning, he leans down into the cockpit, shines his flashlight forward and is rewarded with a vague frantic skittering that seems to come from behind the wooden panel sealing off the front compartment—making it buoyant if the rest of the kayak floods.

Gillette extends his hand and pushes against the panel. The panel slides up. It's on a hinge.

A mouse runs out.

But now the beam illuminates a soggy little mound of urine-soaked debris—a nest. Straw. Bits of paper. Underneath, Gillette can't believe what else he sees.

Is that a folded-up letter?

The throbbing in his skull is back.

The letter is too big for a mouse to have put there.

Which means it was hidden there.

Gillette reaches in and pulls out the soggy letter. The beam shines on smeared blue ink. When Gillette reads the opening lines, his throat dries up, the basement goes warm and his head starts to pound.

"Dear brother," the letter begins, "I tried to do something beneficial. But I fear I made a terrible mistake . . ."

TWENTY

DECEMBER 10TH. 43 DAYS AFTER OUTBREAK.

"*I invented a way to end war,*" Gillette reads.

Sounds crazy, right? Sounds impossible. But imagine that an enemy's tanks can't drive. Their planes can't fly. Their guns won't fire.

That enemy would collapse.

Where to start? With two brothers—you and me, the deacon and the scientist—arguing over the best way to achieve peace? Or with the microbe I designed and built from scratch. And sold to a Cougar competitor?

What competitor? Gillette wants to scream.

I'll probably never mail this letter. I keep writing them and tearing them up. The man I sold my discovery to threatened my life if I talked about it. So I write these letters and hide them. And then I destroy them.

Standing in the basement, reading by flashlight, Gillette has to keep his fingers from crushing the smeared pages. He's shaking, but not from cold.

I wasn't looking for a weapon at first. Just better microbes to clean up spills. I thought, the best place to find new natural oil eaters would be near oil, deep down.

"Under the sea," Gillette says, and turns a page.

Gold at Cornell and Kropotkin at the Soviet Geological Institute both postulate that oil formed originally not from the remains of surface life—plants and animals—as we were taught in school, but by subsurface microbes over millions of years. I thought, if microbes can create oil, they can destroy it.

"Where exactly did you find the microbes?" Gillette says, feeling sweat break out under his armpits.

Bottom line? All life on earth needs unoxidized carbon to live. You. Bacteria. The roses I leave on Dad's grave. On the surface, in open air, unoxidized carbon comes from carbon dioxide, photosynthesis of plants. But in the deep sea there is no light, no photosynthesis. How does life down there get unoxidized carbon? They extract it from hydrocarbons, I believe!

The next few paragraphs are smeared, unreadable. Then,

I believed my new microbe would be found in the vast community living in cool H_2O near undersea oil seeps. Vents down there are passages to the deep earth. They spew forth hydrocarbons for the great mass of microbe-rich slime, a primordial soup undergoing perpetual genesis, forming the base of the undersea food chain.

Gillette turns the page and feels his heart seize up. The ink is badly smeared. It's hard to read. He thrusts the flashlight beam closer.

I got lucky [unreadable] *thirty miles off the coast of* [unreadable]. *What I found was similar to chemolithotrophic sulfur-oxidizing bacteria that Wirsten wrote about, from the Galapagos Rift hydrothermal vents. Part of its genome was identical to a pipe corroder we'd found in an Uzbekistani well. They must be related. But the cell structure of this new microbe was based on sulfur, not carbon. The bug gets its chemical energy supply by oxidation of hydrocarbons. It interacts with metal to produce crystals that would jam machinery.*

His flashlight flickers and goes off. He shakes it. It comes on.

. . . Against crude oil it was a slow eater, unsuitable for cleanups . . . But the bug became voracious in refined fuels. Why? Because it turns out that sulfur in crude oil inhibits its growth. Since refining removes sulfur, with the sulfur gone, the microbe went wild.

The flashlight goes out.

Gillette curses, shakes the flashlight. It stays off. He gropes his way to the stairs. In the living room, the other investigators, having given up the search, crouch by the fireplace or stamp their feet to keep warm. It will be even colder for agents guarding the equipment outside; they are now all sick with flu.

"I need a a flashlight!"

Moments later Gillette stands in a cluster of scientists. Five beams intersect on the paper.

I realized that the microbe I'd brought up would cripple an enemy if inserted into their refined fuel. It was a weapon! But then I had a better idea! What if I could improve it, make it able to survive refining? Dropped into an enemy's oil fields, it would destroy the entire supply!

"You worked at night in the lab," Gillette murmurs. "During the time when Cougar had security problems."

To find the second microbe in the mix I widened my search near the hot vents to include thermophiles, bacteria able to withstand great heat. Those undersea vents spew forth magma sometimes at temperatures exceeding 2,000 degrees. Over millions of years, I theorized, microbes at warmer border areas would evolve heat-resistance. Eruptions would kill weaker ones, leave stronger ones alive, up the survival limit every ten or twenty thousand years.

"Tell me where you found the fucking microbe," Gillette snaps. "If we can find it, we can kill it."

The whole next page is a mass of smeary blue, bled into the soggy page so badly that it is impossible to imagine that the blots ever represented words. The final page is only three quarters full. Gillette's head throbs with frustration. Only a few phrases are clear.

I was astounded when the DNA of all three combined! After all, tetanus is carbon-based and the oil eater was sulfur. Still, there have been instances of species exchanging DNA naturally, across wide gulfs. The resulting spore resisted fantastic heat. Maybe it will prove the panspermia theory, that bacteria originally came from space. That earth's bacterial ancestors survived great heat when meteors carrying them entered the atmosphere. I only knew that I'd created a tool for peace, especially since I'd found the antidote by then.

Gillette says, softly, "Antidote?"

The scientists cluster closer, staring at the page.

"It's unreadable," Agent Saiko groans.

But the next page starts off,

I posed the question to my boss theoretically. I asked, if I came up with something worth billions to Cougar Energy Services, will I get a bonus? That pompous ass said my salary is my reward! After all the years of getting the short end! Well, this time it would be different! I decided to sell my discovery. But to whom? And how to even approach someone? I never did anything illegal in my life and . . .

The writing ends in a mass of smeared ink.

"My God! This is it," breathes Saiko.

Gillette's heart pounds as he fumbles with his phone. Cellular communication is out again. Gillette can't get through to Theresa, but finally finds a spot where the phone works when the group returns to the dance festival grounds.

"Are you sure you can't learn more from the letter?" she asks.

"Maybe forensics can," he says, standing in a clearing by the snowmobiles, and the bungalow in which he sleeps. *The fifty days,* he's thinking, *are almost up.* Ice stalactites hang from pines and roof gutters. He hears an animal snuffling in the trees. His toes have gone numb. His face feels like parchment.

Theresa says, "I'll call the secretary and get back to you. Stay exactly where you are."

The moon is half full, yellow, glowing. He hears, through an open window, an impromptu singalong from the dining hall. Someone plays "Amazing Grace" on a piano. It's the song Chris Van Horne ends sermons with at home.

I miss home.

He calls Marion Street and reaches Marisa, who sounds weak, phlegmy, and tells him that she has a "small cold," which means she's sicker than that, especially if she's home at this hour, not at the church. She says that Les Higuera's sources at ABC have told him that the zoning laws are about to be lifted. She assumes this means that more food and police protection will resume for Marion Street. That resources will be distributed more equally.

There aren't any more resources, he thinks. *It will be a disaster.*

"There's an antidote," he tells Marisa to keep her spirits up, and then call-waiting tells him that Theresa is back. He tells Marisa he loves her and clicks off.

"The president wants you to brief him," Theresa announces excitedly. "We'll send a copter in the morning. Bring the letter. You'll fly straight to Virginia."

"Not Washington?" He feels panic start to rise.

Colonel Novak sighs. "The president has decided that governing from Washington may no longer be tenable. When the zones are lifted, well, you know . . ."

"But my family is in Washington," he says, the hairs on his neck standing. He's thinking that he'll never forget the hideous scenes he's witnessed across the country over the last few days. He knows what's in store for Marion Street, knows that now, right now, he must get home.

"You'll meet the president at Deer Ridge," she says, referring to the deep-cave network built during the cold war to accommodate functioning government for three years if the capital is contaminated. Down there is food, fuel, power, a whole network equipped for comfort.

"I'm going home," he says. "Not to Deer Ridge."

A pause. "Meet him first," she says softly, not ordering him yet. Asking.

"My job is done. I'll send the letter with Saiko. *You* people track Samuelson's research. You don't need me for that. Find the vents. That's a Navy job. Track his money. My family needs me. There's nothing more to learn from me," he says, feeling rage overwhelm him, and the fact that he's been blocked and thwarted since the first night.

He says, "It's supposed to be a deal, Theresa. It goes back to the Constitution. The country protects your family while you fight for the country. I did my part."

"The president wants you specifically."

"Is he leaving *his* family in Washington?"

"You will get on that helicopter one way or the other. In handcuffs, if need be. I won't protect you this time."

Gillette eyes the Yamahas on the snow, glistening blue, gassed up, and the Siglin sleds on which are tied food, jerry cans of gasoline, spare hitches, rope, tools, ratchet straps and zippered waterproof duffel bags. Road maps will be in the saddlebags. He knows the backwoods trails now.

Saiko and the other agents will be fine here. They can call Albany for more snowmobiles. Right now I need to calm Theresa. And figure out how to steal a Yamaha.

"I'll go to Virginia," he lies, as if giving up, realizing with horror that he's using the same bitter tone that Annie spoke with earlier, when she promised to stay away from the zoo. He adds, "I want a leave afterwards."

"You'll get a medal, Greg."

I don't want a medal.

Gillette hangs up, eyes the snowmobiles, brake controls on one side, throttle on the other, like a motorcycle.

Hell, I started out as a motorcycle thief. These things run the same way. The maps will show the way to Hartford. Electrically powered trains go south from there, every morning at eight A.M.*, I overheard on the plane.*

A guard comes out of the dining room, hacking, looking sick, shouldering his M-16. For the rest of the night, agents will be posted out here, guarding the equipment.

First I violated orders by leaving Washington. Now I'll do it by trying to go back.

Gillette finds Agent Saiko in the dining hall, as the singalong is ending. He offers to take guard duty tonight.

"There's no need, Greg. My people'll do it."

"They're all sick. Tomorrow you're back on the road. If your guys don't rest, they'll be worthless in a fight."

Agent Saiko is wavering.

Gillette adds convincingly, "Believe me, after what happened today, I won't sleep anyway. Those guys look terrible. Give one of them a break. I'll come on at two."

Agent Saiko looks grateful.

"Okay. One shift. At four."

After Saiko leaves, Gillette grabs a box of sugar and thrusts it into his pocket. Sugar will disable the other snowmobiles.

He returns to his room and by flashlight writes a detailed report on festival stationery that Agent Saiko can take to the president. He'll hide it under his pillow, with Samuelson's letter, where both are sure to be found.

His letter starts out, *"Dear Mr. President. I regret I must decline to attend our meeting. I hope you understand . . ."*

TWENTY-ONE

DECEMBER 11TH. 44 DAYS AFTER OUTBREAK.

So far this year Washington has been spared the violent storms ravaging the Rockies and northeast. Now the blizzard heads in from the west of the capital, a slowly moving band of violence formed by warm moist air from the Gulf of Mexico colliding with arctic air from the north. In Virginia, people lie frozen to death in homes. Electrical grids are smashed. Wind blasts snap trees in two.

Even before Black Monday, weather this deadly would have set public officials arguing over distribution of emergency money and manpower. Both are gone.

In the capital, at 6 A.M., a first lone flake drifts lazily down toward a man waiting to pass through the military barrier at the Calvert Street Bridge. The dawn light is gray. The breeze smells wet. The flake lands on the black stocking hat worn by "Father" Bartholomew Young.

"You're brave to go into zone C, Father," a young soldier with a Brooklyn accent says, eyeing the collar.

"People are suffering. What else can I do?"

The soldier lets him through, and watches the clergyman trudge north up Connecticut Avenue. He's thinking that the reverend wears the only uniform that is respected on the street anymore. A lone soldier would be attacked, his weapon stolen. A lone policeman wouldn't last out here.

But the black-hatted religious people are off-limits, at least so far. Rabbis. Priests. Reverends. Imams. They trudge from minidisaster to minidisaster; administering last rites, taking confessions, overseeing disinfection fires. Pyres burn in parks. Men hammer coffins inside designated buildings. Bodies will be buried when the ground thaws. Holy men utter psalms to ease the pain of those who've lost husbands, wives, children, lovers.

Father Young has a slightly different role in mind today, though, and rehearses it as he walks.

"They burned my church."

A few more flakes fall. The storm's early warning system is pretty to watch. Snow drifts into burned rooftops, piles of uncollected garbage, a hunched figure dragging a struggling woman—a prostitute if she was out at night—into an abandoned apartment building through its smashed-in basement door.

Father Young goes dazed, horrified.

"Those men burned my church."

The wind carries new urban smells. Furniture and rubber burning, from infected gasoline siphoned and ignited by pyromaniacs. Flesh rotting. And always a hint of something that seems more memory than reality, out of reach. Citrus. Vanilla. Perfume. Detergent.

Ahead, the door of a twenty-four-hour apartment building "nightclub" opens, and out come half a dozen burly, laughing men. The sound of rock and roll shuts off when the steel door closes. A manhole cover clatters into place. Stepping past, Young hears a child screaming below. The screaming stops. He feels eyes on him. He knows he's being watched by guard/observers atop occupied apartment houses. The residents who will stay alive from now on, he knows, are armed with automatic weapons. They've looted camping stores for Coleman lamps, cold weather gear, Sterno stoves.

He tries, "Those men shot my wife."

Soon residents of these buildings will be attacking one another for food. Or because of grudges. Or power grabs. Or sex. Or because they're angry and need someone to hit. The losers will be the ones who resisted looting until it was too late. The ones able to fight with only clubs, sharpened curtain rods, broken bottles, bricks.

"Oh Jessica," he says. "My poor wife."

No, don't overdo it, he thinks. Use no names.

The scene is so different than when he'd first come to Washington. In early October, at dawn, the road had been a showpiece of modern efficiency. Entenmann's bakery trucks had delivered cakes to stores. Rumbling garbage trucks—brightly painted, private carting service vehicles—had made mounds of waste disappear. The glow of headlights had marked the approach of early commuters. Traffic lights had worked. Pigeons (now eaten) had cooed in trees and feral dogs (now eaten also) had fled from him.

Is that a police car coming my way?

Of all the luck. A few working cars still patrol but so rarely that they don't pass for days. Young keeps walking. The headlights slow. He sees four forms inside and hears the squawk of police radio when the window rolls down. He puts on his clergyman smile and leans in before they come out. Below the window, his hand touches his Glock.

A phrase from great-great-grandfather comes to him. Great-great-grandfather, arrested by Turkish soldiers, had written, "I cursed my littleness."

"Hello, Father."

"Officers."

"Mean storm coming, Father. You ought to go home."

"But so many others have no home."

The driver seems worried over Father Young. "It's not just the storm, Father. It'll be a bitch out today, excuse the language." Voice lowered, the man explains that zoning laws will be lifted at 10 A.M. Police and troops will be pulling back soon to protect government buildings. Rioting is expected unless the storm mutes public reaction.

"Dangerous to be out, Father. Want a ride back home, or to the bridge?"

"The street is my home."

They roll one way. He walks another. The snow is soft and tickles his lashes. At 6:30 he starts up the long hill that will crest above Nebraska Avenue near Marion Street. With more light Connecticut Avenue turns from a no-man's-land back into a semisafe thoroughfare. A party of men and women with fishing rods—and saws to cut ice—trudge toward the Potomac. A night hunting party exits an abandoned apartment building carrying wooden traps, probably containing rats. Bundled up people, always in groups, take walks, go scouting, looting, visiting. Four men carrying knapsacks and baseball bats signal him over.

"Want to buy meat, Father?"

"What kind?" The mentor always likes to hear food stories.

The man opens a sack in which lie a half dozen greasy-looking, foil-wrapped packages. A sickly sweet rotting odor wafts out. "Mongolian horse," the man says. "From the zoo."

Father Young stares at the guy skeptically.

"Okay, it's armadillo. But it's good."

The men walk off and he loosens the grip on his Glock.

I don't know where Gillette is, but I know where his family will be, he thinks, heading for the spot.

Apartment buildings have nailed steel signs over their lower windows as protection against bullets, rocks, invaders. Announcements replace windows. SPEED LIMIT 30. DRUG FREE ZONE. MONTESSORI SCHOOL.

Father Young turns off Connecticut by the fire station. St. Paul's Church lies forty yards away, across a lawn. The front door is unlocked. He hears the swell of voices from refugees inside. He's kept tabs on this place for days.

They burned my church, he rehearses, walking in.

And here's the pastor, in the mob, in the main area, bending over a screaming baby. The infant's face is swollen with tears, the head enormous against the shrunken body. The child refuses to eat the watery gruel that its exhausted mother is trying to get into its mouth. Pastor Van Horne turns around to see Father Young standing there helplessly, tears dripping from his blue eyes.

"They burned my church," he says.

"I'm so sorry," says Pastor Van Horne.

"I have food," says Young, unloading coat pockets. Out comes canned Spam and chickpea soup. Canned tuna. Canned plums.

"This is so generous," gasps Van Horne, dumbfounded.

"They shot my wife."

Within a few hours someone from Gillette's family will come here, as they do every day. I can't get on the block so I'll let them come to me.

Outside, the snow falls harder. The wind picks up.

Pastor Van Horne tells Father Young, "Stay with us, friend. This house is open to all." It's 7:15 A.M.

Gillette proceeds east out of Connecticut's Naugatuck River valley an hour earlier, under fading stars, churning on snow-covered Route 44 at twenty-six miles an hour. He's passed the villages of Otis and New Boston, keeping to tracks left by other snowmobiles, so as not to leave one himself. At the old textile mill town of Winsted, Connecticut, he'd cut east toward Hartford. He reaches the midway point, Canton.

I started with a full tank. Hartford is a ninety-minute drive by car from Becket, according to the map. I should be there by seven if nothing goes wrong.

Rural towns are smokeless oases in a sea of white. In the eerie quiet he occasionally spots another snowmobile in the distance. Small shopping malls appear. The drifts get lower. The breeze from the south is brutally cold.

If Theresa sent the copter to search for me, it should stick to the interstate, not smaller state roads.

He left Jacob's Pillow at 4:10 A.M., after assuming guard duty, detaching the sled and disabling the other Yamahas. The engine runs quietly. A blanket over the headlight had killed any glare. By moonlight he'd steered through the forest trails to the state road.

I waved at the civilian guard.

Thirty minutes out of Hartford now, he passes forest and frozen waterfalls reflecting back rising sun. Ahead the air fills with diamond dust, drifting ice crystals. A sign reads, TALCOTT MOUNTAIN STATE PARK.

Keep the speed down on turns.

The cold seeps through his balaclava. The plastic windshield blocks direct wind, but on curves gusts cut behind it, clawing at his face.

Theresa won't lie for me this time.

But with the letter and report in their hands—and fuel precious—he hopes they'll leave him alone for awhile.

Is it a fool's attempt to try for Washington?

I don't care. When the zones are lifted, all hell will break loose.

WEST HARTFORD, says a sign, although the area still looks rural. *I made it,* he thinks as the engine begins coughing.

Then the snowmobile slows. And stops.

Gillette dismounts and opens the engine housing. He's on a long forested curve between blasted-out cliffs. If not for the Hartford sign, he'd swear he's miles from town.

In the vast quiet he peers inside, trying not to think about the train nearing Hartford. He hopes that the problem will be visible, and small. A loose wire. A hose that slipped off, which can be fixed with tools in the saddlebag.

Nope.

He glances at the control panel and fills with horror. He's been on the road for hours, but the fuel gauge still registers full. He taps the gauge. The needle doesn't move. He knocks it with his fist.

The needle drops to zero.

I thought the tank was full. I'll never reach the train station in time.

Gillette starts to walk anyway. Nothing else to do. He sinks into drifts. The snow is powdery, and he pushes through it, his knees getting soaked.

Minutes later he hears a faint growling behind him, of engines coming. Snowmobiles, from the pitch. The riders—still behind the curve—won't see him yet. Are they FBI? Are they looters? He fills with excitement and dread.

Do I risk waiting here for help? If I hide I'll miss the train. What if it's the only train today?

And now he watches three snowmobiles, half a mile back, rounding the bend beneath the small cliffs. The lead driver must see the lone figure ahead, and broken-down snowmobile.

Gillette waits in place like a stalled motorist, an innocent vision of the past. He pulls off the balaclava.

It's too late to run.

Here they come, black-clad riders on black snowmobiles. The riders' faces are hidden behind helmet visors. Gillette sees dead deer on the sleds behind.

So it's not the FBI. These people have been out spotlighting deer.

He feels sweat break out as the vehicles halt, their exhaust rising like breath. The lead driver swings his leg over the seat and walks toward Gillette. The other riders pull rifles from sheaths and train them on Gillette. There are no witnesses, police or cameras here.

If they have fuel, guns and vehicles, they must be black marketeers.

Then he spots a handmade decal on the chest of the approaching rider. "Neighborhood Guard," reads black lettering circling a red, white and blue police-style shield.

The helmet comes off. Gillette finds himself looking into a pair of sea green eyes, veined red as if the man is tired, sick or inebriated. The face is fleshy and middle-aged; large nose and ears, a yellow smile that could be mocking or friendly. Gillette smells oil and whiskey. The man puts one finger to a nostril and blows his nose on the snow. "Just like the old days," he says loudly, to his friends. "Another tourist broke down on Route 44."

Laughter erupts from the other men.

"I used to drive the tow service, Cap," the guy tells Gillette, sounding sober and curious but otherwise unreadable. "What's the problem, pal?"

Gillette hears no threat yet but he knows that half the roadside killings in Africa start off with a smile and offer of help, while bandits appraise the target. He's out of gas, he says. He needs to reach the train to Washington.

"The train?" The guy snorts. "What are you, nuts? Soldiers guard the Amtrak station like it's Fort Knox."

Gillette offers the snowmobile as payment. Fill it with gas and it's yours, he says.

The second driver calls out, "He stole it, Rinker. Anyway, I bet it don't work no more."

The man called Rinker's eyes narrow.

"Put clean gas in it," Gillette says quickly. "It will start right up. You'll see."

Rinker calls back, tells the other two to put fuel in Gillette's snowmobile from a spare jerry can. When the snowmobile starts, they drive it into the bushes, and brush over the tracks. They rifle Gillette's saddlebags, but find nothing inside but tools.

"Okay, Cap. Who's the real owner?"

The shields and logo imply law enforcement, and Rinker wields authority well, but the questions are more curious than demanding, and "Neighborhood Guard" could mean anything. Should Gillette say he's with the government? Or running from it? Or leave that part out altogether?

Gillette looks him straight in the face, and tries a mix of truth and lie. "It's mine. My family's in Washington. The zones are going to be lifted today. I need to be there for my family before everything blows up."

The man's brows rise with interest. "How do you know what's going to happen?"

"The same reason I can get on the train."

"You're telling me you're important. A guy alone, in the middle of nowhere."

"I'm telling you that I need to reach my family."

"Just walk past the soldiers, huh? I couldn't get on that train to Washington. And I offered those guys a lot."

Why did this man need to get to Washington?

Gillette's not about to reveal his special ID unless as a last resort. "I show my license and they check it," he says. "They let me pass after that. I advise the Pentagon. I was in Massachusetts trying to track Delta-3."

"Alone? I doubt it. Where are the other people you were with?"

"I left them. To get back to my family."

The eyes narrow. The man is considering the ramifications of Gillette's pronouncements to his own life. Rinker calls back to his friends, "Cap here says the zones are to be lifted today. He says he *knows.*"

One of the other guys says, "Then we better clear the house out fast, Rinker . . ."

The third guy snorts. "He's lying. He's just a thief."

Rinker laughs. "Who's not, these days?"

Rinker stands back, hands on hips, looking like he's trying to make his mind up about something. His jumpsuit is smeared with oil stains. His face is weather-beaten, shrewd and tough. He won't be rushed.

"Family guy, huh?" he finally says.

"That's right."

"Got a photo of your family, Pentagon man?"

Gillette pulls out his wallet. The sight of Marisa, Paulo and Annie almost draws tears to his eyes. The guy checks Gillette's D.C. license, but also sees the look on Gillette's face. It seems to sway him more than any ID.

"My sister lives in Washington," Rinker says.

Gillette's not sure where this is going. He's aware of time passing. The sun is higher. Maybe the train will be late. Maybe it broke down up north, near Springfield. Maybe it will be held in Hartford. Maybe Gillette's information about the arrival time is wrong.

"If I do you a favor, you do one for me," the guy says.

"Sure. I promise."

"You said that fast, Cap. You didn't even hear what I was going to ask. If I give you something for my sister, you'll bring it to her? She'll meet you when the train gets to Washington. Should be easy if there's no zones, right?"

The guy is staring into Gillette's eyes. He's the human lie detector. Gillette meets the gaze fully.

Gillette says, "She shouldn't go outside alone."

Rinker nods, appreciating the warning. "Everyone makes promises. You a Christian?"

"Yes."

"Which church?"

"St. Paul's, off Connecticut Avenue."

"Liar. There's no St. Paul's in Washington."

"You're testing me. It's there."

Rinker sighs, as if he's made a decision he'd rather avoid. As if he needs to believe Gillette, or has no choice. He's going to gamble on Gillette. Gambling is probably all he does these days.

Like me, Gillette thinks.

"My baby sister's *my* family. I talk to her on the phone every day," Rinker says. "She's sick. I have medicine. She's hungry. I have food."

"I'll get your supplies to her."

The guy sighs. "Some things never change, like me driving people off this fucking highway. Get on, Cap. We'll make a stop first on the way to the station . . ."

"There's no time."

"We'll stop and then see if you can really talk your way onto the train. And Cap? If you lied, if you can't get on the train, or if my sister tells me you didn't come through, I'll find you. People here don't fuck around with Neighborhood Guard."

TWENTY-TWO

DECEMBER 11TH. 8:50 A.M. 44 DAYS AFTER OUTBREAK.

Snow drops more thickly in the capital. Temperatures drop quickly but fiercer winds have not yet arrived. Two inches of fresh snow coats soldiers pulling back from barricades to protect government buildings, in advance of the president's zoning announcement. At the White House, snow whirls off the rotors of the helicopter revving up on the South Lawn. The president and first lady run toward it, evacuating the grounds.

The copter rises and circles slowly, giving the first couple a last painful tour. Snow has begun altering contours below; geometry, familiarity. It's like sand lapping at pyramids, burying time and ingenuity. The copter passes over the Capitol Building, Georgetown, the State Department, the Treasury Department, buildings where legislation was crafted, buildings where secret meetings were concluded and fund-raisers held, blame assigned, credit allocated, threats made. The addresses where the president remembers honest handshakes and misleading ones, kissing a mistress, having a cancer removed, giving his daughter away in marriage, firing his best friend. He remembers peering as an awed eighth-grader from Nebraska through the White House fence, with the first stirrings of political desire.

One more pass over the city, he tells the pilot. I want to see it one more time.

The copter flies over Marion Street, where residents squint up through falling flakes trying to make out the craft above. It's a surprise to hear anything up there.

Then the sound of engines is gone.

And inside St. Paul's Church, Father Bartholomew Young hears an admiring voice at his shoulder say, "You're good with kids."

He straightens and manages a smile. For the last hour he's been tending the sick in the main sanctuary. It's as packed as a Harlem emergency room on a Friday night. Babies scream. Patients moan. The organist tries to distract the sufferers with "Amazing Grace," pumped from two sets of copper pipes extending out like angel wings in the balcony.

"I do what I can," Father Young tells Chris Van Horne. "The kitchen is even out of cat food. Nothing's left."

The electric lights dim but stay on.

The stained-glass window above shows the savior in a red robe, blessing a small boy as biblical-era mothers hold infants in their arms. Glass bunnies and glass squirrels gambol peacefully, amid glass flowers.

The quote set into the window is Matthew 19:14. "Jesus said let the little children come to me, and do not stop them."

"Here come our Marion Street volunteers now," says Chris Van Horne.

Father Young glances sharply toward the arched doorway separating the sanctuary from the rear hallway and double doors to the street. There, a woman and two kids who look like teenagers stamp snow off their boots. He sees a tall, slim black girl and a shorter, curly-haired boy, who looks Latino. The girl argues with the slim blond mother. Both seem upset.

"Ah, the Gillettes," says Chris Van Horne.

"Nice-looking family."

"Marisa? Kids? Say hello to Father Bartholomew Young."

The "Neighborhood Guard" stickers on the snowmobiles, Gillette gathers, enable the drivers to go almost anywhere in Hartford.

I wish they'd go directly to the train, he thinks.

Connecticut's state capital is crippled but functioning. Streets are piled high with drifts, but fewer buildings are burned. People are out walking, or shoveling snow. Gillette even spots parents pulling young children on sleds, making the best of a deteriorating situation. Laughing. Playing.

Hurry, dammit.

Unlike cops in D.C., Hartford's best men outpost on corners, where they warm themselves under blue skies and over oil drum fires, like winos, and wait for 911 calls that will dispatch them to places they can reach on foot.

"They're out of gas, like the troops," calls Eric Rinker over his shoulder as the snowmobile slows before a half dozen cops around a fire. Rinker produces from his saddlebag a bottle of Johnny Walker Red. A cop takes the bottle eagerly. The snowmobiles start off again.

Gillette asks, "Why don't they seize your snowmobiles?"

"God bless Republicans. Our governor says impounding a working snowmobile is taking food from a family's mouth."

Many homes fly red, white and blue "Neighborhood Guard Area" shields. White people wave and smile at Rinker. Blacks they pass scowl, and once, Rinker stops to stare down a couple of Hispanic men—carrying snow shovels and apparently looking for work—who walk away very fast.

I don't want to know more about Neighborhood Guard.

They pull up before a two-story white clapboard house emitting smoke from the chimney, and guarded by a half dozen armed, lounging white men in winter gear who unload the deer carcasses without being asked. Rinker brings Gillette inside. HEADQUARTERS: NEIGHBORHOOD GUARD, reads a banner draped from a window.

"I don't want to miss the train."

"Yeah, yeah. Believe me, they're always late."

Inside, Gillette is astounded to see the living room piled with cartons of food, cigarettes, liquor, spare fan belts, a gasoline generator and boxes of cell phones. Assault rifles lean against a wall. So do racks of cross-country skis, poles, gloves, snowshoes.

Rinker disappears into a back room and returns with a large Timberland knapsack he stuffs with bottles of vitamins, aspirin, antibiotics, a

Ziploc bag filled with gold rings and necklaces. A box of vanilla crunch PowerBars. Ramen noodles. Powdered eggs. Cooked chicken breasts.

Rinker tells Gillette, "My sister lives on Capitol Hill. I called her, told her *you'll* phone when you get close. There are two disposable cell phones in the pack. Use them in case people are tracking you, Dr. AWOL. She'll meet you on Mass. Ave., northwest side, outside the main entrance. Send me a picture of her over the cell phone, so I'll know you got there."

"Got a photo of her?"

"She's smaller than she looks in this shot. She'll wear a red ski cap and red ski jacket. Ask for ID. Put the knapsack in her hands. Only hers."

"You're a good brother," Gillette says.

Rinker pokes him in the chest, hard. "There's also a Walther 9mm pistol, sixteen-round capacity, recoil-operated in there, in case anyone tries to take the food from you. Don't try to find it until you're out of Hartford. Know how to use a pistol?"

"I do."

"Remember, Dr. Greg Gillette of Marion Street. I'll know if you lied, and then I'll find *you*. She's my only family. And by the way, take some food for yourself as a commission."

"I want a different commission."

Rinker's eyes narrow. "You're pushy, AWOL."

"I want a pair of cross-country skis and shoes. If there's snow in Washington, I may need them to get home."

Rinker nods approvingly. "Take your choice. Hungry?"

"Who isn't?"

"I put a roast beef sandwich in there for you."

The Neighborhood Guard stickers—and more scotch—gets them past the barricades blocking Hartford's zone A. The snowmobiles cruise downtown, which must have been fairly dilapidated even before Delta-3 hit. Rundown streets alternate with blocks where attempts have been made at revitalization, at new building.

Reaching the Amtrak terminal, Gillette gapes. It's a scene from the

Russian Revolution. Soldiers on horseback block the entrance like Cossacks. A screaming crowd swells against the line, trying to reach the station. Gillette's going to have to fight through.

"Okay, smooth talker. Do your stuff," Rinker says.

"Move back from the station if you don't have a ticket!" calls an amplified voice. "There's not room on the train for everyone!"

The sharp sound of a train whistle cuts the air, coming from up the line, energizing the crowd.

Gillette gets off, knapsack over his shoulder, skis in one hand, cross-country boots dangling by laces over his neck. At any other time he would look like a vacationer.

"Let . . . me . . . through!" Gillette plunges into the crowd.

He spots the train chugging in on elevated tracks beneath overhead electrical wires. It's moving as slowly as the New Delhi express. Warm air rising off the engine distorts soldiers on roofs. More soldiers sit on a rigged-up cowcatcher, eyeing the tracks, wires, fences bounding the property, VIPs undoubtedly surging onto the platform up there, trying to guarantee themselves a seat.

The station looks ten miles away.

Gillette fights toward the horses, waving his ID. The riders wear helmets with tinted visors. They wield cattle prods on long poles. Horses push against the surging crowd.

People yelling, "Fucking Nazis!"

"This is worse than Saigon!"

He smells animals, sees steam rising off flanks. He almost drops the ID. To the screech of the train whistle someone grabs Gillette's skis. He pulls them back. He's losing ground, being swept back toward where he started. He feels the knapsack being undone off his shoulders. The train slows, up top. Only the rear cars are visible. The train must be boarding. He spins to see a large red-faced woman in a fur hat tugging at the knapsack, and a thin man undoing the straps. Gillette knees the woman, watches her mouth open as she falls back. The man claws at his face.

I'm not going to make it.

Gillette shoves away the man, but two other men spin him around angrily.

"Hey! Leave him alone! Neighborhood Guard!"

It's Rinker and his men, shoving people out of the way, propelling him toward the horses like rugby players driving a ballcarrier toward a goal. A madhouse. A crazy place. A third world capital in the middle of a breakdown.

Gillette's got the wallet out. He's extending the ID, his shouting lost in the din.

A horseman looms. Gillette sees himself reflected in the rider's visor.

The cattle prod jabs toward his face.

Where are those Gillette kids? Father Young thinks.

At 10 A.M., he's finally shaken off Van Horne, the nonstop talker, and left the main sanctuary to search the nursery area for Gillette's kids. He tries the crowded basement quarantine room, kitchen and chapel, auditorium. Even the pastor's office on the first floor, all as packed as the *Titanic*'s corridors on the night the ship sank.

They have to be here. Or did they leave the building?

The damn refugees keep stopping him to ask for medicine, a blanket, blessing, Bible talk. Gillette's wife will be teaching school for the next hour. So his best chance of getting to a family member will be the kids.

But the damn church is so big, filled with Chris Van Horne's endless round of stupid activities. Board games. Math contests. The organist won't stop playing.

Not to mention, it's like Rube Goldberg designed the place, or some medieval monk who loved twisty passageways. The main sanctuary's three exits lead to four staircases, two on each end of the entry hall. Two run up to offices and the choir balcony. Two head down to a basement complex: homeless shelter, kitchen, auditorium, library, connecting hall to the nursery and God knows what else he has not yet figured out. The place used to be a nuclear bomb shelter. Closets and bathroom entrances are surprises, recessed into stone walls. There's no place for privacy. No place to pull a kid out of this nuthouse into the deserted storm.

Are the damn kids moving around in an opposite pattern? Are they behind him? He rechecks rooms he visited ten minutes ago. He turns a corner and finds himself in a spot by a stained-glass hall window he'd already checked. He sees lots of other kids in the Epiphany Chapel, Noah's Ark Nursery, "Shelter for Jesus," and peering at the bulletin board photos of refugees, labeled "New Friends."

But not Gillette's kids.

Great-great-grandfather had called soldiers "sentient puppets on God's stage." *If this isn't God's comedy, I don't know what is,* Father Young thinks.

A tall, lean, beautiful black girl.

A small, curly-haired, olive-skinned boy.

And then finally, when his frustration peaks, comes the accident of victory, a glimpse of two figures disappearing into a supply closet, and a door closing behind them. He pauses outside the door. The argument inside comes through.

GIRL: I don't care what you say. I'm going!

BOY: If you do I'll tell Mom!

GIRL: I dreamt the cubs are alive. Hiding. Their parents were killed.

BOY: Even if they *were* alive, they'd be so starving they'd try to eat you!

GIRL: Cheetahs aren't like that. Cheetahs were pets in olden days. If you tell, I'll never talk to you again.

BOY: Annie, please. *You* lost *your* parents . . .

GIRL: I'm covering my ears. La-la-la-la-laaaaa! You're not my brother. You're some stranger boy, meaner than Teddie Dubbs!

Father Young curses inwardly, hearing adults coming down the steps. He pretends to tie his shoe. When the couple passes, he puts his ear back to the door. The boy is saying, "I'll go with you, then, but *only* if we stay together . . . *only* if you swear we'll be back before dark."

"Paulo, you are soooo great."

The door opens and Young recoils. The kids stand in shock, staring at him, afraid he was listening, that he'll tell their mother—or any adult—what they plan to do.

"I'm lost," Young says quickly. "I can't find the kitchen."

"I'll show you the way, sir," Paulo says, talking fast, thinking he's taking control, distracting the religious man and steering himself and his sister clear of discipline.

Leading Young down the stairs, Paulo glances back at his sister meaningfully. Young takes this to mean that they'll be leaving the church soon.

The kid's cell phone chimes. Being polite and well brought up, Paulo tells Young, "Excuse me," before answering.

"Dad!" the kid says, delighted.

Father Young's heartbeat picks up. They're alone on a stairway. The girl has gone to get ready for the trip.

The boy saying, "Mom gave me her phone. I forgot to give it back to her. You *are*? When? But there's a big snowstorm here. How will you get home from the station?"

Father Young feels a warm glow taking hold. "Luck," the mentor has said, "serves God's purposes."

The boy clicks off. A moment ago he looked troubled. Now he's thrilled.

"My father is coming home," he brags.

"That's wonderful!" Young's face hurts from smiling.

"He's on a train in Connecticut. He works for the government. He's been out of town doing important work!"

"You love him very much, I see."

"He's a hero!"

"And you're the man of the house when he's gone, eh? That's quite a responsibility."

"I take care of Mom and Annie, all right."

"I'd like to meet your dad," says Father Young. "Anyone who raised a fine boy like you is a good father."

"You'll like him. Everybody does except Gordon Dubbs, and he's a thief. Hmmmm. How can I surprise Dad?"

"We'll think of something."

And then another voice, the irritating Chris Van Horne behind them, is calling, "Father Young? There you are!"

"Chris."

The old pastor is beaming, looking red, foolish, disheveled. Van Horne says, "I have a treat for you. Our people want you to give the midday chat today. I confess, I wouldn't mind taking a break from it, my new friend."

Stupid fool!

"What an honor, Chris, but I couldn't dream of taking your place. You do the talk."

"Oh pshaw," says Van Horne. "Today we're talking about the desert, the Israelites wandering and suffering before finding peace. Plenty of fodder there for you, I'm sure."

Say yes, Father Young thinks. *Say anything to get this idiot out of your face. Lie.*

"Of course I'll give the talk."

I'll just slip out and follow the kids, and use their cell phone to call Gillette afterwards.

A good plan.

But then the meddling Pastor tells Paulo, "Let's give Father Young time to compose his thinking. You and I will find Annie. Then we'll sit together during his talk, and after that, a Bible contest! Team red against team blue! Let's see how well you remember your Scripture, young Gillette."

TWENTY-THREE

DECEMBER 11TH. 11 A.M. 44 DAYS AFTER OUTBREAK.

The news gets worse as the train proceeds southward. Cell phones chime around Gillette, during brief periods when reception works. Loved ones are calling, reporting safe zones being broken up around the country. Stories travel through the car by word of mouth, seat to seat.

"What do you mean, my brother stole our food?" a man behind Gillette is saying. "Load the rifle, Sally. Hello? *Hello*?"

Gillette hears about homes of the rich sacked in Boca Raton, Florida. Riots in seaports as attackers flood marinas, seizing rowboats, sailing vessels, kayaks, anything that can be used for fishing. Charleston City Hall is burning. Electricity is gone in Missouri and Oregon. National Guard troops are fighting Army troops on the Vermont–New Hampshire line.

In the train, VIP passengers are crammed by doors, doubled up on seats, and push into the bathroom, where the toilet is stopped up, judging from the smell. Outside, soldiers camped along the tracks give *The Amtrak Patriot* the finger as the train passes. To the west is a tank farm of some sort, a vast holding area for oil contagion. The track suddenly rumbles. The car shakes. An immense orange fireball erupts into the sky from the tanks.

The day of the pyromaniacs is here, Gillette thinks.

The train whistle never stops blowing. The engineer is too frightened to stop, or has orders to keep going. The *Patriot* hurtles at top speed through Connecticut towns where it normally slows for safety.

Gillette uses one of Rinker's disposable cell phones to try to reach

Raines at Fort Detrick, hoping Raines will still help him even if he knows Gillette's gone AWOL. In case Raines's lines are monitored, Gillette punches in the extension of the maintenance office down the hall. He asks the man who answers to get Raines to the phone, fast.

"Who's calling?" The voice goes faint, staticky.

"Clayton Cox and Lewis Stokes."

If Raines doesn't come online in a minute, Gillette will assume the call is being traced and hang up. But at forty-one seconds, Raines's voice whispers, "I guess you figured out they're bugging my phones."

"Any progress on the names? Clayton and Lewis?"

"I ran them individually and together. As a pair, five hundred hits in the U.S. alone. I haven't even started on England, Canada, Australia. Hell, boss. Clayton Cox the Alabama high school principal spoke at a graduation where Lewis Stokes was valedictorian. Cox the wild animal trainer in Tampa answers questions online about his crocodile. Guess the name, boss? The name of the damn crocodile? Homeland Security is tracking the names too, thanks to your tips, but their staff is gutted. And they're still concentrating on the terror angle."

"Any more luck following the money?"

"Sorry."

Gillette asks Raines, "What about Samuelson's research? He found the starter bug forty miles offshore somewhere. Did Cougar records help you figure out which coast?"

Raines sounds frustrated. "In 1999 he's off Australia. Iceland a year later. Mexico. Nigeria. Cougar gave us maps. Some pinpoint undersea vents, but all the navies in the world don't have enough working ships and time to visit every one. We're talking thousands of miles of coastline. Millions of possible spots, looking for a bug you can't see without a microscope."

"Any other good news?"

"Only that you called on this phone. I'm supposed to tell them if I hear from you, and to ask you where you are. Where are you, boss?"

"Maine."

"I'll pass it along. And you won't reach me here after three. They're closing the fort. Moving important people underground. Labs. Research. Nothing key stays up top."

Gillette says, "Can you get home to your wife?"

"Oh, they'll get everyone home. No problem. We're promised all the gasoline in the world for that. They'll even give us a little food as a bonus. Anyway, I'd rather be with my family at the end. Boss, just know you've done more to beat this thing than all those assholes put together. I'm your man for anything you want until three."

"There's still time to figure this out," Gillette says.

"I won't leave my computer until they tear me out of the chair."

The names, thinks Gillette as the train passes New Haven, heads down the coast for New York's Penn Station. The engineer announces there's a blizzard ahead, but says he expects no weather delays until they pass Baltimore, roughly two and a half hours from now.

Clayton Cox. Lewis Stokes. The names of infection.

Gillette closes his eyes and tries to drown out the noise in the car. It's like he's on a third world train. A wind-power exec has started shouting at an Exxon executive, "You people blocked us from getting grants!" The guys start throwing punches. People pull them apart.

Don't think of Cox as a man, Gillette thinks, but an infection. Whoever killed Samuelson is a different type of bug. Infections disguise themselves. Figure out the disguise, find the smoke and mirrors, Larch used to say, and you've got the cure.

Larch-in-his-head waves his index finger at Gillette.

"AIDS sneaks into the human brain by hitching a ride inside macrophages, the blood's immune system enforcers. Streptococcal bacteria, plain old strep throat, elude the body's hunter-killers by molecular mimicry. They dress up like the exact bodily proteins looking to destroy them."

Gillette squeezes his eyes shut, envisions a bloodstream, and white blood cells in it, envisions the disease cell morphing into a faux white cell, then drifting casually toward the killer cells, disguised.

Lewis Stokes is a disease, all right. He's the human face of the Delta-3. *Can I find him by concentrating on the disguise?*

He remembers a village called Thiet, in southern Sudan, where he'd first seen baby Annie, remembers her mother, a tall, sunken-faced woman dying under a tree of Whipple's disease. Whipple's, another trickster, randomly incorporates bits of DNA into the genetic code of its surface proteins, changing appearance. Then Tropheryma whipplei slips smilingly past antibiotics, carrying toxins that kill with rampant diarrhea, anemia, weight loss, inflamed lungs.

Dr. Larch in his head teases him. "You're smarter than a bug, Greg. Or are you? Is a bug smarter than you?"

At least the train keeps working. It glides into Penn Station without incident, takes on a group of miraculously orderly passengers and continues under the Hudson River into New Jersey. New Yorkers adjust to anarchy better than other Americans. They're used to it. The sky goes grayer. Gillette sees a few flakes.

Cell phones stop working at that point, altogether.

He tries to reach Marisa again when they pass Trenton.

And after Wilmington, Delaware.

Finally, a nerve-shattering two hours and twenty-six minutes after the engineer's original announcement, and after passing Baltimore, he hears Marisa on the other end. Union Station is less than thirty minutes away. But snow flies thickly against the window. Drifts grow high outside. Wind buffets the train, which starts to slow.

"I'm almost home," he says.

"We voted to ask Gordon Dubbs if he wants to trade food for heating oil," she says.

"I don't trust him." He's alarmed at the reduced speed.

"Who does? But Grace is sick. The Klines are sick. We have to eat, Greg."

Is the train going to stop?

The engineer answers the question with an announcement. "I have good news for you train history buffs, who notice we're slowing." Although the storm is bad, he says, Amtrak officials hit on an ingenious solution for clearing tracks. They took an old coal-powered steam locomotive out of the Smithsonian, and mounted a plow on front. They found some coal. The plow is clearing tracks between Washington and Baltimore.

"We may even arrive on time," the driver says.

The world breaks into islands. The world descends back to a time before city-states. Social organization revolves around ten-square-foot plots of disputed ground, protected by "Neighborhood Guard."

My family will survive, whatever it takes.

"Where are the kids?" he asks Marisa.

"Listening to the midday talk. There's a new pastor here. Bartholomew Young. A good speaker. A big hit."

"Let me tell you about the desert that the Israelites passed through," says Father Bartholomew, reciting words written by Great-great-grandfather. "The desert is naked, under indifferent heaven. By day, the hot sun fermented the Israelites. By night they were shamed into pettiness by the innumerable silences of stars."

He faces a sea of upturned faces from the raised podium. The audience seeks solace, distraction, wisdom, brevity. The Gillette kids sit on the floor, beside Chris Van Horne. Father Bartholomew's pauses are filled with sniffling. Coughing. Crying. Wind howling outside.

"The Israelites were a starving army."

Paulo is beside Annie, waiting to bolt when the talk is over. But even they respond to his tones and memory, knowing that something different is happening here. What do they sense in him that rivets them? Truth? Suffering? Fate?

"In the desert," he says, "you need direction. You need clarity. You need God."

In his mind he smells the refugee camp. Unwashed bodies and mounds of garbage. If you wanted to eat you had to listen to the men in black and the men wearing skullcaps. To the Bible and the Koran. But the message was always the same. If you want food, listen. If you want water, pretend that what we say is true. To live, you have two choices: Scripture or rifle. Choose your god and bow down.

"I was in the desert once," he says, and from the stupid faces below, he sees that they think he is of them. "I wandered until I found direction. Meaning. A guide."

Each time he says "the good book" he means Great-great-grandfather's memoirs. When he says "the power that saves," he means the mentor, epitome of human kindness and grace.

He had thought it might be difficult to talk, but all he has to do is remember. He'd thought they'd see through him, but the podium is one more vestment cloaking human need.

"We are wandering in a desert now because we worshipped the golden calf," he says, repeating something a British missionary had preached in camp.

"Oil," people mouth in this audience.

"Yes," he says, taking a cue from them.

The organ starts up. People struggle to their feet. Father Young waves Chris Van Horne to the podium and gives him the Bible as the audience breaks into ragged song.

Paulo and Annie are already in the back of the church, pulling on their coats. The rear door opens. The sight of the kids brings to him another old Bible phrase.

Follow the little children.

He seems to be the only one who saw them leave. Their tracks will be easy to follow out there. They're kids. They're not trained. They're Americans, so they're weak.

People reach out to him as he passes, delay him, saying, Great sermon, Father. You have a gift.

He reaches the front door, makes sure no one is watching, walks outside and feels like screaming.

They're out here, all right, pulling away.

Going toward the zoo, as they'd said.

They're all alone in the storm, receding.

But they're wearing cross-country skis!

And Father Bartholomew Young has trained with the SAS, British Special Forces. He knows how to hike in a desert, and live for days without water. He can throw a knife with accuracy and kill a man twenty ways with common household objects. He speaks four languages, and drives heavy machinery. He can strip any firearm in the world and put it back together.

But I can't ski. I never learned.

Father Bartholomew stumbles after them, on foot, cursing, as they pull further away.

Their tracks are easy to follow, at least. I'll catch up when they slow down or turn around.

The end begins with many small acts, around Washington.

A knife raised in an apartment. A State Department clerk robs the cafeteria of bread, while soldiers hold off a mob outside. Staffers left behind in the evacuation siphon heating oil from Treasury Department boilers, and gasoline from White House limos that still work.

People living in zone A—who had been protected hours ago—lock doors, hide food, wait to defend themselves with guns or baseball bats or butcher knives, and hope that the security guards they feed are still in their lobbies, and will act bravely when the time comes to fight.

No more zones, folks, the president has announced.

Think of pandemonium as the new normalcy.

And "limited resources" as the new code of law.

Onward they come, even in the blizzard: the smart ones, the enraged ones, the ones who are hungry beyond caring, pouring over the Calvert Street Bridge past deserted barricades.

Let the other guy die. Not me.

It's the great unraveling of the magnificent city on the hill.

And on Connecticut Avenue, Gordon Dubbs stands at his window, staring across the street at smoke rising from chimneys on Marion Street, which means they have heat over there. Teddie his willing apprentice standing beside him.

War chief and son.

Their own building is freezing, the residents complaining, the pipes freezing up. The blogs—when they work—say that the city is in flames, the president gone, the scientists stymied, troops mutinied. A new era here.

Is it time to attack Marion Street for heating oil?

The Darwin Devil whispers in the ear of Gordon Dubbs, taunts him with a song of evolution. *Survival of the fittest, Gordo.* Who's the fittest? You? Or them? Or someone else in this building, who has more guts than you?

Can I really step over this line? he thinks.

You stepped over it when you killed the old man. Your people need to see you're strong.

"Dad?" says his son. "Are you listening? Basil said there's lots of pussy in the Metro tunnels. He said we should go down there and bring back girls!"

Gordon eyes the smoking chimneys.

Yes. It's time to mount a full attack.

He picks up the phone and calls Gail Hansen. He tells her to come over, that a few cartons of vodka have just arrived.

The *Amtrak Patriot* actually arrives at Union Station, Washington's great white marble elephant, on time, at two.

Passengers break into applause. They didn't think they'd get here. But now that they are here, where will they go? Their meetings are canceled. Snow is a foot deep outside. Gusts top forty miles an hour. The politicians and corporate reps and governor's aides stand stranded in the terminal like air passengers who've survived a crash only to find themselves on a desert island.

Not my problem, Gillette thinks, hurrying through the vast empty marble lobby, past America's now-closed offerings to history: a Starbucks, a Barnes & Noble, a McDonald's.

The place as silent as an Egyptian tomb.

Outside, the view of a plume of dark smoke spiraling from the Capitol dome is terrifying. Then blowing snow blocks it from view. He removes the Walther from his knapsack, having checked the pistol already to make sure there's no round chambered inside. He slides the magazine into the handle. He puts the gun in his coat pocket.

Three figures in red parkas walk toward the station, and he hopes one of them is Rinker's sister. Phones work slightly better around the capital.

He'd called her twenty minutes ago and told her he was coming. She'd said there was no need to meet at the appointed corner. The soldiers had left. Let's meet at the front entrance by the Metro escalator, she'd said.

Now the lead figure waves. Gillette figures the woman, man and boy are a family unit.

"Oh God. Food," she says, when she opens the knapsack. She has Rinker's cheekbones and green eyes. She lifts up a Campbell's Chunky soup can like it's fragile glass. "It's split pea. Look! He sent us tuna fish, Joe!"

She starts to cry.

"Thank you. My brother said you're trying to reach your family," she tells Gillette. "Well, stay out of the Metro. If you go down there, you won't get back up."

She hurries off, hunched over, shielding the knapsack, marching through the drifts with her men.

Gillette punches in Marisa's number. It rings but no one answers. He tries St. Paul's landline and gets through.

"I'm here, baby. I even brought a roast beef sandwich."

"The kids are gone, Greg."

He freezes. His breath frosts. Marisa would never say this if she hasn't checked every possible place where they might be.

"They're not in church, Greg. Not at Marionville. They have my cell phone but don't answer. No one's seen them."

He understands her meaning instantly. "Annie's too smart to go there, Marisa. Paulo would stop her."

"It's been building up in her. I went outside and looked for tracks. If there were any, snow covered them up."

I'll kill those kids, he thinks.

He tries to reassure her. The kids are probably close, he says. Probably they'll show up in a few minutes. But meanwhile, he has to ski up Connecticut Avenue anyway, to reach home. "I'll check the zoo."

"Great. Then *three* of you will be gone."

"Call me when they show up," he says, frantic with worry and rage. "Who saw them last?"

"They were talking to Father Young before his sermon . . ."

"Young?" he asks, straining to hear.

"I told you before. Remember? Bartholomew Young? He showed up this morning? His church burned down."

"*Who* Young? Say it again?"

"*Bartholomew.* Les was kidding around and called him Father Bart. He threw a fit! My name is *Bartholomew,"* she says, imitating the guy. *"Don't call me Bart!* I tried to find him to ask him if he saw them. He must have gone out too."

Gillette stares at his cell phone in horror. Sweat breaks out in his armpits and along his spine. He thinks, *I'm jumping to conclusions.* He thinks, *It can't be.* He thinks, *Of course it can be. This man has anticipated our movements from the beginning, killing families. Of course he'd come after me.*

Should he tell her? She'll panic if he does. But he has to tell her, in case the man calling himself Bartholomew Young comes back to the church.

"Maybe it's *not* him," he says, "but under no circumstances go near the guy. Keep the kids away from him. I'll call the base and try to get someone sent out."

But Gillette is thinking, *Who am I kidding? If he has the kids he won't come back to the church.*

Clicking off, fighting off the urge to get going, he tries Raines's direct line immediately, heart pounding, and gets through. He doesn't care if security listens in now.

"Hi, chief. You still in Maine?"

"Check another name for me. Bartholomew Young," he says, spells it, and then suddenly—maybe it's the panic that does it—something clicks in his brain about the bacteria he'd been thinking of earlier. It's something he's been missing.

Bacteria that disguise themselves take on only partial identities. They adopt only some DNA of their hosts, not all. They don't perfectly mimic enemies.

So he tells Raines, "Break up the names. Maybe we're looking at them wrong. Maybe it's a combination we want, or just last names, or first ones. Run them individually. See if anything clicks."

"The base closes in fifty-five minutes, chief."

"Hook me through to Colonel Novak."

"Are you sure you want to do that, chief?" Raines asks protectively, knowing listeners are zeroing in on Gillette.

"Now!"

He can't stand the waiting. He envisions Paulo and Annie skiing through the blizzard, comfortable in snow as he and Marisa had taught them to be. Pole and glide. He envisions them cruising along Connecticut Avenue.

Then he envisions a lone figure behind.

"Where are you?" snaps Theresa's voice. "How *dare* you—"

He cuts her off, tells her that the man who killed Lyle Samuelson may be at the National Zoo right now.

"The zoo? What are you talking about?"

"He's after my kids."

He puts all his will into his voice, and his logic. "They've known too much about us since the beginning. They targeted families all along. They must have a source inside the investigation. Well, I found the bug, didn't I? I found Cougar, and Samuelson . . ."

"Jesus," she says. "You think?"

"Arrest me after the storm. But right now get people to the zoo."

"I'll try, but there aren't a lot of troops out there at the moment. They're out of gas. Or blocked from coming by fighting downtown. The Metro's not running. I'll do my best. Call when you get there, Greg. I hope you're wrong."

He tries to call Paulo, who doesn't answer, who probably knows it's a parent on the other end. Paulo, goddammit, won't want to face Gillette's monumental ire.

Gillette makes sure his Walther is in his pocket. He digs his ski poles in and shoves off. Left foot forward. Right foot back. Rage gives him energy and frantic rhythm.

Remember the old lyrics?

It's all happening at the zoo.

TWENTY-FOUR

DECEMBER 11TH. 2:21 P.M. 44 DAYS AFTER OUTBREAK.

Push and glide. Push and glide.

Gillette must pace himself, heading up Massachusetts Avenue. If he exhausts himself and gets drenched with sweat, he'll be useless when his kids need him most.

The zoo, he thinks. *I should have known. But you can't watch a teenager all the time.*

Hopefully, Theresa has sent soldiers by now.

His anxiety is a lump beneath his breastplate, a rock in his lungs. Looters leave him alone. They're more interested in zone A buildings. He flashes to the faces of the killer that he saw on the Las Vegas security tapes; unremarkable except for the fact that Detective Hardy pointed them out. Gillette affixes a clerical collar to the faces. The man will be armed. Gillette tries to block away the memory of the FBI man's family in Washington, the dead boy in a Nevada hotel.

Can he really be on Massachusetts Avenue, celebrated corridor of global power? The same route he's traveled countless times with Marisa, to pick up her parents at Union Station? The route he's ridden in cabs, to testify before Congress? He's attended embassy parties on Mass. Ave. more times than he can remember. He and Marisa once even danced at the yearly Halloween soiree at the Observatory, the vice president's residence, near Reno Road. He'd worn a false beard that night, as "Louis Pasteur," the doctor who figured out how to kill rabies. Marisa had put her hair up like Ann Crowe, first schoolteacher in colonial times.

Now the city has turned into some postapocalyptic vision, more quickly than the Rapid Response Team had ever thought.

His cell phone is chiming, but when he sees that Raines is calling—not Marisa—he keeps going.

Some of the plundering around him seems haphazard. In other areas, it's organized by gangs. At Sheridan Circle, looters fan out from the Mayflower Hotel, pulling sleds loaded with supplies. He passes the burning Brookings Institute, the late great think tank, where he'd attended a seminar on "destabilization in the 21st century." There, Ivy League scholars had predicted water wars in the Mideast, flu epidemics out of Asia, nuclear accidents in Eastern Europe. And of course, the reason Gillette was present: suitcase bombs of bubonic plague.

Keep the ski tips out, he reminds himself. Push up hills. Bend down on slopes, to save energy, to eliminate wind resistance.

I never should have left Washington.

The phone keeps chiming. At Dupont Circle, where Massachusetts meets Connecticut, he stops, breathing hard, to rest a moment. He pulls out the unit to answer.

Still no word from Marisa. But this time he takes the call from Raines, who sounds jubilant.

"You did it, boss!"

Gillette stands in the storm, throat raw with terror for his kids, barely absorbing the emotion pouring over the line. Raines checked out the false names, all right, separately, as Gillette ordered. "Want to know why nothing came up before? Because they're all last names! Strung together!"

"How can you be sure?" he says.

"Because they all show up in the same place! They're names from history. British names! Six out of six and that's no coincidence! You're not going to *believe* the link. It's Lawrence of Arabia, the World War One hero."

"Are we reaching for connections here, Raines?"

"No. The only links where all six names consistently appear involve Lawrence of Arabia. They're *people he admired*. He wrote about them!

The Brits are on the line, from London. T. E. Lawrence rode with the Arabs but helped trick them. Helped the Brits and French divide up the Mideast." Raines laughs bitterly. "Oil, get it? He took away oil."

Gillette is stunned. Lawrence of Arabia? All he knows about the man is what he saw in the old movie, and who can tell if that was even vaguely reflective of truth? He flashes to the actor Peter O'Toole leading Bedouin camel troops, attacking a Turkish column of soldiers. He sees O'Toole striding atop a bombed-out Turkish locomotive, while his ragtag soldiers chant his name.

He says, baffled, "What connects this to Delta-3?"

"How do I know? Clayton was chief of intelligence. Cox a staff officer. Bartholomew planned the push on Damascus. Lewis was an Australian who fought for Lawrence, with Stokes. Punch in the names and up comes Lawrence's book, *Seven Pillars of Wisdom.* Hell, it's like an American naming himself after presidents. Lincoln Washington. Wilson Adams. Only a Yank would do that. Can you see an Al Qaida fanatic naming himself after dead British generals? I don't think so! Could Delta-3 come from Britain?"

"I need you to do something else for me."

"Everyone's going crazy here. I can stay another two hours. Homeland Security is dropping the Arab angle—well, they'll never *entirely* drop it but—"

"Find Father Bartholomew Young in Washington. Go to D.C. police records. Churches. Utility records. A telephone number. A credit card. *He's here.* And get hold of a Las Vegas detective named Duane Hardy. He's got videos of the guy."

"Boss, this is one place where the stupid zoning law helped for a change. There are databases of everyone who lived in zone A or B. HS needed to figure out who to move out. They're on this already. They started looking when we linked up the names."

Gillette resumes poling, pulse racing. He turns north up Connecticut Avenue. He prays as he moves. Please God, protect my children. Take me, not them. Don't punish my kids if I should have stayed with my family.

Nearing the Calvert Street Bridge, he sees figures struggling in drifts, like wraiths. Then the snow clears for a moment and he stops, shocked, because standing ten feet ahead, in the middle of Connecticut Avenue, is a leopard. Alive. Free. A leopard in front of a broken-down Metro bus.

What's that in its mouth, Gillette thinks.

A human arm, in an olive-colored parka sleeve.

My kids don't wear that color. Annie hates it.

The animal's breath steams. The leopard regards him as if out of some peyote vision. They stare at each other. But the smell of sour meat is real, and the odors of dung and wet fur. The eyes are green behind gauzy snow. Small ears press back on a roundish head that seems more powerful up close than through the bars of a cage. The tail swishes.

Gillette works the Walther from his pocket slowly, as if the creature knows what it is. He wonders if 9mm bullets can stop it. He shouts at it. "My children!"

That's an adult's arm, he thinks with relief.

The green round eyes blink.

Enraged, he screams, "My children!"

The leopard turns away casually, toward the incline to Rock Creek Park. It pads through drifts and is gone. Gillette had been a curiosity to it. Nothing more. Certainly no threat.

How many other animals are out here?

Pole and glide. Christ. A leopard.

Three quarters of a mile to the zoo, he estimates.

Please God. Let me find them alive.

Father Bartholomew Young puts his Glock 9mm pistol back into his parka pocket. The looter at his feet sprawls face up, where he'd been backpedaling. His arms are spread outward. In the stained snow lie dozens of eyeglasses.

Not food. Not weapons. Eyeglasses. Wire ones and rose-tinted ones. Horned-rimmed glasses. Lenses of all kinds. Armageddon arrives and one nut hoards eyeglasses. Had the guy been an optometrist? Maybe he'd realized that vision in the future will be a precious commodity to sell.

All I needed was your snowshoes.

But the idiot had tried to flee.

Father Young pulls off the straps binding the Yukon Charlie's Back Country model 825s to the guy's boots. He takes the Velcro-secured gaiters too. He unstraps the poles and strips off the gloves and wool-lined Saranacs, both warmer than his thin leather London Fogs, which he discards.

Off we go on snowshoes, faster, and even faster!

This is much better than walking in shoes!

The tracks that the kids laid are deep and steady, running up Connecticut Ave. like a rope attached to their feet. Father Young's legs throw up snow like paddlewheels. His leg muscles are powerful. He uses the poles to push.

Even if they're ahead of me, they'll turn around and come back the same way. So we have to meet.

Traveling Connecticut Avenue is like taking a river in a jungle country. You stay in the middle. You don't venture near shore. He follows ski ruts down the long incline to Yuma Street, past dead cars with their bumper stickers—where snow has blown off back windows—naming a rogues' gallery of people who might have slowed oil dependence and didn't. Bush. Clinton. Bush.

Down another hill, past another strip mall. And voilà, he's at the zoo! The ski tracks turn left, past the stone lions, snow-blasted shrub garden, "Asia Trail" and gutted guardhouse.

He stops. He listens for the voices of teenagers.

Come out, come out, wherever you are.

"Have some vodka, Gail. You'll feel better."

Generalissimo Gordon Dubbs, lord of 5110 Connecticut, a man who feels himself growing into his destiny, flashes his handsomest smile at the terrified woman bound by duct tape to a Windsor chair. The abandoned janitor's basement apartment is his new "interrogation center," modeled after his time in Iraq. The bedroom is cold and empty, the bulb bare, and he's laid out tools on the wood floor.

Gail Hansen's face is streaked with tears. She's shivering and sweating and not looking so rich and superior now. Snot runs onto her once-stylish blouse.

Teddie says, "Dad?"

"Not now. Watch and learn. Gail? How many security cameras are on the block, besides the ones on both ends?"

"Shove your fucking vodka you know where, Gordon."

She looks away wildly, at the ceiling, walls, beach photo, into the living room with its secondhand furniture. Anywhere but at him. Gordon yanks her head up by the hair. Her scream arouses him. Teddie catches his breath.

"How many more did you say, Gail?"

"Only those two."

"Dad?" Teddie says.

"I've told you before. It's impolite to interrupt."

The strong will survive from now on, he thinks. The Marion Street types will feed the predators. Teddie better learn that. If he's squeamish, he needs to get over it.

Gail blubbers, "They know I'm here."

"Oh, we both know that's not true. You've been sneaking over here for weeks. They think you're locked in your house, like usual. The snow will cover your tracks, you drunk."

Dubbs extends a highball glass half filled with Three Czars vodka. Let her smell it. She wants it. Let the boy see how to work addicts. You dangle the treat first.

"Drinkie, drinkie, Gail. Which house has the security monitor?"

"How can you let the boy watch, you pig!"

He balls a fist, hits her in the face, feels the satisfying crunch of cartilage snapping, soft flesh splitting. It arouses him even more. She grunts as her head snaps back. She's hyperventilating. No one's ever treated the woman this way. She starts to scream, gags, chokes, starts weeping. Teddie steps closer. The boy, Gordo notes, smells of cologne even though he hasn't started to shave.

Little man!

She gets out, "The monitor is in the Higuera house."

That's better. Gordon shows her a hand-drawn diagram of the street which he made. "This house?"

A nod. A sob. "The women watch it, upstairs."

Good. She's volunteering information without being asked now.

"How many guns on the block, Gail?"

Teddie stares into Gail's face, fascinated and flushed and actually excited. It's homeschooling. Science lab.

"Bob has a shotgun," Gail tells him. "He's a Marine."

And the way she whispers "Marine" is how some people say "God." Like "Marine" means "archangel." But I was in the Army, Gordon thinks. Cops are tough too. I've taught my people to fire weapons. One ex-Marine is no match for us.

That jarhead hit me.

"Dad, her nose is squishy," Teddie observes as she says, "Both guards have handguns. One apiece. There was another gun but it broke." Then she whispers, "I'm . . . so sorry," and he realizes that this last part is addressed to the neighbors she betrayed, not to him. She's dribbling like an infant. Disgusting.

"Do the guards stay put? Or walk around?"

"Walk . . . around . . ." She slumps now. All will has left her. It is as if will holds the human skeleton together, and its loss can melt solids, change physics, alter souls.

If we take out the guards, only the shotgun will be left.

"What other defenses are there, Gail?"

"I don't know. I'm not on that committee. Please don't hit me again."

Committee? He laughs. The world collapses and Marion Street forms committees.

"Which committee are you on, Gail? The drunk committee? The I'll-do-anything-for-booze committee?"

He hears Teddie chuckle.

The overhead bulb flickers but stays on. No doubt about it, power outages will be rampant soon, and that will make Marion Street's cameras wonderfully blind.

The feeling that comes to Dubbs now is of incredible power. He wonders if this is what it felt like to be a medieval lord, unconstrained by law, absolute ruler of peasants, like in films he used to watch.

A line from *Duke, Devil and Destiny* comes to him.

He steals the line, and tells Teddie, "All things are possible for the strong and daring, my fine young son."

There will of course be alliances to make later with other groups. Agreements carving up areas of influence. Trade deals. Assistance pacts. More powerful people will rise and require fealty, as kings did from dukes in olden days.

But the key will be surviving the first round of slaughter, to live until the top dogs arrive. The key will be supplies, heat, reputation, fighting back.

"Was that so bad, Gail? Have a sip."

In the police he'd had power over street people, but even when dealing with the lowest retrograde, there had been rules to follow or avoid. Here there are no rules. He can make his own. He can do whatever he wants with her. But before he does, is there more he can find out? Some benefit he can obtain that he has not yet considered?

"I hate myself," she says.

Leaders, Gordo knows, must be scientists of human behavior.

"Dad?" Teddie asks.

"Now you may speak, son."

"Can I try it with her too?"

Gordon grins. *There's* a benefit he hadn't considered.

Good boy, Teddie. Show me your stuff.

He's following them on snowshoes, Gillette thinks, looking down at the markings in the snow in horror, having reached the entrance to the zoo.

He digs his fingers into his palms. Two sets of cross-country ski tracks head between the stone lions ahead, overlaid by the wide oval marks of steel-clawed snowshoes. That means the snowshoes passed last.

I can't pole forward and hold the gun at the same time.

He leaves the Walther in his pocket. He heads into the zoo. Is someone watching from the shrubbery flanking the entrance? The guardhouse where the door swings back and forth in the wind? The bunkerlike visitor center coming up on the left, embedded in a man-made hill?

Has that leopard come back?

The ski tracks—unlike the snowshoe tracks—go into the visitor center but also come out. Meaning that Father Young had realized that the kids had moved on by the time he got there. So he'd made up time following tracks receding into the zoo. Gaining on the kids.

The snow is almost blinding. Gillette follows arrows pointing toward the American prairie exhibit and the giant pandas. He finds himself at the cheetah enclosure, a small hill ringed by a low fence and moat, all going in and out of vision in gusting snow. Two wooden boards have been laid over the moat as a ramp for hunters to get in, or escaping animals to get out. He sees no animals.

"Cheetahs are losing their race for survival," reads a sign beside a photo of the cubs, showing their furry round heads, serious expressions, cute black facial markings flowing like mascara from nose to jaw.

Annie had sat for hours at dinners, parroting information on the signs. "Daddy, in Renaissance Italy noblemen carried cheetahs to hunting fields on horseback. In Moghul India, Akbar the Great kept cheetahs at court."

Fucking baby cheetahs, he thinks, spotting a zebra carcass—butchered cleanly by humans—on the far side of the enclosure.

His cell phone chimes and cursing, he grabs it, and turns the volume off, hoping that the sound hasn't carried. There's no point in answering. Any information he needs at the moment is written in the tracks in the snow.

I'll kill him.

He poles forward. The zoo seems bigger than he remembers. Paths meander and diverge. There are forested areas thick with red and pin oaks. An upper and lower level of exhibits. A concrete reptile house.

What if the tracks he follows double back, and will bring him face-to-face with the man he's after?

The man who is after me too, he thinks.

I have to warn the kids. I have to get him to come for me. To leave them alone.

He halts to think, outside the panda house, the concrete bunker with a zigzaggy ramp leading inside. He's visited the zoo so often with Annie that even with the snow blowing, he envisions his surroundings. Is there something he can use here? Some way to get the man to come for him?

He conjures up the nearby zoo plaza and "Panda Store" filled with stuffed animals. The popcorn wagon up there. The public restrooms and the mini–ranger station where panda tickets are sold.

That's it! That's it! Inside!

Thirty seconds later he's looking into the hutlike ranger station, with its door askew, shelves and desks looted, minifridge open, and bits of fur, animal turds, feathers and blood on the floor. He unstraps his skis to get inside. A ripped ranger parka hangs on a peg. Graffiti has been spray-painted over the park service logo: "EAT ME." He jerks back when he hears a sneeze above him, and looks up to see a white owl staring down from a beam.

Maybe the kids have been captured, he thinks, reaching for the silvery microphone, praying that his idea will work. The kids must be alive, he tells himself, because if Bartholomew Young wants to use them against me, he'll want them functioning, able to talk, at least for a while.

Loudspeakers, he knows, are scattered on light poles throughout the zoo. Now he must send his parental will out through the boxes, through the electronic void.

He flicks switches on the console. Nothing happens. He jiggles the switches desperately. He drops down to look for loose wires or connections. Oh God, let me reach my kids, he thinks. Let this fucking thing work.

Walking in snowshoes is like walking in deep sand.

It floats. It grips. You have to lift your legs high, Bartholomew Young thinks.

He can see from the fresh tracks ahead that he's close to the Gillette children, even if he can't see them yet. The way the snow pulls at his feet reminds him for a moment of the first time that the mentor had sent him on a mission, after saboteurs had blown up a company pipeline in Algeria. The attackers had demanded money, or they would strike again, they said.

He'd tracked eight men through the desert for a week, from the sandy part to the dry mountains, and finally a wadi where they lived in a cave and thought themselves safe. There they built campfires, roasted mutton, sang songs, boasted about their skill with explosives, and their ability to strike the oil lines at will.

Ninety-two seconds was all he needed.

After that, the mentor's property was left alone.

The zoo is a smashed-up city. He's passed a man in a parka wearing eagle feathers on his head, and carrying a steel crossbow, living out a hunting fantasy amid wreckage.

Another time he heard snorting and had been surprised to see a pygmy hippo waddling from a clump of bushes on tiny feet, confused, shivering, snow on its wedge-shaped head. He was astounded that it had lived this long.

But the hippo was a lucky freak. Basically this place is less zoo, more butcher zone. Blood smears walls. Dung and meat smell is everywhere. The zoo is an exercise in the kind of natural selection that involves guns.

And now the blowing snow parts for an instant, and the man calling himself Bartholomew Young sees two smaller figures ahead. He'll be in hailing distance in a second. Despite his irritation, he has to admit to feeling admiration for these two, for the girl's dedication to orphaned young animals, and the brother's fierce loyalty to the sister. Hell, they're adopted. And Gillette, whatever his faults, did a good thing by adopting orphans.

"Your mom sent me to get you," he'll tell them.

As the snow closes in again he calls to them but gets no answer. They must not have heard over the keening wind.

"Your mom is pretty mad at you," he'll say.

There they are, on the path!

He's about to call out again when a burst of static, an electric tremor ripples through the air of the zoo.

"Annie! Paulo! It's Dad!" an amplified voice booms.

Father Bartholomew Young stops in his tracks.

"Leave the zoo the back way! Use the park! Stay away from Father Young! He's dangerous!"

You don't know the half of it, Young thinks, turning away from the kids without regret. Whatever else Gillette will accomplish today, he has just saved his children.

"Don't try to find me! Call your mother at the church!"

Keep talking, thinks Young, eyeing a loudspeaker on a light pole, trying to remember the last place he passed a zoo map. The map will show the locations from where Gillette will have access to the intercom system.

Gillette's voice booms, "I love you! Get out!"

TWENTY-FIVE

DECEMBER 11TH. 3:49 P.M. 44 DAYS AFTER OUTBREAK.

Gillette lets go of the microphone reluctantly. He must leave the shack before Bartholomew Young arrives. The urge to keep warning his children is overwhelming. Suppose one last message makes the difference between life and death?

Outside, light has begun fading. The silence is profound now that the booming of the public address system has stopped.

Gillette steps out into thickly blowing snow. His plan is crude, *but visibility is so bad it might work,* he thinks. He plants his skis in a drift piled against the ranger station. He drapes the spare ranger parka over the skis, creating a scarecrow. Anyone coming up the path will now see the "shoulder" of a man partially concealed behind the shack. As the scarecrow's head is behind the building, there's no need to use a hat.

Come after me, he prays. *Leave my kids alone.*

The Walther feels heavy and inadequate in his hand. He circles around the back of the shack, keeping low, crosses the rear of the small plaza and steps through the shattered plate glass window, into the panda store. From here he should have a view of anyone crossing the plaza. His footprints, hugging buildings, will be invisible to anyone looking for him.

Crouching behind the cash register, peering out, Gillette is aware of hundreds of black button eyes fixed on his back. Stuffed apes. Stuffed otters. Stuffed penguins.

The wind in the trees goes, *Ooooooh.*

Maybe I should go looking for Bartholomew Young.

Then, coming up Olmsted Walk, Gillette sees something moving. A vague low shape in the gray glides toward the shack. His gun comes up.

His breath catches. Apparently the baby cheetah has grown since Gillette last saw it. It's almost adult-sized now. Thin, but taller. Shoulder blades rotating. The cat never even glances his way.

What do you know! Annie was right, he thinks with parental pride, tracking the animal past the hut, turning back and cursing inwardly, only now seeing the human figure sneaking toward the ranger station, hugging the front of the panda shop. Gillette's keyed-up state causes him to move too quickly. He swings the Walther left but the man must hear or sense him, because he throws himself sideways an instant before Gillette fires. Shots explode across the plaza.

The figure scrambles away on all fours, like an animal, and Gillette thinks, *Maybe I hit him.*

But then a voice calls to him from the half-buried area of picnic tables and bushes on the far side of Olmsted Walk. The man does not sound in the least bit injured.

"I need to speak with you, Doctor."

The voice is mild, reasonable, unaccented, and floats like snow out of the gray.

"Sir? I have your children."

He's horrified but he doesn't answer. The man is trying to fix his position. Is he lying? The taste erupting in Gillette's mouth is like rusty iron. He feels as if a baseball bat has started tapping against his sternum. It's just his heart.

"That Paulo! A gutsy boy," the voice says.

The voice seems to be moving now, left to right, but the snow blows too thickly to see the man. Gillette resists the urge to fire. How many shots has he squeezed off? Did he fire twice? He'd started out with sixteen rounds in the magazine. Had he had a full magazine? If he removes the magazine to check it, Young might hear the click.

"That Annie of yours is a stubborn girl, Doctor."

Gillette moves also, staying in the souvenir shop, easing ten feet left to lie in broken glass behind overturned shelves. Stuffed giraffes lie by his face.

"Poor kids! They're tied up in the snow. The drifts will cover them up."

Gillette envisions his kids smothering. The man wouldn't do that, he tells himself, wouldn't just tie up kids and leave them in snow. It would be inhuman. But he remembers the photos of the FBI man's slaughtered family.

The voice calls, "I'll make you a deal, sir. Come out. Talk to me. Then I'll let them go."

I bet you will, Gillette thinks.

The voice keeps moving. "You're thinking about those other families, I know. But *those* times I was supposed to distract people. *This* time is for information. I'm not an unfeeling person. I honor my commitments. I'll ask some questions and then I'll shoot you. See? I tell the truth about things. Great-great-grandfather only killed when he had to also. *I'll let them go.*"

Great-great-grandfather? Gillette thinks.

"What kind of father are you?" the voice calls.

Gillette's mind works frantically. *If he needs to ask questions are we close to figuring out Delta-3?*

The voice is muffled, losing patience. "If I have to come over there, the offer is void."

But as long as the man keeps talking, Gillette knows he's on the other side of the plaza. He risks lowering the gun and pulling out a cell phone. He punches in Marisa's cellular, hoping he'll get through to Paulo and the kid will answer. His hands are shaking. He hates the loud beeps the unit makes. If Paulo doesn't pick up this time, he'll have to give up, have to assume that the kids have been taken. There's no choice. He hears ringing, and a clicking. He's gotten through!

You don't have my kids! he exults.

But then the world collapses because the voice that comes over the line is the same one from across the plaza.

"Sir, throw your weapon out, across the walk. Come out of the store, hands up. This will only take a few minutes. It was smart of you to call."

Gillette's will dies away. His energy is gone. Who was he to think he ever could have outsmarted professionals. Standing, he throws the Walther across the plaza. He doesn't hear its soft landing.

He calls, "I'm coming out."

Maybe the man will keep his word. Maybe there's some spark of morality there, some code of conduct that enables Father Young, Clayton Cox, whatever the hell the guy calls himself today, to live with the horrible things he does. Even the worst murderers have personal codes.

Please, God, let my kids still be alive.

He steps over broken glass. The snow blows harder, and drives at his eyes. He walks onto the plaza.

Goddamn zoo, he thinks.

The man's voice says, "I'll keep my word, sir."

The figure stands up, fifteen feet away, in plain view.

But then an amplified voice booms out suddenly from the direction of the zoo entrance. It's a loudspeaker.

"Dr. Gregory Gillette! This is Captain Robert Arnett of the United States Army! Gregory Gillette! Come to the main entrance!"

"I wouldn't answer, sir," the figure advises calmly, moving forward as if no soldiers are close.

Gillette sees the man's hand outstretched and the gun. If he moves the wrong way, Young will fire. Maybe the soldiers will save my kids, he thinks.

"Come along," Young says, gesturing toward the rear of the zoo calmly.

"Colonel Novak sent us!" the loudspeaker calls.

The man before Gillette seems no different than any other man. He's just a man in a parka. The face looks older than the one on the Las Vegas video, but the shape of the head is the same. It's the same man.

The loudspeaker voice seems to be coming closer now, down Olmsted Walk, as troops advance into the zoo. The voice seems to seek out Gillette like radar.

"Your children are safe! They're with us!"

Whaaaaaa?

Bartholomew Young looks closely into Gillette's eyes. His pistol has a long tube screwed in front. He shrugs. "They got out of the zoo. You did well, sir. They're fast on those skis. The boy dropped his phone in the snow and I picked it up. You can't trust teenagers with valuables."

Gillette can hardly breathe.

"Marisa taught them to ski," he says, aware of how stupid that sounds, trying to delay, talk, think, breathe.

Gillette sees the blue eyes under a stocking cap blink. Otherwise the face is as expressionless as one on a stuffed animal. Gillette waits for the bullet. But he's happy. His kids are safe.

"Do you know who he is?" the man asks.

Gillette bows his head. This man must not learn anything from me, he thinks.

He hears a metallic click, imagines a trigger easing back. The damn soldiers must be close.

Gillette speaks quickly. "You need me. You need information, right? Everyone else who knows it is gone."

Gillette feels the gun pressed to his temple, steady and cold through the hat. But there's no shot.

The man asks, "Do you think your children are safe? They're not safe. They're only safe at the moment."

Gillette looks up.

The man says, "If it's not me, someone else will come. If it's not you, we'll ask your children or wife. There is only one way for you to stop it. With answers."

Gillette says boldly, "We know who he is."

The man pauses, peers at him closely, trying to see the truth. "That boy is not your natural son. He is not of you, yet you would die for him. Do you want him to die for you?"

The pressure of the steel silencer increases. Gillette smells garlic. The man says, "The name, then. Say it."

Gillette hears a new sound now, not electric but it sounds like . . . snorting? From a *horse*?

"The name, Commander! I promise I'll stay away."

Gillette says nothing.

Suddenly the man is gone and, sure enough, a horse-drawn sleigh is coming, and in the sleigh sit soldiers.

He means to come back, Gillette thinks, starting to shake.

The sleigh that reaches him is a great big one from the Smithsonian. An open rig once used for Christmas rides and 19th-century picnics, drawn by two large black horses.

The conveyance is completely at odds with the modern weapons and equipment wielded by the men inside.

Gillette points out the direction in which Father Young fled. The snowshoe tracks lead toward the giraffe and elephant houses and acres of forested zoo.

"Your kids are safe in the other sleigh, sir."

The Rangers spread out on foot, following the tracks in a beater's line. Without snowshoes or skis they look as if they're pushing through surf. But they seem tough, competent, angry. Their captain explains to Gillette that a second team is heading into the zoo from the rear, to trap Father Young.

"No chance he'll get out. Here come your kids, sir."

Disbelief gives way to release. In the second sled, the kids are dwarfed by their troop escorts. Paulo looks fascinated. Annie knows she's in more trouble than ever before in her life. Gillette hugs them, squeezes them, then starts to shout.

"How could you lose that phone? It's not a toy. I've told you a thousand times! You don't take care of things."

Paulo breaks into tears.

"I'm sorry, Dad." Now he looks pale.

They'd never even seen Young. They'd had no idea he was near. They'd heard Gillette on the loudspeaker and fled instantly, skiing straight into oncoming soldiers, they explain as the captain's cell phone chirps.

"For you, Doctor."

He expects Marisa but it's Raines again. Gillette can barely concentrate on what the man is saying.

My kids are alive.

"Boss? We found his apartment on the list!"

"List?"

"I told you! All zone B landlords submitted lists of tenants! Bartholomew Young, occupation clergyman. He rented an apartment on Q street, near Dupont Circle."

"Any other Bartholomew Youngs on the list?"

"None. HS is on it, trying to get the FBI to the apartment. They'll send guys on foot if they have to."

The kids look scared finally. What almost happened has sunk in.

"Hold on for a moment, Raines."

Dupont Circle, he knows, is close, a few minutes away.

"Captain? I need to get to Dupont Circle. I'll take three of your men and a sleigh," he says, showing the ID.

The officer thinks he's afraid at first. "Oh, we'll have him in a few minutes. No worries. You're quite safe here, sir."

Gillette gets back on the phone and tells Raines, "Give me the exact address of that apartment."

Paulo grins, realizing that he and Annie get to come on a mission. There's no time to get them to Marion Street. Gillette warns the kids that they're on "thin ice." They'd better follow instructions exactly at that apartment, better not take even a single step without his permission. Do they understand? Do they promise?

Paulo loves this. One minute the kid is terrified. The next he's having the time of his life.

Yeah! You betcha, Dad!

To Paulo, it's just like CDC's annual take-your-kids-to-the-office day.

Bartholomew Young churns downhill in the blizzard, hearing the Rangers seeking him ahead and behind. Visibility is terrible so they'll have to use their helmet mikes. Wherever he goes, he'll leave tracks. *If I climb a tree the tracks will show it. If I double back I'll run into them. Continue ahead and I'll hit oncoming troops.*

The soldiers are like game beaters on the mentor's estate, driving foxes toward hunters. Which makes Bartholomew the animal just now.

He's afraid, of course, and as usual emotion brings the words of Great-great-grandfather to him. "We no doubt enjoyed rare moments of peace and forgetfulness, but I remember more the agony, terrors, mistakes."

The loudspeaker never stops. "Bartholomew Young. You will not be harmed!"

Gillette had said, "We know who he is!"

The voice orders, "Put your hands over your head!"

He's in thick woods, churning along flat ground piled with deep snow. He realizes that the odds of surviving are terrible, but his best chance of getting out of here lies in attack.

Great-great-grandfather had been captured once by Turks, who raped and beat him. Young will not let himself be caught. He'll die first.

He unfastens the snowshoes and wedges them beneath the back of his parka. The parka, zipped tight, will keep them from falling out. He'll need them later. Now he drops to the ground, burrowing into the white powder.

The shock of the cold is tremendous, but he wills himself not to feel it as he belly-crawls forward. The snow is so deep that he almost disappears. The soldiers will not be looking down, but ahead. Also, they wear boots, not snowshoes, so they must struggle to move.

Snow runs down the back of his collar.

I must be almost even with the oncoming line.

And sure enough, materializing on either side through trees he sees two forms, moving in tandem. Bartholomew and the men form three points of a contracting triangle. The booming announcements mask the sound of movements. The soldiers will pass within twenty feet of him on both sides. If it weren't for the snow cover, all he would have to do is wait until they pass. But if he does that they will spot his tracks and turn.

He fires the silenced Glock. *Pffft!*

The man on the left slaps at his neck. A faint gurgling sound reaches Bartholomew. He hopes the man's mike was not turned on, and he's already turning toward the second soldier even before the first one is down.

Pffft! Pffft!

The man on the right doubles over. Dead on the spot.

They'll assume I'm in front of them, he hopes.

He hears shouting now. The troops know they're being fired upon. They're taking cover, cautioning one another to hold their fire until they see the target. They don't want to shoot their own comrades, the men coming toward them. That's the problem when you trap an enemy between close lines. That's the vulnerability that he plans to exploit.

The snow falls thicker. He crawls ahead. In minutes, the soldiers will find his trail, see where he wriggled forward and made his break. But it will be too late. He's through the line. He's reached the back end of the zoo, the fence, and Rock Creek Park and Harvard Street.

The soldiers left no rear guard here. They used everyone for the assault.

And now he hears firing behind him. What are they shooting at? An animal? Each other? Gillette?

Bartholomew Young, free man, remembers the face of the doctor, close up.

Was the doctor lying? Do they really know who the mentor is? He must find out for sure.

If it's true he must warn the mentor.

The pain comes hot and hard in his left thigh, even before he hears the shot. It drives him into the snow.

TWENTY-SIX

DECEMBER 11TH. 4:56 P.M. 44 DAYS AFTER OUTBREAK.

Father Young's block looks like a war zone, Gillette sees as the sleigh pulls up. Townhouse doors are smashed. Windows are broken even on upper floors. The street is deserted at the moment and he hopes whatever destructive human wave rolled through here is gone.

He dreads splitting his escort up, but orders one of his men to guard the sleigh. Everyone else heads into the building. The agitated horses snort and stamp in two-foot drifts, perhaps sensing that their bodies would feed a dozen people for weeks, if hungry ones come along.

Maybe I should have left the kids at the zoo.

But he'll not leave them again. And he needed to check this apartment.

I don't want them anywhere near Bartholomew Young.

They halt inside the foyer, listening for looters. The building is silent, so maybe the looters are gone. Gillette tells the kids to wait here with Corporal Arnold McKenna, as the boy introduced himself. The foyer is safer than staying outside. It will also be safer than going upstairs.

Especially if Young booby-trapped his apartment, Gillette thinks, remembering the terrorist prep videos he'd watched at the Pentagon, on the Rapid Response Team.

Paulo argues, "Aw, Dad. I can help."

"Do it by listening to me for a change."

Private Duane L. Pettigout—an impossibly thin, pimply kid—accompanies Gillette up the steps two at a time. The higher up they get, the worse the damage. Banisters are smashed. Graffiti drips in wet red paint down white walls. "EAT THE RICH!" Human shit lies on the beige carpet. But the building is warm. Zone B had heat.

The FBI should have been here by now.

Apartment 4C lies at the end of a corridor where every door is smashed in, the jambs splintered.

No worries about a booby trap, at least.

Lights flicker in the hallway. Power will be failing tonight, Gillette thinks. Peering into 4C, Gillette knows that his chances of finding evidence here just dropped.

The apartment is savaged, a diorama of destruction. Furniture smashed. Shelves knocked off walls. Couch slashed.

Wait for the FBI? Or go in?

New looters may show up soon.

He wishes he wore body armor, wishes there were more soldiers along. He's sick with apprehension for the safety of Marisa, and can only pray that the wreckage he's witnessing is less severe outside zones A and B.

That my people are safe.

"Pettigout, wait in the hall. Check in every couple of minutes with the guys downstairs."

"Want me to help search?"

"I don't know what to tell you to look for."

They whirl at a crunching sound—glass underfoot—to see a large and well-dressed blond woman standing in the doorway of the adjacent apartment. Thick makeup smeared. A chic black pants suit on, and matching high heels. Pearls around her fleshy neck. In the flickering light she looks terrified. She's a resident, not a looter, Gillette sees.

"I hid when they came," she whispers.

She could be anywhere from forty to sixty years old, her face swollen with tears. It strikes Gillette that over the last few weeks, he's only seen fat people in zones A or B.

"They had bats and axes. I heard them breaking things. I hid in the bathtub. I was sure they would find me."

"Who lives in 4C?" Gillette asks.

"Father Young. He's gone. Everybody's gone. Joel said I'd be safe but I can't find him. He took our money and Mister Ted! Can I come with you?"

She breaks into a fresh round of tears.

"Yeah, you'll come when we go." Gillette has no idea what to do with this woman. Take her home?

"Did Joel send you?" she says hopefully. "Did Joel have the cat with him? Mister Ted hates to go outside."

"Wait in your apartment, ma'am."

She seems about to protest, as if she fears he'll disappear. But she backs out of sight. Gillette hears, faintly from outside and blocks away, a single burst of automatic weapons fire.

Pettigout eyes Gillette uneasily. "We was fighting house to house before, sir. Them mobs is nasty. We don't want to be trapped here if people come back."

"The FBI will be here any minute," Gillette says.

"Maybe. But I heard troops was dispatched to defend FBI headquarters. I heard was some tough fighting there."

Gillette feels his heart racing. Even a trained forensics team would have a terrible time isolating evidence here. Dozens of strangers have rampaged through these rooms. Snow blows in. Shoe prints are everywhere.

Gillette spots emptied cans and boxes by the open refrigerator and understands what had fueled the mob's ire.

He had food.

Young must have had *lots* of food, judging from the empties. Empty cans of peas, corned beef, corn, tuna. Empty boxes of granola and raisin bran. Empty cans of frozen OJ. Empty olive jars and cheese wrappers, licked clean. Bread wrappers without a single crumb inside. Penne boxes torn open, as if whoever found them ate the pasta uncooked, to make sure that they got to eat at all.

"Hoarder," reads graffiti painted over the sink.

Why hasn't the FBI arrived?

From the hall, where Pettigout is monitoring his helmet communication system, the kid calls to Gillette.

"Arnie sees people across the street, in the park, sir. Arnie says people are coming back."

Gillette keeps his ski gloves on, to keep from leaving fingerprints that would further confuse the forensics team. *Where are they?* It seems impossible that looters would have overlooked anything of value here. Gillette steps into the center of the room, determined to search.

All the lights go out.

For an instant he fears that Bartholomew is responsible. Then he realizes that power has failed.

He hears the woman in the next apartment start screaming. "I can't stand it anymore!" He hears the crash of pots striking the wall. Is she throwing a tantrum? Thrashing around in panic, looking for candles?

"Pettigout. Give me your flashlight."

Apartment 4C had seemed bad enough before, but now, by flashlight, it's worse. He stands pivoting, trying to stay calm, hoping that some object will trigger a thought, memory, answer. A Delta-3 carrier lived here. He slept here and hopefully left messages, drawings, hints, DNA. He peered out at the world from these windows. He measured success or failure each day from this room.

Okay, Bartholomew, Clayton, or whoever you are, what have you left for me to find?

The beam crisscrosses debris. He tries to imagine this madhouse reconstructed. The overturned coffee machine and toaster on the kitchenette counter look new. The open suitcase on the floor must have contained the clothing strewn around the room; shirts size medium, pants waist 33, length 32. The flashlight beam glints off the wreckage of a laptop computer.

Maybe the hard drive still works.

He closes it and wraps it in a shirt, to protect it, and prevent himself from being cut. He throws in unlabeled computer disks too. They may at least carry fingerprints. He feels like a junk collector, not an evidence collector.

"Arnie counts ten people in the park, sir."

Gillette finds another flashlight, gives it to Pettigout and instructs him to keep trying to reach Marion Street and the soldiers at the zoo by cell phone. Reaching the soldiers, he is to ask if Father Young has been killed or captured. Reaching Marion Street, he's to say that Gillette and the kids are safe, and will soon be home.

I haven't heard any sound from next door for a few minutes. The woman has quieted down.

"Sir, Arnie counts eighteen people now."

These days, wrecked apartments are more common than clean ones. Gillette finds only plastic dishes and pots in the kitchenette cabinets. He spots books and videos amid the bed stuffing and ripped pillows on the rug. The flashlight beam lights up titles. *The Godfather. Crash.*

"Arnie says two guys are coming across the street."

He wishes he knew what was inside the computer. Because he's found no photos. No notebooks or drawings. No maps. Phone numbers. Names.

Ten more minutes pass.

"Arnie says the men want to buy our horses."

Gillette sees old Dr. Larch in his head.

"Did you check everywhere?" Larch asks.

"In twenty minutes? No one could do that."

"No one will ever check if the looters come back!"

Gillette now hears snatches of raised voices from the street. Sell us *one* horse, someone is saying. Why should you have *two* horses while our kids have nothing to eat?

The soldier down there shouts, "Get back or I'll fire!"

In the bathroom he finds no prescription medicines, but he bags the toothbrush. If he gets lucky, it will have Bartholomew's DNA. The linen closet contains linens, nothing more. Back in the living room, Gillette stands on a table to give himself a fresh perspective. Searching by flashlight means that 99 percent of the apartment is always invisible. But the beam accentuates clarity in the 1 percent he sees.

He opens books and flips pages to see if anything has been wedged inside. He slides the videos from their sleeves to make sure nothing is hidden there.

Pettigout is right. We have to leave.

And then, as he's about to go, Gillette becomes aware of what the flashlight beam illuminates like a spotlight, on the floor. The book looks old, small, bound in black leather, like a Bible. Gold lettering on the spine shines up at him. Gillette sees the title. His mouth goes dry.

Seven Pillars of Wisdom. By T. E. Lawrence.

Gillette reaches down and picks up the book. On the cover he sees a pair of crossed scimitars in gold leaf, and beneath it, "The sword also means cleanness and death."

Raines said all the false names are in this book.

He opens the front flap and shines the light on an inscription written in blue felt pen on the title page.

"No better relation is better than a prudent and faithful friend," he reads out loud. "Benjamin Franklin." And "No one can serve two masters. Matthew 6:24."

"Is this your book?" he says out loud. "Or did it come with the apartment? What are the odds of finding this particular book here by coincidence? Pretty small, I'd say.

"Hmmm. British publisher. 1927. But felt-tip pens didn't exist then so this book was a gift, later on. The giver wrote the inscription. Is the giver important to you? Is he the 'master' you serve, Bartholomew Young?"

"Sir?" Pettigout warns sharply. "Arnie says the people in the park are dividing up."

Gillette feels as if an electric current runs through the book into his hands. It's a link. It's hard to thumb pages with gloves on, but he does it crudely, positioning the book in the beam. The pages stop moving as if of their own volition. He's looking at a drawing of a man.

It looks like the face at the zoo.

What in God's name?

Because the man in *this* drawing is dead. This drawing was made over eighty-five years ago, when the man was in his twenties. The man in the drawing has shorter hair than the one in the zoo, cut in the manner of the early 20th century. It's longish on top, brushed left, short on the sides, boyish and clear of the ears. The face is youthful in the lack of lines, smooth arch of chin and neck that have not become brittle or fleshy with age. But there is depth in the stare. And torment. The left eye partially wanders, perhaps seeing a different reality. The right eye looks straight ahead as if clearly acknowledging every brutal truth on earth.

He said something about a great-great-grandfather.

Gillette looks at the name of the man in the drawing.

Could it possibly be true?

"Pettigout, let's move!"

Gillette takes the book and other evidence. On the way out he stops at the apartment next door, and calls into the dark for the woman to come along. There's no response. He steps inside, assuring her that he's a friend. This apartment is bigger, and impressionist landscapes hang askew on walls. He swings a door open and pushes into the kitchen. The smell of gas is overwhelming. Antique crockery is smashed beside an overturned hutch.

"Holy Christ."

The flashlight beam crosses the kitchen and stops on a pair of shoes that at first seem to be standing on tiptoe. The beam travels past the ankles and legs to the oven. The woman is wedged in, tight.

She can't be dead. It's impossible after so short a time. But when he pulls her out the face is blue, the eyes bulging, the tongue protruding. She does not respond to first aid.

Heart attack.

He and Pettigout head downstairs to collect the children and get out.

The drifts outside have deepened. Gillette sees that in the small park across the street, flashlight beams cluster. Forms mill about like ghosts. He can hear the nervous horses snorting in the alley beside the building, where the driver must have moved them. He urges the kids to hurry.

"You're the one who held us up," Paulo says.

They reach the sled. The driver grips the reins and tries to soothe the horses. The man knows animals. Perhaps this particular soldier grew up on a farm.

"Step right up, sir. Don't mind those people."

"Take us to Marion Street."

After all, Gillette reasons, the FBI never showed up. FBI headquarters might be as much of a wreck as this townhouse. He hears gunfire from the direction of downtown. There's no way he's endangering his kids any more today.

No way I'm going anywhere right now except home. I'll give the evidence to the troops at the zoo, if they're still there. Or I'll call Theresa.

They pull away from the alley as people run toward them from across the street. Gillette covers the kids with his body. Pettigout fires into the air and, screaming, the people fall back.

Annie tells Gillette over and over that she's sorry. Sorry she went to the zoo. Sorry she took Paulo. Sorry she made him come after her. Sorry she worried Mom.

"It's all right."

Gillette's last view of the receding crowd shows forms converging on the townhouse, like jackals charging a corpse.

A computer. A toothbrush. And an old book. Is it down to that, he asks himself?

Evening is coming on.

He hopes that Bartholomew Young is killed or captured.

Delta-3 rules the city, region, earth.

TWENTY-SEVEN

DECEMBER 11TH. 6:02 P.M. 44 DAYS AFTER OUTBREAK.

"We are surrounded by enemies," says Gordon Dubbs.

He stands facing his subjects from atop a desk in the living room of apartment 2D, from now on to be called the "great hall." Couches have been replaced by folding chairs. Walls are hung with looted art; medieval scenes; a poster from his favorite film, *Duke, Devil and Destiny;* a cheap print of jousting French knights; a German lithograph called *The Feudal Lord and His Whores.*

"Our enemies want to take what is ours," he says.

In the front row sits Teddie, proud of Dad, and looking older since his interrogation of Gail Hansen, and their disposal of the battered body forty minutes ago in Rock Creek Park. All residents are present except Lester Gish, on duty in the defense room, monitoring the security screens. The people here carry the kind of grudge at the world that Dubbs knows how to exploit.

"Soon we will have to deal with gangs from other buildings. AWOL soldiers. Refugees. Criminals who will want our food. Marion Street has oil. Have they offered to share? No. They hate us. Tonight is justice," Gordon says.

On the wall, or "operations board," he's drawn a large map in black marker. It depicts the eight houses of Marion Street, the cars blocking each end and the alley out back. Dots show security cameras. Stick figures are guards.

"The oil is in this corner house."

Dubbs uses a pool cue to point out locations. He's laid out booze, cheese, olive tapenade, smoked oysters, Vienna sausages, crackers and cakes.

"The first group will go in quietly at the far end of the street, surprise the guard and disarm him. You'll have AK-47s. He'll have a handgun. We don't want to waste bullets. Don't shoot unless you have to."

Basil Prue is nodding, liking the word "shoot." Frank the ex-bouncer looks afraid.

"Our second group will do the same thing at the other end of the street. That will give us two of their three weapons and two hostages. When they learn we have hostages, they'll hand over that other weapon. Because what's one fucking shotgun against our arms? Meanwhile, the third group will go in through Gail Hansen's backyard, come out mid-street and take control. I'll lead that one."

Dubbs the future warlord knows that not all of these people are ready yet to take the logical end step: full elimination of enemies. That will come later.

"Remember, our firepower is overwhelming. They'll give up or run. But if they open fire, shoot."

Jovita, the six-foot-two Montana girl and ex-chauffeur, looks ferociously excited. Dubbs adds, "We'll make *them* move the oil here."

It'll be prisoner labor, he explains. Let those yuppie pansies experience pain for a change. That's the punishment for hoarding. Let 'em haul sleds, drag cans, lug plastic jugs, wrap up oil drums in quadruple blankets and roll them, inch by inch, up the hill. Let them sweat. They *stole* the heating oil, didn't they, looted it, and the penalty for looting is a firing squad, right? Hell, even the president said that! In war, once you beat an enemy, they're prisoners! Either way, tonight's the night to hit Marion Street because they won't expect anyone to be out.

Frank Gerard asks uncertainly, "What about witnesses?"

Dubbs laughs. "Witnesses? Witnesses hide in the best of times. I patrolled blocks in D.C. where people would get shot in the middle of the afternoon and the neighbors don't want to get involved. No one's going to see anything."

Witnesses will tell people, don't fuck with Dubbs or he'll kill your ass. We need witnesses.

Frank the ex-bouncer raises his big hand again.

"But what if the police come back? You're talking about shooting people, Gordo. Those other times were self-defense. This is murder!" He looks around, bewildered. "Are you people serious? We looted too! We're the same as them!"

Dubbs climbs down from the desk and strides across the room. Without hesitation he smashes the cue into Frank's skull. The man cries out, toppling. No one moves to help Frank. They seem shocked or fascinated. Frank is covered with blood, moaning and lying in a fetal position when Dubbs resumes his position atop the desk.

"It's survival. Not murder. It's what you do from now on if you want to stay alive. It is," Dubbs says, remembering a legal term from his police days, "exigent circumstances."

Which basically means you do what you want.

Jovita's hand goes up. "I think we should wear uniforms, like a real army," she says, as Dubbs rehearsed her. "We found those fatigues in the army-navy store. Why not wear them tonight?"

Dubbs nods as if this is the first time he's heard the idea. He explains that in battle, it is necessary to recognize your own people immediately. The audience seems receptive, still in shock over Frank. Dubbs wins the vote.

Then Teddie, on cue, suggests that ranks be assigned so during the attack there will be a chain of command.

They vote yes on that too.

From now on, Dubbs says, he will be commander. *No more calling me Gordo,* he thinks. Jovita will be a colonel, Basil a major. Everyone else will be a captain, and Lieutenant Teddie Dubbs the new aide-de-camp.

Maybe one day, these people will call him "My Lord."

"We're all officers in this army," Dubbs remarks. The joke breaks the tension. Everyone laughs.

He's already got a coat of arms in mind for the building, crossed AK-47s against a medieval shield. But that proposal can wait until the celebration, after they've taken Marion Street's heating oil.

"We'll go in when the power fails tonight," he says. "We'll wait until their cameras are out and hit 'em hard. And tomorrow, when the city's buried and temperatures hit zero, in here we'll put on music . . . eat some fine food . . ."

Teddie raises a hand. "What if the power stays on, Commander?" Gordo had explained to him earlier that the word "Dad" should only be used in private from now on.

"Then we'll go in at one, when they're asleep."

After the meeting he and Basil will chain Frank to a radiator in the "detention apartment." Tomorrow he will dispose of the rebel, after shooting him with a gun they seize from Marion Street. Just like the old days in the police, when Gordo used to plant weapons on suspects.

He wraps up his speech. "Tonight, the people who laughed at us all our lives will be getting what they deserve. Tonight is a crossroads. A week from now, people who do the smart thing tonight will be on top, and everyone else," he says, sweeping his arm toward the window, blizzard, future, "will be out there."

"When I was a kid, I loved watching science fiction movies," says Private Duane Pettigout.

The sleigh proceeds up Connecticut Avenue, over the Calvert Street Bridge, at 6:30. Darkness has fallen. Gillette hears the horses panting. They must raise their legs high to push through powder on the road.

"Them movies had it wrong," says Pettigout.

Gillette's frozen fingers keep working the cell picture phone. He keeps trying to reach Raines or Marion Street. He punches redial. He hits it again.

"Them movie viruses always killed *people* . . ."

The soldiers face outward into the blizzard, for defense. Gillette and the kids sit on the floor so as to be less visible.

"In them movies, people who survived had plenty of food, and cars filled with fuel. But Delta-3 wiped out the supplies, not the people, so we got to fight for what's left."

The sleigh slows, passing the zoo. Gillette had hoped that soldiers would still be here. But they're gone, and whatever tracks they made are blown over.

Pettigout says, "I'm sure he's caught, sir. Or dead."

"Keep going, driver."

Maybe all phones are out, he thinks, and all towers. Maybe tonight is the end of power. Our machines turn into useless trinkets. Our technological magic survives as whispered stories of the past. Our greatness becomes a song of memory for starving children. We denied dependence until it ripped its mask off, and then it was too late.

The struggling horses pull them up the hill toward the fire station. Buildings flow by, ghostly, through snow. Power goes on and off in sections, creating a slow-motion strobe effect. Gillette feels as if he's been away from home for years, not weeks. They see no one outside. Tonight's looting is mainly inside former zones A and B. Easy pickings were stripped from zone C weeks ago.

Gillette can't believe it when they reach St. Paul's. The arched doors appear as if through a snowy tunnel. He glimpses the high bell tower. He'll try to use the phone again, from there.

Will Marisa be here?

The kids come inside with him. He sees instantly that conditions have deteriorated while he was gone. The faces that turn to him are devoid of curiosity, more like skulls. There's no organ music. No one plays games. The smells could kill.

"Marisa went home," Chris Van Horne says, as he wipes the brow of a pregnant woman. She's hacking, sick.

Gillette climbs the circular staircase to the bell tower. At least there is heat in the church, he thinks. At the top he extends the phone out into the elements, like a divining rod, and punches redial for the hundredth time.

This time he hears ringing at home.

Once . . . twice . . .

Dammit. I finally get through and you're not there.

He tries Raines and gets through.

Gillette explains what he found at the apartment. He'll keep the computer, disks and toothbrush on Marion Street, until after the storm. But he wants to send Raines a picture of the book title page. Is that possible?

"Click away. We have no more food here, but great technology."

"Check the publisher. Year of publication. Inscription. There's a symbol in pencil on the inside flap, a seller's mark, I hope. The book is British. The price is in pounds. I don't think it's a coincidence that it was in the apartment. I think our Lawrence wannabe carries it around."

"Carr*ied,* not carr*ies.* The soldiers got him. His tracks led into a burning house. He didn't come out. He's dead."

"They're sure? They saw the body?" Gillette says hopefully.

"No, but they stayed close for an hour. They questioned people. They did perimeter searches. They never found his tracks leading away. They say he burned up."

Gillette experiences some relief, although he wishes that the man had been seen dead, or captured, so he could have been questioned. Another chance to track Delta-3 is gone. He tells Raines, "Try to find the book's seller or buyer. Maybe the Brits can help."

"Well, Britain's in better shape than us, boss. They've got more nuclear power. More electric trains. More food to eat, with all that coastline. And they've got coal."

Gillette clicks off, tries Marisa, gets no response. Typical, he thinks. I can call a fort thirty miles off, but can't get through two blocks away.

Downstairs, he gathers the kids and goes outside. He tells two of his soldiers to stay at St. Paul's, to defend against looters. He tells them to shoot one of the horses so the refugees can eat.

"But these are government horses," says Pettigout.

"No. They're this week's food distribution."

The remaining horse lacks strength to pull the sled alone. So they unhook it and lead it across Connecticut and down Nebraska. After a few minutes comes a shot behind them as Gillette envisions the other horse put down.

"Stop right there!" Bob Cantoni's voice comes from behind the bulky form of a barricade car.

Gillette shouts, "Who do you think you are? A Marine?"

"Holy shit! Greg? Is that you?"

He hears his best friend announcing over the comm system that Gillette has returned. They hug in the storm as neighbors stream from houses, bundled like Eskimos. They've shoveled paths on the block, "defensive corridors," he hears Bob explain. Marisa stands in front of him. She's got tears in her eyes. Her lips are cold and delicious. Then he feels the kids holding him too, feels emotion welling.

Marisa says, "You're home."

"I brought us a horse," Gillette says.

He doesn't have to explain further. He sees from the famished look in their eyes that they understand. A horse, he thinks. The Cantonis like to bet on horses at Laurel racetrack. Lisa Higuera rents riding horses with Annie, in Rock Creek Park. Paulo, at four, would endlessly ride the mechanical mustang in front of the drugstore.

Gillette thinks, *This animal saved my children. It brought the troops to the zoo. Without this horse I never would have found Bartholomew's computer, book, brush.*

"We'll butcher it in my backyard," he says.

He hopes the man who threatened his family is really dead.

At the same time, Father Bartholomew Young lies in an abandoned first-floor apartment on Columbia Road, off Harvard Street, nursing his wound in a corner. The building has been stripped by looters. The wound—like the one that the Turks inflicted on Great-great-grandfather in winter 1918—throbs painfully. But he's found a bottle of rubbing alcohol. The alcohol stung when he applied it. Now, sitting with legs splayed to allow him a view out the window, to see if soldiers are still after him, he uses a ripped-up cotton shirt to tightly field-dress the wound. The bullet has passed through.

I must get to Gillette. He's the last reachable person who can tell me if they've traced the mentor.

In memory he's back in the zoo, shot, his thigh burning as he glances down to confirm that no blood is dripping a trail. He's unsure whether a sniper shot him, a random bullet hit him, or whether the soldiers have realized he passed them and turned to chase. Do they have infrared? Can they spot him from heat? Is he even now a red blob on someone's scanner, partially hidden behind an oak while laser beams crisscross the trunk?

When no other bullets kick up the snow, he decides to gamble on a lone shooter. He can't stay in one place.

I guess I'll know if it's a sniper then.

He crawls out from behind cover. His leg burns when he moves. No more shots come. He reaches the exit fence.

Then from a ridge behind, a random gust of wind carries a soldier's excited voice.

"I see his footsteps!"

The pain grows worse as he starts to run. He limps through powder, over a small footbridge and across Rock Creek Parkway toward Harvard Street. Snow blows into his mouth.

His knees work like plows. If he stops to strap on the snowshoes he'll make better time. But snowshoe tracks will be unmistakable to the soldiers. He must wait to use snowshoes, must convince his pursuers that he's wearing only boots first.

And no fleeing into Rock Creek Park either. There, his tracks would be the only ones in the snow.

Reaching Harvard Street, he heads up an incline toward a fiery glow, into what eight weeks ago had been some prime D.C. real estate. A wooded suburban-style block in the Adams Morgan neighborhood.

Now half the houses look deserted. Others are in flames. Pyromaniacs are siphoning infected fuel, he thinks.

Gillette will have described me to the soldiers. I need new clothes and new footprints. I need to stop the bleeding before it leaves a trail.

He needs, in other words, a house.

He makes for a pair of burning brick colonials. One is engulfed by flames. But the fire in the other is smaller. Silhouetted, looters stream in and out, a disorganized mob.

No one notices one more looter. He lurches over the body of a dead man in the doorway, who clutches a blood-soaked nine iron. The old man wears house slippers and a sweater, not a coat, so he must be the owner, must have tried to defend his property. Looters bump past, frantically grabbing valuables. A small boy on the kitchen floor scoops peanut butter from a jar with his fingers. A woman screeches, "I saw it first!" Two men fight in the upstairs hallway, in the glow of flames illuminating a shoebox spilling out crumpled wads of cash.

Grabbing a flashlight off the floor, he heads for the master bedroom, which is emptying as looters carry off jewels. In the closet he finds a pair of men's boots. They're size eleven instead of his own size, nine. The pattern on the soles is zigzaggy, not linear. The boots will leave different tracks in the snow.

Ah. The boots are dry and warm.

His feet are freezing.

He rifles the bathroom for med supplies, but he's gotten here too late. The medicine cabinet hangs open, its shelves devoid of anything except cosmetic products: shampoo, hair color, mint mouthwash.

He looks out the smashed window and experiences a thrill of fear. Far down the block he sees the jerky glow of flashlights. The soldiers will soon reach the house.

Maybe the looters will distract them.

Back in the hallway, the fighting men still roll on the carpet, clawing at each other's faces. Fire has driven everyone else downstairs. Father Young pulls out his silenced Glock and without hesitation shoots both men in the head, angling the shots upward so blood will spray away from their coats.

The grunting stops. The hallway is silent except for the crackle of fire. Father Young begins to cough. It's getting hard to breathe.

The first man's parka, sprayed with blood, is useless.

But the second man's thick peacoat will suit him fine.

Young's throat is burning and his eyes are tearing. The world dissolves into sulfur and rubber stink.

Moving quickly, he strips off the dead man's coat. The fire is spreading up the walls as he switches stocking caps, green for blue.

As the last person out of the house, he turns to see fire shooting from the rooftop and windows. Smoke drifts up, ash falls down. Men struggle to keep a woman from rushing inside, probably to try to save one of the men whom Father Young just shot. She's screaming out the name, "Roger." Then the soldiers' flashlight beams converge on the crowd and an amplified voice calls, "We won't hurt you! We're looking for a man . . ."

Father Young shouts, "They shot Roger! Soldiers! Run!"

Everyone here knows about the shoot-to-kill law.

The crowd scatters, all except for the crying woman, who runs into the building, foolishly.

Father Young, limping away, carrying his snowshoes, leaves size eleven tracks in the snow.

And now, in the apartment, he stands, grimaces with pain, makes his way to the building exit and dons the snowshoes.

I can't go back to my apartment for the book. Gillette knew my name. He probably sent the FBI there. And I can't reach the mentor by phone in this storm.

He must get to Gillette tonight somehow, for answers.

The great World War One hero Thomas Edward Lawrence, his ancestor, he knows, refused all honors and decorations after coming home from battle. He changed his name to J. H. Ross, and T. E. Shaw, to hide himself from the adoring public. He died believing he had failed his best friends.

I won't fail my friend, Bartholomew Young thinks, stepping out into the storm. *I'll protect him.*

It's funny, he thinks, because 1918 was a snowy year for Great-great-grandfather, just like this one is for me.

For twelve years he's carried the mentor's gift as a talisman, personal bible and lucky piece. He'll miss the book. But the loss will not endanger his mission. Anyone finding the volume will assume it belongs to the rented apartment. No one should associate him with names in the book. The inscription is generic. The book is simply gone.

Limping down Columbia Road, toward Connecticut Avenue, he recalls the last meeting with his mentor, at the man's estate, months ago.

The cliffside walk by the ocean. The waves booming in limestone caves beneath. The wind out of the oil fields to the north. One of the mentor's arms around his shoulder, the other sweeping out to encompass the future of "balance restored, proper people in control."

The mentor going on about new empire. The mentor envisioning glorious peace in a formerly haphazard world.

"But success depends on keeping the secret, and then giving only selected people their oil back, at a price," he had said.

Marion Street is a forty-five-minute walk from here. The Glock is in Father Young's belt. He chews his last Chuckles—licorice flavored—as he moves.

His mentor had said, *"Afterwards, when we bring you home, you will have fulfilled your ancestor's mission."*

Afterwards you'll live here, with me, he said.

Tears come to the eyes of the onetime beggar boy.

He recalls Great-great-grandfather's admission, at the end of his book. After all the words about battle had come the moment when the great T. E. Lawrence—killer of Turks, commander of men—had confessed, "My strongest motive throughout had been a personal one, not mentioned here, but present to me always."

Afterwards I will live like the mentor's son, he thinks.

Afterwards will come filial love.

I will have a real name.

At least it's a white Christmas. Well, almost Christmas.

Hark the herald angels sing.

At 9 P.M. the meat comes out from two kitchens, two houses, on platters carried to the Higuera house down shoveled paths. Conversation ceases as food arrives. Knives and forks come up. Eyes glitter, Gillette sees.

Alice Lee is dead, he thinks. Grace Kline is sick at home, cared for by her mom. Gail Hansen is locked in her home, probably drunk. Pettigout is on guard with Julie. Marisa is on security-monitor duty. *I am happy to be home.*

The faces are thinner and grayer around the tables. Mostly the eyes have changed. Even in the brief period Gillette has been home he sees that Bob Cantoni has become edgy, more alert. Les is listless, a scarecrow. Neil Kline, the enviro-lawyer, keeps snapping at people. And Joe Holmes is obsessed with logistics. How to build a barricade. How to store the uneaten meat of the horse.

The steaks are hacked chunks carved with kitchen knives and roasted without seasoning. But it's the most delicious meal Gillette has ever eaten. Water tastes like wine. The fat on the bones is manna. "Don't eat too fast," Judge Holmes advises, although she doesn't seem to be doing it herself. Gillette—who has dined better than any of these people—feels the proteins saturating nerve endings, sating needs that have been growing for weeks.

"Okay, Greg. Tell everyone what you told me," Bob says as the electric lights flicker, plunge off, go back on.

He tells the tribe everything. He explains about Cougar Energy Services and the letter in Massachusetts. He tells them about the murders. When he utters the word "antidote," their eyes get larger. They want to believe, but know that belief may bring disappointment.

He tells them that Father Young was involved with the Delta-3 infection. The cure may be close, or Young wouldn't have come for him. Too bad the soldiers didn't get to question him. But the evidence from the apartment may help.

"You think Cougar created the bug?" Les asks.

"I think Dr. Samuelson sold it to someone he thought was friendly. A government or business. Someone who knows oil, sitting back and waiting to claim that they discovered the antidote. Samuelson quit Cougar four years ago. So he sold Delta-3 around then. Why not release it right away? Because the buyer needed time to set up a distribution network that wouldn't be traced back. The opening came when Cougar announced its new bactericide program and took on Tangier clients. I figure our bad guys already had people in place at other pipelines and oil fields. Maybe they even infected their own supply first so no one blames them."

"But who are they? Or which country?"

Gillette says unhappily, "I don't know."

Neil Kline snaps. "Truth is, you don't know any of this is real. It's just your theory."

"That's right."

"There may not be any antidote at all."

"That's true too."

"Which means when this fucking horse meat runs out, we're back where we started."

"Don't curse in front of the kids," says Judge Holmes.

"We can't have cursing during the apocalypse, can we?"

Neil storms out, into the snow. He starts his guard duty shift at ten, four minutes from now.

After dinner Gillette and Bob walk the street, reviewing security precautions. Gillette has brought back two additional firearms—Pettigout's 9mm Beretta and Pettigout's M-16. Bob figures they can supplement the guards.

"I wish the soldiers had seen Young's body," Gillette says.

The storm has tapered off but the sky is overcast. Gillette estimates the actual snowfall at two feet, with drifts double that. Floodlights shine down on swept-off car tops—so guards can see—and snow trenches and razor wire atop the fences in back, along the alley.

"We'll double guards tonight," Bob says. "I'll stay upstairs in my house. Pettigout goes in Les's. That covers the whole street, not just the ends. We'll watch the backyards through the windows."

"I'll take a turn too," says Gillette.

"No, you'll sleep with your wife. There's plenty of time for you to take guard duty later."

Gillette grins. "*Semper fi.* Thanks."

"You're on leave tonight. But keep the Beretta."

There are so many things to talk about with Marisa but the most important thing at the moment is to not talk at all. She's upstairs in their heated bedroom when he gets home, candles lit. She's wearing the black lace nightie. She looks thinner when he pulls back the covers. Less flesh on the thighs. Shoulders more hollow, like her cheeks. But the eyes are as blue and exciting as the first time he gazed into them, as a student, at a party, years ago. He smells toothpaste on her breath. She must have saved the toothpaste until he got home, so it wouldn't run out.

He had not thought he'd get to do this again. They touch gently and then passionately. He lets her help him forget, for a little while, what is outside the house.

Microbes can't do this, he thinks, afterward. They split apart to create. They don't come together. They spend their lives alone. How can creatures like that win out?

They fall asleep entwined. In his dream it is summer. He's at a Nats game with the kids, high up in the stands at RFK. Paulo and Annie jump up and down, with other fans, shaking the stadium. It's July Fourth. The hot dogs taste spicy. The Nats catcher hits a homer and fireworks go off.

No, not fireworks. Gunfire, he realizes, hearing the quick loud bursts of automatic weapons.

The audience starts screaming. Even the players are fleeing, little dots down on the field, falling down.

I must protect my children, he thinks, horrified, opens his eyes and the dream goes away, but not the shooting.

The shooting comes from outside.

He's here, Gillette thinks, reaching for the Beretta.

TWENTY-EIGHT

DECEMBER 12TH. 1 A.M. 45 DAYS AFTER OUTBREAK.
30 SECONDS BEFORE ATTACK.

As a sergeant in Iraq and a cop in D.C., Gordon Dubbs used to feel trepidation before going into action. He experiences a bit now. Pressed against the rear wall of Gail Hansen's townhouse, cold and wet, he directs his troops by comm system, looted from the spy shop, toward the car barricades on both ends of Marion Street. He grips his AK-47. The hostage seizures, coming in seconds, will signal the opening of tonight's campaign.

Power failed throughout northwest Washington ten minutes ago. Entry through Gail's back fence had been easy. Snow here is thigh-high but soft.

Nine . . . eight . . . seven . . .

In his mind he sees the block as if from the air, the shoveled paths in the street. Dubbs, Lester and Prue ready to burst into midstreet, to command the invasion. He envisions the yuppies in their warm beds, asleep.

Did I think of everything?

Five . . . four . . .

Will they be surprised!

His chest tightens and expands with flutter, but from adrenaline, not fear. Gordon's no coward. He's filled with too much rage. Violence has always been his outlet.

"Go," he whispers into his mike.

He hears Jovita's low, threatening voice over his earpiece, from behind the left-side car barricade.

"Put the gun on the car, slowly. Hold up your hands."

Excellent. Just what I told her to say.

Except then he hears two unexpected sounds, over the comm system and also in the night air, as they are so loud.

A man's enraged voice shouts, "Help!"

And a single gunshot sounds.

They weren't supposed to fight!

As if that first shot was a trigger, lots of shooting erupts from *both* sides of the street. Idiots! Marion Street was supposed to surrender. His own people were supposed to save bullets, use stealth.

He shouts into his mike, "Stay behind the barricades!"

Goddammit, why don't people listen?

Gordon turns to Prue and Lester, who have gone wide-eyed under their balaclavas. They didn't expect opposition either. But Gordon has faced it before.

"Follow me," he orders. "And do what I say!"

Gillette leaps from bed naked, grabs the Beretta and yanks on pants and unlaced boots. Marisa is already out of the room, bare feet pounding on the upstairs hallway, naked cave-mom heading to save cave-kids. He hears her shout for the kids to keep down. Downstairs he throws his parka over his bare chest, opens the front door and peers out.

His heart slams against his ribs. Holding the Beretta in both hands, he crouches against the doorjamb, keeping low.

Power is off but the moon is out. The snow looks purplish. At the far end of the street, a body lies on the ground by the Cantoni Honda, and flashes of light correspond to the bursts of automatic weapons. He recognizes the two-tone parka worn by the inert person in the snow.

Oh damn. It's Neil Kline.

No bodies lie on the other end of the street, so he hopes Les—on shift now—is unharmed, but from behind the Gillettes' Subaru come more muzzle flashes. *They're using my car as a shield,* he thinks, enraged. He hears the rushing whine of bullets. Shingle and brick chips fly off the

Higuera house. Attackers are shooting up the place. Is Pettigout inside? But then snow starts bursting up everywhere. The attackers are shooting wild.

"The guard dropped his gun!" an attacker shouts on the Subaru side. "I have it!"

Gillette's questions come in a rush.

It can't be him. It's not one guy. Is it a gang? Looters? Do they know about our oil?

And then he sees attackers in Army jackets—soldiers, he thinks—wearing balaclavas, rushing around the Subaru.

Are they mutineers, like in Vegas? Are they here for the evidence? Father Young said other people would come.

He drops off the front porch to land in the small gap between hedgerow and house. Snow spills inside his jacket and boots. It's freezing. He shakes snow off the bushes to see better. Gillette spots three ninjalike figures rushing down the center of the street, in the shoveled path. They will pass before him like tin shooting gallery targets. He raises the Beretta with both hands. The sight sways. His ungloved hands shake with adrenaline. He squeezes and squeezes. The firing adds to the din. One figure topples. A second throws his rifle onto the snow and doubles over. He looks like he's screaming but Gillette can't hear over the sound of his firing. Now two soldiers are down. The third, surprised by Gillette, throws a rifle away and is running, falling, crawling back toward the barricade.

I drove them off on that side.

Gillette hears more screaming from the other end of the street. All the firing comes from that side now. He stays shielded behind the bushes, moving toward the fight. It's a goddamn invasion. Images come to him in pieces. Two teams of attackers are visible through the bushes, one rounding the Honda, another appearing from behind Gail's house. They must have come through her backyard.

One man by Gail's house shouts, "Get back 'til we find the shotgun!" Gillette recognizes the voice. He realizes who the "soldiers" are. They're not soldiers at all.

Dubbs.

And at that precise moment comes the booming blast of a shotgun. One of the men by the Honda goes down.

Boom . . . Boom . . .

A second guy is down. *Bob must be firing from upstairs in his house.* And Gillette sees rapid bursts from an upper window of Les's house also, across the street.

It's Pettigout.

The attackers at the Honda are fleeing. One tries to crawl away and crumples. Another leans to help the injured one, and—*BOOM*—is suddenly propelled down as if by an invisible hand. The figure does not get up either. A third invader is draped over the hood of the Honda.

Now only Dubbs and one man are standing, pressed against the side of Bob's house, shooting at Pettigout.

Goddamn you to hell!

Gillette bursts from the bushes, firing, charging.

One of the men goes down.

The last attacker is backing toward Gail's house and retreating around it. He swings his weapon toward Gillette and fires. Gillette hits the ground, rolls, pulls the trigger. He's out of bullets. But the attacker is gone. And when Gillette reaches Gail's backyard, the fence door is open and the man has fled. Gillette slams and locks the door.

It's hard to hear. His blood rushes in his ears. He looks down at himself. His open parka is soaked with snow. The fabric is partially shredded by bullets. His chest is wet and heaving. But he is not injured. He is astounded.

Are they really gone?

Gordon Dubbs stumbles through the snow-filled alley, replaying the quick demise of his grand plan. Minutes earlier, he'd reached the front of Gail's house to watch helplessly as attack discipline fell apart.

No one was doing what he'd told them. His people were rushing the street on both sides, overconfident, knowing the guards on both ends of the block were disarmed.

Where's the goddamn shotgun? he thought.

He watched Jovita lift into the air and fly backward, the soles of her boots up.

Boom . . . boom . . .

Still, we outnumber them, he thought. We have firepower. When this is over, heads will roll at home.

Boom . . .

Another attacker went down. *The shotgun,* Dubbs thought, *is in the Marine's house, next door, almost right above us.*

"Follow me!" he hissed at Lester and Basil, leading them across a snow strip to the side of the Marine's house.

It should have been easy then, because all they had to do was stay pressed against the wall until they reached the front door. They were hidden from view from the Marine up there. Once they got in, it would be three against one.

But then why had Prue fallen? He couldn't be shot. The Marine couldn't see down here. Shit! Who shot him? Dubbs thought as Lester began pirouetting like a ballerina, rifle dropping in slo-mo. Blood spraying from his mouth. The eyes panicked as he jerked like a spastic puppet, and from the ripping hiss of bullets in the air Gordo realized with shock that Lester *was being hit by automatic weapons fire.*

But no one who lives here was supposed to *have* an automatic weapon. Gail had said there were only three guns on the street!

Gordo realized from the flashes across the street that another shooter was in that house.

He fired back, jammed in another magazine.

And then geysers of snow erupted near *his* feet, in a conga line coming closer. He dove out of the way as bullets smacked brick. His rage was so blind that he leaped back up and sprayed the building. He was good. Accurate. He'd won shooting medals in the police. He could see the line of bursts his own bullets made on the house opposite. Into the window. No more firing came from the window then.

One down!

But Gordon, turning back to the street, thought, in shock, *Nononono!*

Because his army was destroyed. Impossible!

Uniformed bodies lay by the Honda and the Gillette house. None of Dubbs's people remained standing.

How can disaster happen so fast?

And now here came fucking Gillette like a madman, bursting out of bushes and firing yet another weapon. How many damn firearms were over here anyway, Dubbs thought, backing away, spraying the street. It's not fair, he thought. Gail lied. She said there were only three weapons. She couldn't have held information back. The plan would have worked except for that bitch. There was nothing wrong with the plan. It was a great plan. Maybe she didn't lie. Maybe those assholes knew she was talking to me, so they fed her false information.

Get out of here now! he'd thought.

And so the great Commander Dubbs had raked the Marine's window to keep the man back. He'd reached the alley as the top of the fence slats blew out toward his face, and now he tells himself as he runs in the alley that he will not be beaten by these people. Gordon has known setbacks. His whole life is a setback. He retreats through backyards to the dark safety of burned-out Ingomar Place.

There's no more shooting. From the direction of downtown come thumps and explosions and the glow of flames. Once again, other people, lucky ones will be triumphing tonight. Making this new world their own.

Me and Teddie. We have food. We can find new help. We have weapons. We'll learn from our mistakes.

And then the lights around him in buildings go on.

Light!

Somewhere out there, even during the evil time, an engineer works in a power station. A switch has been thrown, a wire fixed, a connection jury-rigged, and for one more night let there be artificial light.

It's like a baseball stadium, when all the seats are illuminated. Lights flood the ruins downtown. Lights glow in the Oasis, and on Marion Street. Floodlights shine out again, except for an area where their defense system was shot up during the fight.

From the wreckage of the first house that was destroyed when the jet fell, weeks ago, Gordon draws into his lungs cold air. He's survived. Peering out, he sees in the distance Gillette's white face looking over the fence. Gillette holding one of the captured AK-47s. Like he's looking for someone to shoot.

Fucking Gillette and his fucking street, life, oil, friends. Every time the average man gets ahead, gets breathing space, the Gillettes step in and take it all away, Dubbs knows with fury.

Well, Dubbs thinks, aiming the AK-47, *I'll be back.*

"Excuse me," a voice says politely. There's a tap on his shoulder. "What are you doing exactly, sir?"

Dubbs whirls and looks up. He's so stunned he doesn't move. The man standing over him smiles as if they're chatting at a vegetable stand, discussing the price of an onion. How did he get here so silently? Dubbs even recognizes the guy. He's seen him outside St. Paul's. He's a religious man, a priest maybe. Why is he here?

"Don't do that," the guy says.

The guy bends down so fast that only at the very last instant does Gordon see beyond the smile. He recognizes the look. It's unmistakable. It is a million years old. It is what he saw sometimes as a policeman, except back then, men with eyes like that were in leg irons, or a witness box, or there were other cops along to help subdue them.

"I need to talk to him," the man says.

And the short, happy reign of Warlord Dubbs is over.

Gordo lies in his looted uniform, neck broken, in snow.

TWENTY-NINE

DECEMBER 12TH. 1:30 A.M. 45 DAYS AFTER OUTBREAK.

Neil Kline sprawls by his guard post, riddled by so many bullets that his clothing is in tatters. Blood drenches the snow around him, looking like wet tar, gleaming on bone shards and shell casings. Neil clutches his pistol where he fell.

Gillette still hears shots in his head.

Of course he fired at them. He was too angry to do anything else. His shot warned us. He saved our lives.

The neighbors venture out of their houses. Chris and Grace Kline run toward Neil, once the husband, once the father. Gillette and Bob kneel beside fallen attackers, slide off balaclavas and try to identify faces, making sure these people are no longer threats.

Are any attackers still alive?

The three corpses beside Neil were killed by shotgun blasts. One body, Gillette sees, is that of a young woman.

Two more attackers lie in front of Gail's house, killed, judging from the stitched wounds across their chests, by Pettigout's automatic weapons fire from across the street.

Gillette recognizes these two. They were Dubbs's friends, from the Oasis. The men whom Annie feared but could never identify as the attackers of her friend.

"Where's Dubbs?" Bob says.

"Someone got away, out back. Maybe it was him."

Around them the neighbors move swiftly, shocked but functioning, drilled by Bob in emergency procedures while Gillette was gone. They gather up fallen weapons. They check each other for injuries. They look for the missing, and make sure the barricades remain guarded, and that the perimeter behind the homes is still intact. Through his horror, Gillette is proud of them.

In front of Gillette's house lie the two men he shot, one unconscious and pumping blood from the groin. The other man clutches a knee, screaming for a doctor.

The groin wound looks mortal, Gillette thinks, but before he can check further he catches sight of Les hobbling up the street toward him, arm around his wife, Lisa. Alive!

Les sits down heavily in the snow, head in his hands.

Oh no. Not him too.

Gillette's not about to tend to the enemy over friends. He orders Eleanor Holmes to help the wounded attackers, to try to stop the bleeding. But the judge just stands there, looking down at them with flat, hard eyes.

"They're accessories to murder, Greg."

"Do it for yourself, Eleanor. Not for them. For how you'll feel later."

"We wouldn't have a later if they'd won."

Gillette's eyes drop to the long, wood-handled knife Judge Holmes grips in her hand. Joe generally carves the Marion Street turkeys with it, on Thanksgiving.

He wants to argue with her more. But he doesn't. It's not like the police will come and arrest these men later. Not like Marion Street will give them food. It's not as if the quality of mercy hasn't become self-destructive, or due process still requires cooperative strangers. Maybe the judge is right. There are no other doctors here. There's no medicine or ambulances, operating rooms, sutures, bandages. It's been years since he even watched surgery up close.

"Don't leave me with her," the wounded man sobs.

Walking away, he calls over his shoulder for someone to get into Les's house and find Pettigout. The young soldier had been stationed there.

The wounded guy—the conscious one—must know what's about to

happen. He screeches, "You're a doctor!" Suddenly his scream goes high and steady and when Gillette glances back, Chris Kline has joined Eleanor, wielding her favorite nine iron. She's sent Grace into the house. The women's arms rise and fall. By the time Gillette reaches Les, the screaming stops.

"Yo. Greg. I think I hit one of them, but I dropped the gun when I ran. I was out of bullets."

Les sits on his haunches, biting back pain, leaning back as if to get away from his trembling left leg. Dark spots glisten on fabric. His face is bone white. He seems disgusted that he lost his gun, but proud that he fought. For the first time in weeks, Les sounds animated.

"Les, you saved your family tonight."

"It's funny. I covered stories in Somalia and Kosovo. I did a month in Iraq. Where do I get shot, huh? Marion Street!"

"You took on three guys with automatic rifles with a pistol. You get the Marion Street silver star, Les."

"Where's Neil? I don't . . . see him. Is . . . he . . . okay?"

Les is breathing in gasps. He unzips his parka and his hand flutters to his throat. His eyes grow dazed. His face starts to slacken.

Gillette shakes Les gently. The eyes refocus a little. Gillette says, softly, "Stay with me, Les."

Annie has arrived, holding a pair of sewing shears. When did she start looking like an adult? She hadn't looked this way at the zoo, only hours ago. But her face is calm, resolute. There will undoubtedly come a reaction later, but right now she's focusing on what needs to be done. Gillette's proud of her. "Take the scissors, Dad," she says. "You said to cut away clothes when there's an injury."

"Help me get him into his house! Les, can you stand?"

"I can . . . beat you in a race anytime, old buddy."

Gillette spots Bob Cantoni leaning out of Les's bedroom window, across the street.

"Pettigout's hit! He's unconscious, bleeding badly!"

Gillette calls out instructions for Bob to put pressure on the wound, and try to slow the bleeding. He leaves Les with Annie and Lisa. He finds

Pettigout upstairs, sprawled against a corner wall as Bob tries to butterfly off the flow of blood with his hands. The boy's right shoulder joint looks shattered. The cephalic vein spews blood. The deltoid muscle is shredded and filled with bone shards.

"Help me get him down flat," he tells Bob.

"He was out when I found him."

This kid fought for us and I will save him.

Gillette cuts the fatigue shirt off and notes that the wound comes from a single gunshot. He rolls the kid over. There's no exit wound so the bullet's still inside.

I hope it didn't jump around in there.

"Find Les's tool kit. I need needle-nose pliers. A box cutter. Dental floss or a sewing kit. Strong thread."

"Why dental floss?"

"And have someone start a fire in the fireplace. Heat up the poker."

He probes the wound with his finger and Pettigout cries out and starts to thrash. Gillette has no anesthetic to give him, to dull the pain.

Extract the bullet. Clamp the vein. Then cauterize.

When did Marisa get here? She's helping him press down on the wound, and staunch welling blood. The kid groans and sweats.

"I found all your stuff," Bob says, rushing in.

Gillette uses the box cutter to widen the incision. He inserts the needle-nose pliers into the wound as Marisa pours water on it to clear away blood, so Gillette can see. The pliers hit bone. He twists the prongs. He draws out the pliers. He's extracted the bullet.

Raines said that people will be coming from Fort Detrick. I'll call and ask them to bring med supplies.

He reaches in again with the pliers, finds the torn vein and clamps it off to stop blood loss. He tells Marisa to wind the thread around the handles, tightly, to lock them in place. Handles jut from Pettigout's bare shoulder.

Marisa, finished, makes it out of the room and into the hallway before she throws up.

But the bleeding has stopped. I won't risk cauterizing. I'll wait for the antibiotics.

Now for Les, he thinks, exhausted.

"Marisa, call me if Pettigout starts bleeding again. Bob, come with me."

It's suddenly hard to see. The hallway is blurred. Something is wrong with his vision. No, nothing is wrong except that he's crying. It doesn't stop him. It simply goes on, involuntary. Tears course down his cheeks as he reaches Les in the living room to find that Annie has slit Les's trousers, and washed the wound.

"How's Pettigout?" Les asks with concern.

He's watching, grimacing. Marion Street's new tough guy. Bob tells Les, "The reason you never got shot overseas is, chickenshit reporters stay far back from the action."

"Chickenshit Marines are too dumb to know when to leave."

Gillette cleans the wound and dresses it, using ripped-up towels as bandages, finishing as Joe Holmes arrives to report that he went looking in Gail's house for liquor, to use as a painkiller. He found Gail gone and dozens of empty bottles in view around the house. Bottle after bottle. Gail had been afraid to throw them out and have her neighbors see the extent of her need.

"I bet she got the booze from Dubbs," Joe says.

Les grimaces. "You think she's over there now?"

Bob Cantoni says quietly, "Let's get them."

The three men regard each other.

"We have their weapons," Bob argues. "Dubbs ran away. They have medicine. Food. We can't just sit and wait for them to regroup. They'll be back sooner or later."

Les actually nods. "Bob's right," he says.

"Well! Look what it took to get you two to finally agree," says Annie, which breaks the tension, actually produces laughs. But it's short-lived.

"No more fighting tonight," Gillette says. "That place is as well defended as we are. If we go over there, the same thing will happen to us."

"Then tomorrow we scout 'em out," Les says firmly.

Outside, the floodlights are back on. The security cameras all function but one. Gillette's been working on patients for hours. He's about to drop.

Joe Holmes and Marisa will take guard duty. Annie and Paulo will sit up with Les and Pettigout, and wake Gillette if things go bad. Tomorrow they must try to bury Neil, who has been carried into his house. And they will have to dispose of the attackers' bodies somehow.

"We need our doctor fresh," Bob tells him. "Sleep."

Gillette doesn't think he'll be able to, but he's spent. He drops off almost at once in his own bed. The house is warm. The sheets still smell of sex. Could he actually have had sex with his wife tonight? It seems like months ago. At least the enemies are gone. He should be more relaxed. But he's not relaxed.

In his dream he's in Sudan again, in a *tukul,* a conical mud hut. Sun drifts in through the small oval entrance. On the dirt floor lies a dying Dinka man, tall, emaciated, coughing, struck down by resistant tuberculosis.

And Larch is there too, even though the doctor had been long dead when Gillette received Sudan assignments.

"Red snappers," Larch says, kneeling beside the sick man, eyeing Gillette sideways, like this is another test.

Gillette knows that red snappers are a CDC term for tuberculosis bacteria, slender rod shapes under a microscope when viewed with acid-fast red stain.

In the dream, red snappers swim in the mud walls like fish in an aquarium, wriggling as if the poison inside is trying to burst out.

"Vacuoles," presses Larch. "Tell me about vacuoles."

Gillette thinks, Vacuoles? Vacuoles are how the body kills invaders. They're capsules that surround germs. Special watery compartments in human killer cells. They contain toxic secretions that keep humans safe.

"Vacuoles are where resistant TB hides," he says.

That's right, says Larch. TB gets into the body. The hunter-killer cells find it right away, surround and absorb it and think they've killed it, but resistant TB is still alive inside the vacuole. It waits. It waits until the carrier weakens from age, or disease, or tiredness. It can afford to wait because the carrier thinks it's dead.

Gillette asks Larch, "You mean he's still alive?"

Larch says nothing. He just shrugs. When he stops asking questions, you know your answer is right.

"But he went into the burning house and didn't come out," Gillette tells Larch. "The soldiers said so."

The red snappers on the wall become Delta-3 microbes, swimming, glowing, multiplying as if mud walls were oil. Gillette tells Larch, "You're saying that D-3 uses our own defenses to protect itself. What are our own defenses? Army. Police. Oil people. The last ones we would ever blame for an attack."

The microbes morph again and assume human shape. The little bugs in the wall have all become the man from the zoo.

And now someone must have entered the hut from behind, even though there is no doorway there. Because someone is tapping Gillette on the shoulder. He wakes up. The tapping is real, he sees with dread. Bartholomew Young stands over the bed, bends close. Gillette feels the silencer against his temple, as a hand covers his mouth.

Vacuoles.

"Is he safe?" the man whispers.

His hand tastes like blood. Is the man bleeding? Gillette's pistol is in the drawer of the night table. It is too far away to reach. But he must try to reach it. Because this man will kill his family when they return.

"Dr. Gillette?" The hand over his mouth loosens.

It doesn't stop, he thinks. It will never stop.

Dr. Larch is back, still whispering, irritated, waving his pipe to make a point. "TB changes its name, don't you see, Greg? It says, I'm a vacuole. Don't refer to me as TB anymore."

"Dr. Gillette!" the man hisses, pressing the gun harder against his skull.

And Gillette thinks, The names!

"Call me Lawrence," he whispers, to keep talking, to buy time, to try to think and rattle the man.

The pressure against his temple lessens. In the moonlight Bartholomew Young's eyes grow wide. The breath has no odor, just heat, as if he is pure energy, and came out of the earth, like oil. But the name had an effect.

"Lawrence . . . Allenby." Gillette adds a second name from the book, offering associations, not answers, dangling possibilities and squeezing himself into a different appearance to try to live. He must adopt the guise of knowledge. He wonders, *Why didn't anyone see this man come over the fence?*

The man is stunned. Whatever he expected, whatever he prepared to hear, this utterance was not it. His features seem, for a moment, disconnected. Gillette sees a jumble of emotions. Like Picasso is rearranging the face, stacking feelings and deciding which one should go where.

"Great-great-grandfather, you said," Gillette says, remembering the inscription in the book, trying to tantalize and distract and move the man back so he can lunge for the drawer.

But Gillette can't move yet. If he does, the man will kill him. If he even tenses he'll break the spell. Only his mouth may move. Which guess will keep him alive?

"The computer—" he begins.

Bartholomew Young relaxes. Gillette has said the wrong thing. The computer in that apartment was a trick to fool people. The guy has left false clues all along.

And then like a bullet, a thought comes.

"The DNA," Gillette says, remembering the evidence.

The man stops.

"We know who he is, and who *you* are."

The gun withdraws. So the man is on file somewhere.

Young exhales painfully and Gillette realizes he has been injured.

Get him to glance away.

The man stares into Gillette's eyes, waiting for more. Mere hints will not suffice. Gillette goes for the safest ruse. He starts off, "We checked all the databases and—"

It's not right. *Databases was wrong.*

Gillette quickly risks, "In Great Britain . . ."

The man freezes. The eyes turn sad. "Yes," Gillette nods. *"There."* And

Young steps back, but just a few inches. Gillette needs him distracted to have even a minimum chance to lunge for the drawer.

The man grimaces. He leans to the right, so the injury must be on the left side.

"Ah," Young says. And that almost inaudible exhalation, that pause is like a small incision. An opening in skin, and opportunity for bacteria to slip in. Because a microbe needs only the tiniest entry point to wreak havoc. A puncture so small, human eyes can't see. Gillette must be a microbe and exploit the wound.

Suddenly, quietly and quite astoundingly, the man begins reciting something. His voice sounds scratchy in the dark, almost sad. Gillette sees sweat on the brow. Is it from revelation, or pain?

"I loved you," the man whispers.

Gillette thinking, *What the . . .*

"So I drew these tides of men into my hands . . ."

It's a poem. Is it from that book?

". . . the seven-pillared worthy house," the intruder says, ending the recitation. "So you know."

"Yes."

"He was a great man."

"Actually, that's the part we don't understand," Gillette says. "I mean, we understand that *you* think he was special, but really. A criminal? A pedophile? A person who spent years in the worst prisons?"

Silence. Then the man hisses, "What are you talking about?"

"Look, the last thing I want to do is make you mad but I'm curious. We just couldn't figure this part out. Why so much pride in someone so perverted that . . ."

"T. E. Lawrence was a hero!" the man snaps.

"Oh? Oh! I see now! You think that *he's* the one you're related to? Of course. Now it makes sense."

The gun jams swiftly back against Gillette's head. The man's hand is trembling. But is this from pain or shock? In the moonlight, Young's expression is part rage, but also panic. Gillette only knows that somehow he has attacked the very core of the man. He has found the vulnerable point.

He must put all his will into what he says next. Gillette reaches back and summons up the gang kid who put on a false face to enemies, the CDC doctor who lied his way through jungle roadblocks to reach victims, the father who showed his children confidence when he was fearful, the bluffer who tricked his way out to Nevada.

"I'm sorry," Gillette says. "But whoever told you that you were related to Lawrence of Arabia was simply not passing along anything even remotely close to the tru—"

"James Fitz-Barr would not lie to me!"

"Loyalty," Gillette says boldly, remembering the inscription he'd found. "Special friends. Is Fitz-Barr the man who gave you that book?"

Bartholomew Young looks like a block of pale wood in the darkness.

"Is he the man," Gillette presses, feeling close now, needing to know, "in England?"

But it doesn't work. Gillette has made his mistake and he knows it instantly from the quick transformation of the face. The killer is back.

"England? Did you say *England*?" the man says. And then the man smiles. It is an utterly relaxed expression. The man has what he came for. He knows what he needs to know.

"You almost had it right. All you had to say was Scotland. You were close. You were ve-ry good. You almost did it, but Sir James would never . . ."

Gillette sees the finger on the trigger tighten. And then a tremendous blast lifts the man from behind and flings him forward, into the air. Mouth open and fixed and surprise fills the eyes as the body lifts over Gillette like a stuffed dummy. It crashes down on him, driving out air.

Am I dead?

The weight—the body—is being pulled away. It's hard to hear, the blast was so loud. Someone is shouting. A face swim's close. It's Bob Cantoni, yelling, "Are you all right?"

"The answer is in Scotland," says Gillette.

"I saw him come over the fence. I had to move slowly. I couldn't let him hear me."

Gillette finally gets through to Detrick from the church tower, an hour later. He orders the duty officer to wake Theresa, who comes on the line sounding wide awake and worried. Out the window, in the distance, all of downtown seems to be on fire.

"Are you all right?" she says. "And your family?" She sounds more like a friend than a commander.

"I have a name!" Gillette says. "James Fitz-Barr."

The copter can't land in the deep snow, and on the narrow street, so it hovers fifteen feet up while Gillette climbs the rope ladder with the evidence in a knapsack on his back. It's 7 A.M., December 12th. First light.

Raines is inside. The copter lifts off and Gillette sees his family and neighbors getting smaller, gathering up the medicine and food parcels that the crew dropped. The copter heads south toward Virginia, skirting smoke columns rising downtown, fires still raging, and over debris and bodies on streets and rooftops and Metro entrances and on Potomac River ice, left there after last night's battles. Small bands of stunned citizens stand in the white.

"I hope the president has food," says Raines.

"Tell me what you found out."

"Sir James Fitz-Barr is an oil guy. Some high muckety-muck British lord who owns an oil and fluids company. Ex–British Intelligence. He ran their petroleum security division before taking over the family business."

"And the book I found?"

"The Brits figured out the name of the seller, from the mark inside the cover. Old Edinburgh shop. But the owner can't be located. The shop burned down. And even if they find the guy, he may not have a record of the buyer."

"Fitz-Barr was the buyer," Gillette says. "No doubt."

"Well, word is, the Brits swear up and down that he can't possibly be our guy. Old family. Patriot. Connected. Old school pal of the head of

British Intelligence. The Brits think we're making up charges again, like against Iraq after 9/11. They say we're barking up the wrong tree."

"What does the president say?" asks Gillette, getting mad again, thinking, "here we go again."

Raines smiles. The sun is up, and Virginia, below, is a white ravaged wasteland. *We have fifty days to figure this out,* Os Preston had said that first night at the Pentagon. Well, the time is just about up, Gillette knows.

Raines says, "The president wants to know what *you* think, boss. I get the feeling you'll call the shots this time. What do you want us to do?"

THIRTY

DECEMBER 15TH. 2 P.M. 48 DAYS AFTER OUTBREAK.

Sir James Fitz-Barr, ex–deputy chief of British Intelligence, petroleum division, CEO of Deep North Oil and head of his ancient family, sits in the comfort of his Cliff House library, sipping scotch, watching a replay on BBC-TV of Washington in flames. These are the final shots taken before communication failed in that doomed city.

Here's the White House surrounded by American troops, firing at their own people. It's like seeing one of those old movies about the fall of Rome.

Ah well, he thinks, swirling ice in his glass. *They were always a fractured society. Take away the money, oil and cars, and they go at each other's throats.*

Here's the face-shot of the American president, hiding in a bunker in Virginia, assuring the world that the great colossus will rise again, that the oil bug will be defeated, that civilization will find a way to survive.

Oh, it will survive, but you people are finished. Another three days and we'll release the antidote, he thinks, shuts the set and opens the library curtains, so he can see the magnificent rain-swept grounds.

You blew it, America. Now it's our time again.

He gazes out the huge bay window of his nine-hundred-year-old ancestral home. He's planning press conferences to announce the discovery of an antidote to Delta-3 in a subsea hot vent, by his research ship, off Iceland last week. The ship had been one of a half dozen European science vessels running on clean fuel from national reserves.

What he won't announce, of course, is that the murdered American scientist Samuelson found the antidote years ago and sold it along with his lab-created Delta-3 bug. The man had simply showed up one day, a disgruntled employee of a Nevada fluids company, a heaven-sent gift.

"We got lucky in the end," Fitz-Barr will announce. "Nature taketh but giveth back. Delta-4 neutralizes the effect of Delta-3 when applied to infected oil. It's a natural enemy," he'll say, giving the exact same explanation that Samuelson had given him. "The two microbes evolved in adjacent vents. The application of the antidote bug will neutralize the effect of the hybrid. Then Delta-4 will be killed by refining. Within weeks, pipelines should be running. Oil fields should be clear."

After we make a few changes as to who owns them.

Outside it is raining, and the usual cold wind blows from the direction of his offshore fields between Scotland and Norway. Fields that he had ordered infected with Delta-3 a month before the night that the Imam in Pakistan—on secret retainer for years—made his prediction of disaster. The blind faker going on about the west collapsing—as he'd been instructed by Fitz-Barr's agent—before Fitz-Barr had made sure that both Imam and agent were eliminated. It was easy to do. After retiring from Intelligence, he'd maintained by proxy contact with freelance wet workers in strategic parts of the world.

We'll ask for 10 percent of all profits. We'll deny the antidote to the Saudis, Iranians, Iraqis, Venezuelans. Let them give us back control.

The last two months have been comfortable enough at Cliff House and the local village, four kilometers away, on another bay. Clean fuel is stored here for his Land Rover, motorboats and village fishing boats. Electric power in this area comes from a nuclear plant. Seafood is plentiful. Sheep provide meat. His villagers are loyal, if ignorant of his intentions, and he's been sharing provisions with them. They've blocked access to the peninsula to hungry refugees from the cities. In fact, Fitz-Barr's security staff is out on the main road today, kilometers away, mopping up after repelling an attack last night by outsiders trying to reach the flocks.

His comm system is top level. His guests—all out shooting game today—are other members of the "Strategic Oil Council," as the secret group calls itself. They've been waiting out the emergency, talking politics, planning a new world and playing gracious host to the crews of disabled ships that periodically drift helplessly into the coves.

We're like the seven sisters, the original big oil companies who helped run the world.

He wonders what has happened to the beggar boy—the blond man sent to the U.S. to keep track of what the Americans know. He hasn't heard from him in days. He was concerned at first for his own safety, and fearful when the first disabled ship arrived, but now it is the norm. Nothing bad has happened to him. Besides, his connections at Whitehall are strong. If any action were contemplated against him, he'd get a call first. A polite inquiry. None has come.

If the beggar boy has been killed, that means I won't have to do it. I'd rather not do it. But he's the last connection to me. He had a good life. I gave it to him.

Fitz-Barr regards himself as a "human engineer," a master of psychology. You must know the needs of your people, his father used to say, or create new ones that you can control. Always appear to be a friend. Give people their dreams and they will do anything for you.

We're bred to lead, he thinks.

Lawrence of Arabia indeed! What a laugh!

The poor kid. The boy had been a social experiment. It had been like training a dangerous dog, after the kid, a needy stray, followed him in the Amman bazaar and saved his life that first day.

I have such an eye for talent!

The sad, filthy, pathetic, no-name beggar boy whose whore mother had apparently told him lies about his background—and to whom Lord Fitz-Barr had given meaning with the brilliant lie about DNA. That and an old volume he'd purchased in Edinburgh for the middling sum of thirty-one pounds. And for this illusion he'd received in exchange the kind of unbridled fealty that his ancestors—the approving faces on his wall—had demanded of yeomen since before nation-states existed at all.

We take care of them. They take care of us.

The blond man has been the most loyal, most deadly individual that Lord Fitz-Barr ever hired. A natural assassin. Mozart with a knife. A genius at illusion, who for months before Delta-3 was released crisscrossed the world, eliminating anyone who knew too much about the bug.

Now Fitz-Barr rings for a mug of tea from his housekeeper, dons a waterproof anorak and steps out onto the vast lawn with his pet Alsatian. He strides toward the cliff walk, out for his daily constitutional and private view of his isolated bay. But when he gets there he sees that out on the bay water is roiling, a submarine is surfacing.

Must be one of our nuclear models, he thinks, intrigued. *Or maybe it's one of the last vessels still out there, and it broke down from infected oil.*

Sailors on the conning area of the submarine seem to be preparing a small inflatable for the trip to shore. Perhaps they are out of provisions and hope to stock up in the village, although why land here instead of in the village? he wonders. Fitz-Barr will walk down to speak with the sailors. He enjoys hearing eyewitness accounts of damage caused elsewhere by Delta-3. And he enjoys the gratitude of stranded sailors.

He strolls on, down the zigzaggy limestone cliff path, lost in memory, reviewing the history of his special group.

The Strategic Oil Council was established in 2001 shortly after the Al Qaida attack on the World Trade Center and Pentagon, and the Iranians began developing a nuclear bomb. Western nations passed useless U.N. resolutions to try to stop it. Fitz-Barr knew that in a few short years, if trends were not reversed, fanatics, using oil money, would have the power to trigger Armageddon.

So the SOC would institute the hard measures that governments would not take, to regain control of oil.

And when that scientist from Cougar approached me, it was as if God had given approval to our group. We bought Delta-3 and began planning distribution. We waited for the proper moment.

Down on the rocky beach, he sees, the sailors have landed, formed two lines and trot toward the cliff walk. There is purpose in their formation, and in their rapid progress through the icy rain.

Days before that Imam gave his speech, I had my own equipment infected. The antidote is stored on this property, and in strategically protected locations on other lands I own.

The sailors round a bend, closer now, and he sees with surprise from their uniforms that they are not British sailors, but Americans.

They're probably stranded, far from home, like so many others.

Fitz-Barr goes to meet the sailors in a good mood. He'll act concerned. He'll share provisions. Maybe sailors are ill, and he can send for the village doctor. The Americans will be able to relate to him some truly splendid eyewitness disaster stories.

I'll invite their captain to dinner, he thinks.

It never occurs to him for an instant that he might be in danger. After all, Americans are allies. They have not sent troops to Britain ever, in history, except as helpers. Fitz-Barr is secure here. His family has been secure in this place since before the time of Robert the Bruce.

Even if I were to be implicated at this point, there's no proof. And once the antidote is released I'll be unassailable.

The dog starts to bark as the men come into view again.

"Easy, boy."

The men slow at the sight of Fitz-Barr, who greets them with a genuine smile. Their captain, at the lead, looks to be a trim and healthy forty-year-old. Fitz-Barr expects to see urgency in the expression, or need, at least, but the cold rage on the black face freezes him in place.

"Are you James Fitz-Barr?" the man demands, holding a photo, looking from it to Lord James's face.

Fitz-Barr experiences a twinge of doubt. The man has not addressed him as Lord. Or even sir or mister. Then Fitz-Barr recovers. The captain is simply boorish. These insufferable people act as if they control the world, even when their Congress is burning. Fitz-Barr is not about to sink to the level of this man. He plasters on his finest host-of-the-manor expression.

"I am he. Who—may I ask—are you?"

"Take him," the man orders his sailors. "Fast."

Heart slamming in his chest, Fitz-Barr turns to run. He hears a shot. The dog is down. The invaders are actually grabbing him and putting handcuffs on him, on his own property. It is absolutely impossible.

"This is outrageous," protests Fitz-Barr as they propel him down the cliff walk toward the beach. The villagers are too far away to help. These thugs are actually *snatching* him like he's Saddam Hussein or one of those tinpot dictators that America periodically aids and then arrests!

The captain snarls, as they hurry along, "Where is the antidote?"

"You are breaking the law!" Fitz-Barr cries.

"Yeah. Quite a rash of that going around."

THIRTY-ONE

DECEMBER 22ND. 55 DAYS AFTER OUTBREAK.

The crowd lining upper Connecticut Avenue breaks into cheers when the food convoy appears, their cries as loud as when plows cleared the road yesterday. And much louder than when the president returned to the White House two days ago, passing overhead in a helicopter. Gillette stands in the front row, arms around Paulo and Annie, breath frosting. Les Higuera's frustrated voice comes from behind.

"Are you going to tell me what really happened in Virginia or not? We both know the official story is bullshit."

Paulo smiles at Les. "You cursed. You owe me a dollar."

Truck after truck rolls past, their escort troops waving like liberators. The president wants the world to see the U.S. capital functioning again, so services are returning to the city quicker than elsewhere. But supplies are also beginning to flow to the rest of the nation, as shown constantly on again-uncensored TV.

"The first pipelines were announced clean yesterday," a man behind Gillette is telling his kids. "The president released the entire strategic oil reserve. Some president! More of an asshole. He'll never be elected again."

Gillette notes wryly that the most prominent sign that things are getting better is the fact that Washingtonians are talking politics again. But the city is in shambles. Recovery will take years. Federal buildings were burned. Thousands of refugees must be accommodated, shops and homes rebuilt, entire transport fleets replenished with undamaged cars and trucks.

Delta-3 can be killed now. But destroyed engines cannot be rebuilt.

Trials of looters are being conducted around the clock, Gillette also knows.

He's to report back to CDC tomorrow, at his own request, to work on medical needs around the country. He's happy to be off the antiterrorism team and back to concentrating on human disease.

"Did the British and Americans really team up to arrest Fitz-Barr?" Les presses, more the pesky journalist than best friend. "My reporters heard that an American sub landed in Scotland and seized the guy. That he was put in front of a firing squad at sea by CIA guys, *a bluff,* and he broke and told everything: the formula, the location of reserves of antidote, names of conspirators. We heard that only afterwards did the president and British prime minister cook up the story about cooperation."

Gillette lies. "I wasn't there when the decisions were made."

Temperatures hover close to zero, but no one seems to mind. Gillette eyes his kids with love. If seminal events mark the formation of every generation—Pearl Harbor, 9/11, Delta-3—then his children, he knows, transformed, are watching the passing vehicles with new sensibilities. Vast appreciation for the supplies, to be sure, but Gillette also notes how their expressions flatten out at the sight of chemical exhaust rising from the trucks.

Already—in reopening schools, Marisa said—teachers plan assignments on alternative forms of energy: solar, nuclear, hydrogen, wind. Yesterday Paulo had come into Gillette's study holding a book titled *Power!* "Dad, do you know that the first solar energy house was unveiled in 1948? That's sixty years ago! How come no one ever did more with solar energy after that?"

Gillette had told the boy, "Maybe you'll explain that to me someday."

Les squeezes closer in the crowd. "Greg, you were with the president. In the room. If I break this story I'm back on top. Deep background. Tell me."

"There's nothing to tell," Gillette says, flashing back to that horrible morning in the exact re-creation of the president's Oval Office, 150 feet under a mountain in Virginia. The dimensions of the room the same as in the White House. The carpets, bookshelves and plush chairs identical, and the presidential seal woven into the deep blue pile. But instead of windows looking out on a rose garden, banks of TV screens had been set into the walls. Gillette will have nightmares for the rest of his life reliving what he'd seen on them as he sat with the president, chief advisers and the chairman of the Joint Chiefs of staff.

Tehran and Oakland burning. Soldiers cooking human bodies in Buenos Aires. Riots in Hong Kong. Cholera outbreak in Australia. Military arsenals seized in Wyoming and near Shanghai. Starved astronauts floating in the space station. Thousands of penitents and flagellants flooding St. Peter's Square, the plaza by the Wailing Wall in Jerusalem and the Blue Mosque. Farms and ranches overrun by attackers from cities. People clawing at each other on piers in Tokyo and San Francisco, fighting for a few feet of space from which they could try to catch a fish. People throwing explosives into harbors, and swimming out to gather up edible sea life floating on top.

Oil fields set on fire by vengeful mobs.

Gillette had sat with a king's-eye view of a world convulsing in late-stage petroleum withdrawal.

"Seize Lord Fitz-Barr. Patch it up with Britain afterwards, if you can," Gillette had advised when asked.

Now Les grimaces from his wound and says, "If you won't tell me the truth about Virginia, at least tell me exactly how Delta-4 works."

"That will take a while to know," Gillette answers truthfully as the last truck passes. "All we're sure is, it kills Delta-3. Then Delta-4 is destroyed by refining. Inject it into dirty oil fields and it removes infection. Inject it into infected pipelines, flush out the whole thing, and you can use the pipeline again."

The crowd begins to break up. They're left eyeing 5110 Connecticut across the street—abandoned, as it turned out, after Dubbs's failed attack on Marion Street. The building had been looted by Gillette and his neighbors, and they'd found most of Dubbs's food stores left behind.

We donated half the supplies to St. Paul's. We gave a quarter of it to neighbors on Ingomar and Jennifer streets. And we kept 25 percent—enough to last three months—for ourselves.

"Let's go see how Raines is doing, moving into Gail's house," Gillette says, turning toward home. "She had no relatives. It'll be months before probate court decides what to do with that house. Maybe we can figure out a way to keep it for Raines."

A police car, one of the first back on the road, passes and Officer Danyla waves at them.

When they reach Marion Street they find that the Raines family has joined the other neighbors in Gillette's house, for a pre-Christmas dinner. The house smells of sizzling meat, spices and apple pie. Bob Cantoni wears an apron. Marisa is putting out pinot noir bottles and Hi-C cans on long tables in the dining and living rooms, and Lisa Higuera is setting out china plates. Judge Holmes sits in a corner, staring out the window, distant and depressed since the night of the attack on Marion Street. She's declined to adjudicate looter trials.

"I'm going to quit the bench, Greg. I'm unfit to be a judge," she says. "And I'm going to turn myself in for what I did to those men. They deserved a trial."

"You're not turning yourself in, Eleanor."

She bursts into tears.

Paulo sits with Grace and Chris Kline by an extra place set for Neil—in his honor.

Raines and his wife, Elizabeth, are helping themselves to hors d'oeuvres laid out: Ritz crackers and olive tapenade. There's a Christmas tree up, a trimmed hedge cut from Gail's backyard. Tiny lights blink. Paper angels sit in the green tree.

"Too bad Colonel Novak couldn't come," Raines says, offering Gillette a cheese spread and cracker combo. In civvie clothes, he looks even bigger.

"She's spending the week at the fort, with her people who can't get to their families. But she said she'll come for dinner one night when she's in town."

Marisa appears in the kitchen doorway holding up Paulo's old musical triangle from second grade. The high steady dings it makes—like a chuck wagon cook's signal—have been part of a Marion Street tradition for years. Block dinner is served!

"You're not ever going to tell me what really happened, are you?" Les asks.

"Leave the mayor alone," Bob snaps.

"Before we start eating, Pastor Van Horne wants to say a few words," Marisa says. "As if just 'a few words' are possible for him, ever."

It's good to hear the neighbors laugh, Gillette thinks.

"But before the pastor does that, I want to make a toast," Marisa says.

She raises her glass. She turns toward Gillette. He feels the love directed at him and sees the people he values most in the world standing to face him, holding up their wine, or water, or juice. Annie's eyes tearing. Paulo grinning proudly. Bob and Les standing together. Even Judge Holmes extending a glass. They all knew this toast was coming. More than the president, more than the country, this was what he fought for, he thinks. *This.*

"To the mayor of Marion Street," his beloved wife says. *"The Mayor of Marion Street!"*

"He killed the bastard and he came home."

ACKNOWLEDGMENTS

I would like to thank the following people who helped in the creation of *Black Monday.*

In the sciences, I am deeply grateful to Dr. R. L. Folk of the geology department, University of Texas, for sharing his knowledge and theories relating to nanobacteria; to Dr. Howard Shuman and Dr. Jonathan Dworkin of Columbia University, for information they provided about DNA, gene sequencing, microbiology and spore forming bacteria; to Dr. Gloria Coruzzi at New York University; and to several scientists who work for the U.S. government but asked not to be thanked in public. You know who you are.

Thanks to Alvin Natkin, formerly of Exxon and former president of the International Petroleum Industry Environmental Conservation Association, for sharing so much information about oil refining, distribution and chemistry.

To Renee Ridzon, MD, formerly of the CDC, finally, a story!

In Washington, D.C., thanks to R. Kenly Webster, Jim Grady and Bonnie Goldstein.

In North Carolina, to Philip Gerard. And in London, to Bill Massey, for advice, readings and friendship.

Many thanks to Marysue Rucci, Gypsy da Silva and Emily Remes at Simon & Schuster, Patricia Burke at Paramount Pictures, and to my beloved agents at ICM: Esther Newberg, Josie Freedman, Chris Earle and Kari Stuart.

Thanks to Ted Conover, Bob Denley, Bert Kimmel, Ken Levy, Gary Schoolsky, Sue Reiss and to Wendy the Plumba.

I'd also like to acknowledge two superb books that were helpful in the writing of *Black Monday.* Thomas Gold's fascinating *The Deep Hot Biosphere* sets forth new ideas of where oil and gas came from. And T. E. Lawrence's *Seven Pillars of Wisdom,* a beautiful and captivating read, is the autobiography of the famed World War One hero, Lawrence of Arabia.

Any mistakes in *Black Monday* are mine.

ABOUT THE AUTHOR

R. Scott Reiss lives in New York. Film rights to *Black Monday* have been optioned by Paramount Pictures.